AF323323

Statistics and Modeling of
Regional Climate
Variability in China

**Qingdao–Hamburg Cooperation in
Oceanography and Climate Studies
Since the 1980s**

Statistics and Modeling of Regional Climate Variability in China

Qingdao–Hamburg Cooperation in Oceanography and Climate Studies Since the 1980s

Edited by

Hans von Storch
Helmholt-Zentrum Hereon, Germany
Ocean University of China, China

Delei Li
Institute of Oceanology, Chinese Academy of Sciences, China
Laoshan Laboratory, China

NEW JERSEY · LONDON · SINGAPORE · BEIJING · SHANGHAI · HONG KONG · TAIPEI · CHENNAI

Published by

World Scientific Publishing Europe Ltd.
57 Shelton Street, Covent Garden, London WC2H 9HE
Head office: 5 Toh Tuck Link, Singapore 596224
USA office: 27 Warren Street, Suite 401-402, Hackensack, NJ 07601

Library of Congress Cataloging-in-Publication Data
Names: Storch, H. v. (Hans von), 1949–   editor. | Li, Delei, editor.
Title: Statistics and modeling of regional climate variability in China : Qingdao–Hamburg
    cooperation in oceanography and climate studies since the 1980s / Hans von Storch, Delei Li.
Description: New Jersey : World Scientific, [2025] | Includes bibliographical references and index.
Identifiers: LCCN 2024002508 | ISBN 9781800615809 (hardcover) |
    ISBN 9781800615816 (ebook for institutions) | ISBN 9781800615823 (ebook for individuals)
Subjects: LCSH: Marine meteorology--China. | Marine meteorology--Mathematical models. |
    Climatic changes--China. | Climatic changes--Mathematical models.
Classification: LCC QC994.6 .S73 2025 | DDC 551.509164/6--dc23/eng/20240310
LC record available at https://lccn.loc.gov/2024002508

British Library Cataloguing-in-Publication Data
A catalogue record for this book is available from the British Library.

Cover photograph of Hans von Storch by Dr Andreas Pohlmann, reproduced with permission.

For any available supplementary material, please visit
https://www.worldscientific.com/worldscibooks/10.1142/Q0466#t=suppl

Desk Editors: Nandha Kumar/Rosie Williamson/Shi Ying Koe

Typeset by Stallion Press
Email: enquiries@stallionpress.com

Dedicated to
Ocean University of China

Foreword by Jürgen Sündermann

This book documents significant progress in the understanding and numerical simulation of climate-related physical processes in shelf seas. Choosing the Chinese coastal waters as an example, the complex interactions of atmospheric and oceanic dynamics, from the global to the regional scale, are analyzed and simulated within a hierarchy of coupled models. The results are significant for understanding the anthropogenic climate change and a sustainable management of our natural environment. At the same time, they manifest the value and benefit of joint international research efforts.

Our findings are based on almost forty years of German–Chinese cooperation starting with the International Union of Geodesy and Geophysics (IUGG) conference in 1983 in Hamburg where we first met our Chinese colleagues. We realized that we shared strong common interest in our research activities and could jointly achieve new insights into ocean, climate, and environmental research.

Our cooperation started with a personal invitation by the Shandong College of Oceanography to visit Qingdao, to give a series of lectures and to discuss possibilities for further joint activities. Together with rector Professor Wen Shengchang and his research group, we developed a long-term strategy for cooperation in shelf sea research. Our initiatives have been generously supported on both sides by the relevant state offices. So we could integrate further research groups. On the German side this included the GKSS Research Center Geesthacht (now called the Helmholtz-Zentrum

Hereon), the German Weather Service Hamburg, and the Hamburg University of Technology, and on the Chinese side, the First and the Second Oceanographic Institutes in Qingdao and Hangzhou, the East China Normal University in Shanghai, and the South China Sea Institute in Guangzhou.

The cooperation has been always perceived as important and turned out to be very successful — on both sides. Besides the scientific aspects, I felt the human contacts, the private invitations, and the individual traveling care to be very generous and just as useful. I have been impressed by the high theoretical standard of my Chinese colleagues which was barely known in the West. The still underdeveloped computing facility was balanced by original approaches in theoretical oceanography. The combination of our capacities then led to new joint projects and results in shelf sea research and an exchange of scientists.

A first highlight in our cooperation was the project "Analysis and Modelling of the Bohai Sea Ecosystem (AMBOS)" in the years 1997–2000. It included two field experiments in the Yellow Sea with the Chinese research vessel DONG FANG HONG *R/V Dong Fang Hong* (with a female German cruise leader). My following personal distinction with of the "Friendship Award" bestowed by the Deputy Prime Minister of China in the People's Hall of Beijing documents the general recognition of our fruitful cooperation.

I am deeply satisfied that Hans von Storch and the Chinese scientists from Qingdao (such as "my" doctorands Xueen Chen and Jiang Wensheng, and the new generation Delei Li) have continued and extended this joint research line. In a time of global climate change, international scientific and societal cooperation is needed to maintain a healthy (and peaceful) environment. This publication contributes to this endeavor.

Jürgen Sündermann
Hannover, September 2023

Foreword by Xueen Chen

Over the past four decades, the collaboration in marine sciences between institutions in China (primarily the Ocean University of China (OUC)) and institutions in Germany (primarily the University of Hamburg and the Helmholtz-Zentrum Hereon) has evolved into a remarkable testament to the power of international scientific cooperation. This enduring partnership has borne witness to significant scientific achievements in coastal oceans, facilitating an extensive exchange of knowledge and fostering strong relationships among scientists from diverse backgrounds and cultures.

As Professor Sündermann has outlined in his preface, the story of the Sino-German collaboration began to unfold during the 18th General Assembly of the International Union of Geodesy and Geophysics (IUGG) in Hamburg in 1983. It marked a pivotal moment when Chinese oceanographic scientists, emerging on the international stage, engaged in discussions and established vital contacts with their German counterparts. This initial encounter laid the foundation for a flourishing partnership, particularly between the Shandong College of Oceanography in Qingdao (SCO, later OUC) and the Institute of Oceanography (IfM) at the University of Hamburg.

Scientific exchange between OUC and Hereon culminated in the symposium "Coastal Oceans: Interdisciplinary Scientific Prediction and Management" in Qingdao in 2014. This gathering highlighted the importance of coastal issues and led to the signing of two significant memoranda, reaffirming the commitment to collaboration.

The "Chinese and European Coastal Shelf Seas Ecosystem Dynamics — a Comparative Assessment" (CHESS) project, driven by Xueen Chen from OUC and Corinna Schrum and Wenyan Zhang from Hereon, has been conducted from 2020 until 2024. It aims to gain insights into complex marine ecosystem dynamics and their responses to climate change and increasing human activities by comparing and assessing different nearshore continental shelf marine systems in China and Europe. The CHESS project has been reviving and further strengthening bilateral cooperation and communication in the field of marine science between China and Germany.

One of the hallmarks of this collaboration is the mentorship of students who have gone on to become leading scientists in their own right. Distinguished individuals, such as Daji Huang, Wensheng Jiang, and Xueen Chen, received their Ph.D. degrees in the 2000s from the University of Hamburg, making significant contributions to our understanding of Chinese coastal seas. Professor Hans von Storch, among others, has played a pivotal role in fostering the growth of young researchers. Several students have earned their Ph.D. degrees under the guidance of Professor von Storch and myself, contributing to a new generation of oceanographic/climatic experts. Currently, ongoing collaborations continue to nurture the development of promising scholars under the CHESS project.

The chapters presented in this anthology represent the culmination of years of dedicated effort and interdisciplinary Sino-German collaboration in the past decade. It encompasses a wide array of research areas, from regional dynamical downscaling to the internal variability of marginal oceans, from the variability of polar lows in the North Pacific to the climatology of wind powers in the Bohai Sea and Yellow Sea. The research has also explored the critical intersection between climate variability and extreme events, such as storm surges, and their impacts on coastal regions. The anthology offers valuable insights into various facets of coastal oceanographic and climatic phenomena and underscores the importance of international cooperation in advancing our understanding of the ocean and climate.

As we gaze into the future, the partnership between Chinese and German institutions in marine sciences promises continued scientific excellence. The challenges posed by marine ecosystem dynamics and their responses to climate change, sea-level rise, and extreme events necessitate collaborative efforts on both sides. This cooperation remains poised to address pressing questions, expand our knowledge of regional climate dynamics, and find innovative solutions to the issues that lie ahead.

In closing, we celebrate the legacy of four decades of collaboration, salute the achievements of the past decade, and look forward to a future filled with shared discoveries and continued partnership in the pursuit of scientific excellence. Together, we are forging a path towards a better understanding of our environment and its vital role in our shared global future.

Xueen Chen
Qingdao, September 2023

Contents

Introduction

This anthology showcases the outcomes of the decade (of 40 years in total) of collaboration between scientists from Qingdao and scientists from Hamburg. During the initial phase, which lasted until roughly 2010, Professors Jürgen Sündermann from Hamburg University and Wen Shengchang were the primary drivers of this collaboration by laying the groundwork for understanding the dynamics of China's marginal seas, and the capability to build and run hydrodynamical models of Bohai and Yellow Sea. Later, in the last 15 years, Professor Hans von Storch from the former GKSS Research Center (now Helmholtz-Zentrum Hereon) located near Hamburg and Professor Xueen Chen from the Ocean University of China built a productive cooperation. The anthology focuses on the last decade and provides an overview of the methodologies, reconstructions, and potential future climate changes in the Shandong region and its neighboring marginal seas.

For this book, we present 12 original scientific articles which significantly contributed to the existing knowledge base. They have been presented in a new format but are otherwise unchanged.

1. The Qingdao–Hamburg Cooperation

1.1. *The First 30 Years*

The 18th General Assembly of the International Union of Geodesy and Geophysics (IUGG) in Hamburg, Germany, from 15 to 26 August 1983, marked the first time that Chinese geoscientists became visible in the Western world following societal reforms in the People's Republic of China. As a result, German and Chinese oceanographers had the opportunity to establish numerous contacts and engage in discussions during the event. This led to the development of a lively correspondence between various German and Chinese marine research institutions, especially between the Shandong College of Oceanography in Qingdao (SCO) and the Institute of Oceanography at the University of Hamburg (IfM). Upon invitation of Professor Wen Shengchang (文圣常), Professor Jürgen Sündermann visited Qingdao in the summer of 1986 and delivered a series of lectures at SCO on the modeling of physical, chemical, and biological processes in the sea.

In November 1986, Professors Wen Shengchang and Jürgen Sündermann signed the first official memorandum on scientific cooperation between SCO and IfM (Figure 1), which was supported by the State Oceanic Administration of China (SOA) and the German Research Foundation (DFG). The partnership between the University of Hamburg (UH) and the Ocean University of Qingdao (OUQ), which succeeded SCO and preceded the Ocean University of China (OUC), continued to develop and included other marine disciplines, such as meteorology, biogeochemistry, hydrobiology, field measurements, and remote sensing. Additionally, scholars from both institutions exchanged knowledge, and young Chinese scholars studied in Hamburg. The Joint Sino-German Governmental Commission on cooperation in marine research and technology facilitated regular meetings where project proposals and experiences were shared. Over the years, collaborative efforts were also established with other oceanographic institutions in China. The partnership between OUQ (OUC) and UH continues to thrive with support from SOA and DFG, and leading scientists from both institutions have been visiting each other almost yearly. Notable individuals from Qingdao, such as Jiang

Fig. 1.   Jürgen Sündermann in Qingdao in 1994, centre. To his left, the retired rector of OUC Professor Wen Shengchang (文圣常), to his right, the director of the Institute of Physical Oceanology, Professor Feng Shizuo (冯士筰).

Wensheng and Xueen Chen, obtained their Ph.D. degrees from UH, and each developed models for the Chinese coastal seas.

On 21 April 1989, Professor Sündermann was appointed as an Honorary Professor of OUC in Qingdao (Figure 2), and his continued partnership was acknowledged during an academic ceremony in Qingdao on 5 August 2004.

The joint project "Analysis and Modelling of the Bohai Sea Ecosystem" (AMBOS) was the most significant achievement of their cooperation. The project took place from 1997 to 2000 and included two ship cruises of *R/V Dong Fang Hong* in the autumn of 1998 and spring of 1999. The outcomes of this project were published in a Special Volume of *Journal of Marine Systems* (Vol. 44, 3/4, 123–254, 2004), with guest editors J. Sündermann and Feng S. (冯士筰).

Most of the individuals who initiated this collaboration on both sides are now retired; Professor Wen S. died at the age of 101 years in 2022.

Fig. 2.   Certificate of Appointment of Jürgen Sündermann to become a Honorary Professor at the Ocean University of China — 19 April 1989.

1.2. *The Last 10 Years*

After retiring, Jürgen Sündermann recommended to Professor Hans von Storch that he should participate in the established cooperation between OUC and Hamburg. As a director of the Institute of Coastal Research in Geesthacht and an affiliated professor at Hamburg University, this academic exchange was a perfect fit for his scientific interests. He collaborated with Professor Xueen Chen (陈学恩) in Qingdao to build a dynamic cooperation with Ph.D. students, sponsored by the China Scholarship Council (CSC) (Figure 3).

Hans von Storch has been a frequent visitor to China, not only to the OUC in Qingdao but also to the Institute of Oceanology Chinese Academy of Sciences (IOCAS) in Qingdao,[1] where he serves as a member of the Scientific Advisory Board. He holds the position of guest professor at OUC (Figure 4) and was actively involved in academic exchange programs.

---

[1]The cooperation with IOCAS began in 1992 with the visit of Cui Maochang (崔茂常) to the MPI in Hamburg, which resulted in two papers, one in English and one in Chinese, on downscaling sea level variations.

Fig. 3. Ph.D. students from Qingdao, from top left: Xia Lan (夏兰), Feng Jianlong (冯建龙), Li Delei (李德磊), Zhang Meng (张萌), and Tang Shengquan (唐声全), Lin Lin (林璘), Chen Fei (陈飞) studying with Xueen Chen (陈学恩) and Hans von Storch (bottom right; Photo by: Andreas Pohlmann) at HZG.

At the turn to the 20th century, Qingdao was under German administration, then named Tsingtau. As part of this administration, regular weather observations were made and (via the Transsiberian railway) reported to Germany. These data were stored in Hamburg in

Fig. 4. Inauguration of Hans von Storch as guest professor at Ocean University of China in Qingdao in 2013.

the "Seewetteramt", the Hamburg affiliation of the German Weather Service. They were almost forgotten, until Gudrun Rosenhagen, the then head of the Seewetteramt, found them. Upon the initiative of Professor Fu Gang, the original colonial German documents, covering eleven years (1898–1909) of weather observations in Tsingtao, were handed over to the Municipal Meteorological Bureau of Qingdao. On that occasion, a symposium was organized in Qingdao (Figure 5).

Another highlight of the OUC-HZG cooperation was the symposium "Coastal Oceans: Interdisciplinary Scientific Prediction and Management" held in Qingdao on 20–24 October 2014 (Figure 6). The symposium focused on coastal issues and brought together scientists from China and Germany as well as Sweden and Poland. On that occasion, two memoranda were signed.

The first, named the "Four Seas" (四海), was initiated by the leading scientists Hans von Storch, Detlef Schulz-Bull, Jiang Wensheng (江文胜), and Luo Yongming (骆永明). It recognized the similarity of Bohai/the Yellow Sea and the Baltic and North Seas' systems (Section 1.3). The work reported in this anthology may be understood to large extent as contributing to the goals of "Four Seas".

The second was a "Memorandum of Understanding" between OUC and HZG. It was signed on behalf of OUC by vice-president Li Huajun (李华军) and on behalf of HZG by Hans von Storch (Figure 7).

Fig. 5. Ceremony on the occasion of transferring the original weather data collected by the German colonial administration on the Qingdao weather service on 8 April 2014.
Speakers: Gudrun Rosenhagen; in the front row: Professor Fu Gang (傅刚, Executive Vice Dean of Graduate School of OUC), Wolf Rosenhagen, Hans von Storch, Wu Tiejun (武铁军, Assistance Mayor of Qingdao City Government), Professor Hu Yongyu (胡永云, Peking University and Vice President of Chinese Meteorology Society), Dr. Shi Yuguang (史玉光, Director of Meteorological Administration of Shandong Province), Dr. Gu Runyuan (顾润源, Director of Meteorological Administration of Qingdao City).

Fig. 6. Participants of the symposium "Coastal Oceans: Interdisciplinary Scientific Prediction and Management" in Qingdao on 20–24 October 2014.

So far, six students received their Ph.D. degrees under the supervision of Hans von Storch, Xueen Chen (陈学恩), or Jiang Wensheng (江文胜), namely Chen Fei (陈飞), Xia Lan (夏兰), Feng Jianlong (冯建龙), Li Delei (李德磊), Zhang Meng (张萌) and Tang Shengquan (唐声全); presently, Hans von Storch is working with Lin Lin (林璘)

Fig. 7. Signing the memorandum of understanding of OUC and HZG on 20 October 2013. Right: OUC vice-president Li Huajun (李华军), left: Hans von Storch.

on the presence and significance of hydrodynamic noise in marginal seas. On the German side, several scientists joined the effort, in particular Beate Geyer, Frauke Feser, Matthias Zahn, Ralf Weisse, Leone Cavicchia, and Peter Bisling. The articles in this anthology are taken from the work of these Chinese students and their German supervisors.

1.3. *The Four Seas* (四海)

In the framework of the Sino-German Cooperation in Marine Sciences, the symposium "Coastal Oceans: Interdisciplinary Scientific Prediction and Management" took place in 2014 in Qingdao, China. On that occasion, the joint perspective of the "Four Seas" (四海) was discussed.

"Memorandum on Sino-German Coastal Research Dealing with
Bo Hai and Huang Hai as Well as Baltic and North Sea Regions"

Hans von Storch, Detlef Schulz-Bull, Jiang Wensheng (江文胜) and Luo Yongming (骆永明) Co-signed by Bian Changwei (边昌伟, OLUC), Chen Xinping (陈新平, SOA-NMHMC), Chen Jianfang (陈建芳, SOA-Second Institute of Oceanography), Xueen Chen

(陈学恩, OUC), Ralf Ebinghaus (HZG), Kay Emeis (HZG), Holger Freund (U Oldenburg), Peter Fröhle (TU Hamburg-Harburg), Fu Gang (傅刚, OUC), Jürgen Jensen (U Siegen), Li Keqiang (李克强, OUC), Li Xiaoming (李晓明, CAS Institute of Remote Sensing and Digital Earth), Li Yan (李琰, SOA-NMSDIS), Li Qiang (李强, Tsinghua U), Liu Dongyan (刘东雁, YIC-CAS), Liu Sumei (刘素美, OUC), Markus Meier (SMHI, Norrköping), Mu Lin (牟林, SOA-NMSDIS), Kai Myrberg (SYKE, Helsinki), Thomas Pohlmann (U Hamburg), Beate Ratter (HZG), Gudrun Rosenhagen (DMG), Gerald Schernewski (IOW; EUCC), Torsten Schlurmann (U Hannover), Corinna Schrum (U Bergen), Sun Jun (孙军, Tianjin University of Science and Technology), Jürgen Sündermann (U Hamburg), Wei Hao (魏皓, Tianjin University), Jörg Olaf Wolff (U Oldenburg), Xiang Rong (向荣, CAS-SCSI).

1.4. *Coastal Science as Own Research Field*

Coasts are those parts of the ocean which are significantly influenced by the neighboring land, be it in terms of currents, winds, temperatures, sea ice, salinity, suspended matter, specific species, and anthropogenic substances; but coasts are also that part of the land which is significantly influenced by the neighboring sea, be it opportunities for transport, risks related to flooding, tsunamis, storm surges, and where the points of departure such as shipping, tourism, and oil and gas extraction are found. Not surprisingly therefore, the science in coastal dynamics and management extends far beyond oceanography. Certainly oceanography is an important component of the interdisciplinary mix of scientific efforts dealing with coasts, but other fields such as meteorology, coastal engineering, land use planning and management, freshwater hydrology and ecology, climate, sociology and cultural science are also needed for constructing holistic views of the subject of coasts.

What makes the issue of coasts special is not only the presence of the "other" counterpart, i.e., the sea versus land and land versus sea, but also the presence of an often dominant factor modifying or even constructing the coast — humans who use the coasts in various ways, conditional upon temporally and culturally varying preferences, who make use of opportunities and try to deal with often great dangers. The German coasts, there is a saying according to which "God created the ocean, but the Frisians created the coast" (the Frisians are people living at the Dutch and German North Sea coast). Because of both, the culturally conditioned values and preferences, and the different geophysical, morphological and ecological set up, the various coasts of this world are very different, and face very different challenges, risks and opportunities. As a consequence coastal science is fragmented into regional research communities; in some quarters coastal sciences is considered as a mere variant of oceanography, and in others it is essentially coastal engineering, and in still others human geography.

The challenge is to bring together these different communities, disciplines, challenges and concepts together, not only within the scientific community but also in a trans-disciplinary effort by combining the real world, in which the coast is subject to competing practices and decisions, with the body of scientific knowledge.

We are convinced that a deepening and a sustainable cooperation between German and Chinese coastal scientists will hold the potential for significant progress in coastal sciences, not only in the two countries but also beyond. We expect synergistic progress, in terms of better understanding of processes, subsystems, the holistically framed system comprising natural dynamics and human interference, and perspectives conditioned by different societal values and preferences, when we consider in parallel, in particular in a comparative mode, the two North German marginal seas of the North Sea and the Baltic Sea, as well as the two Chinese marginal seas Bohai and Huang Hai.

This progress will also have a management dimension, an instrumental (monitoring) dimension as well as issues of the built environment (coastal defense, ports, shipping lanes, offshore activity).

1.5. *Four Seas* (四海)

We focus on the "Four Seas" (四海) not only because many of us are familiar with these seas after many years of scientific analyses, but also because of their societal significance with a broad range of important modes of uses, from oil and gas, nearshore aquaculture, offshore fishery, wind energy, shipping and other industrial activities to tourism and natural preservation. Both systems are made up of one sea which is more enclosed, and another one which is half open to the world ocean. In both systems major ecosystems have their home, but they also serve, or served, as dumping sites for onshore activities. A difference is the fact that the North European seas have many different national coasts while in case of the Chinese seas only a limited number of international partners share responsibility for the marginal seas.

We should also mention the emotional value of the Four Seas for the people living at their coasts — which certainly has an influence on the culture of the regions.

Within the "four seas" concept, we cover issues of hydrodynamics and ecology, of coastal climate of observational projects, the construction of monitoring systems, and the various dimensions of human usages (fisheries, oil and gas, natural reserves...). Other issues, which deserve attention deal with past changes, possible future drivers in changes, the construction of scenarios, the challenge of detecting (changes beyond natural variations) and the attribution (of plausible mechanisms). A particular challenge will be to map similarities and differences of regional societal values, decision procedures and stakeholder networks. Finally, we welcome systematic efforts for assessing the body of scientifically legitimate knowledge and for describing to "what extent this knowledge is based on agreement or is contested."

2. Added Value of the Cooperation

The value added from the collaborative effort over the past ten years is illustrated by 12 original scientific articles, which contributed to the existing knowledge base.

Most of the fundamental challenges in global climate science and global change science have been addressed, even if some significant issues still persist in understanding the overall functioning of the climate system, including the role of clouds and radiation, system stability and the potential for abrupt and irreversible changes (referred to as "tipping points"), and the simulation of a complete ice-age cycle. However, these matters are primarily of academic interest and no longer have a significant impact on contemporary climate policy and management, as far as "mitigation" concerns. This is because nations have already committed to complete de-carbonization in the next few decades, as outlined in the Paris Agreement.

For a considerable period, the concept of "adaptation" received minimal attention, primarily due to concerns that emphasizing adaptation might support a public perception that mitigation efforts were less urgent. For instance, Al Gore argued in his book *Earth in the Balance*, that adaptation would be a misguided strategy (cf. Stehr and von Storch (2023)), because it would lead people to view mitigation as less needed, that the climate change challenge could be better dealt with adaptation. In those days, arguing for the need of adaptation was in the Western context as arguing as advocating for "abatement *or* mitigation" as alternative instead of the more obvious "adaptation *and* mitigation" approach. It is evident that adaptation is necessary regardless of the circumstances since climate change is already occurring. Even if the global community successfully achieves the goals set in the Paris Agreement, further changes will transpire, necessitating adaptation measures.

The need of knowledge for adaptation is quite different than that for mitigation — the latter addresses global issues, while adaptation is needed on the regional or even local scale, where the hazards emerge. Thus, one has to be able to understand climate variability as well as global climate change scenarios regionally. The appropriate technique is named "downscaling" — either by constructing statistical models or hydrodynamical models. The history of this now matured techniques (Chapter 1), and the generated added value (Chapter 2), for the case of East Asia, by applying such techniques, is discussed in the two reprinted articles in Part I.

During the investigation of regional climate variations and changes in East Asia, a new scientific challenge emerged, namely the issue of unprovoked regional variability, referred to as hydrodynamic "noise" in our field's terminology (Part II). The term "unprovoked" indicates that in ensembles of simulations, where only the time of initialization is shifted or when identical setups are run on different computers, the resulting outcomes differ (Geyer *et al.*, 2021; Lin *et al.*, 2023). Hence, these variations cannot be attributed to specific external factors or details of the initial conditions. This internal variability was initially observed while attempting to downscale statistics of meso-scale eddies in the South China Sea (Chapter 3). Two subsequent papers explored the relationship between the intensity of the noise and model grid resolution (Chapter 4), as well as the scale-dependency of the signal-to-noise ratio, which was found to be significant for large scales but negligible for small scales (Chapter 5). Interestingly, similar variability emerges in the shallower regions of Bohai and the Yellow Sea, albeit with reduced noise intensity when tides are activated (Chapter 6).

This line of analysis emerged somewhat unexpectedly. Initially, the first study (Chapter 3) aimed to develop a downscaling model for regional ocean dynamics, but all efforts proved unsuccessful due to the apparent random behavior of meso-scale eddies. Consequently, this transformed the study from a downscaling investigation to a new exploration of regional dynamics. The Stochastic Climate Model introduced by Hasselmann (Hasselmann, 1976) emerged as a significantly useful concept when dealing with this new challenge.

Storms play a significant role in the regional climate of various areas, including the Bohai/Yellow Sea region and the North and Baltic Sea region, as described in the "Four Seas" concept (see Section 1.3). Consequently, studies have focused on investigating North Pacific Polar Low storm variability (Part III) on decadal time scales. Drawing inspiration from Polar Low studies in the North Atlantic, Polar Lows' emergence and variability for past decades was quantified and scenarios for possible future developments were developed (Chapters 7 and 8).

In Part IV, four articles demonstrate the combination of concepts developed in Geesthacht, and data and models prepared in Qingdao. These applications aim to analyse significant impact variables both in the past and potential future scenarios. Specifically, they focus on estimating potential wind power (Chapter 9), and low-level jets (Chapter 10). Also hazards associated with storm surges are investigated — along the coast of China (Chapter 11) and studying the specific case of the Bohai region (Chapter 12).

In summary, the outlined collaboration serves as a remarkable example of effectively combining potentials, capabilities, and talents from diverse scientific and cultural backgrounds. The cooperation showcases the success achieved through scientific collaboration, including the exchange of visits by scientists. This approach not only strengthens relationships but also fosters understanding and friendship across cultural boundaries. Given the current confrontational political climate, such soft methods are highly valuable and essential.

References

Geyer, B., Ludwig, T. and von Storch, H. (2021). Limits of reproducibility and hydrodynamic noise in atmospheric regional modelling, *Communications Earth & Environment* **2**, p. 17, doi:10.1038/s43247-020-00085-4.

Hasselmann, K. (1976). Stochastic climate models. Part I. Theory, *Tellus* **28**, pp. 473–485.

Lin, L., von Storch, H. and Chen, X. (2023). Seeding noise in ensembles of marginal sea simulations — the case of Bohai and Yellow Sea, *Advances in Computer and Communication* **4**, pp. 70–73, doi:10.26855/acc.2023.04.001.

Stehr, N. and von Storch, H. (2023). *Science in Society: Societies, Climate Change and Policies*, World Scientific Publishing Europe Ltd., doi:10.1142/q0399.

https://doi.org/10.1142/9781800615816_0002

Part I

Downscaling

This Part I is focused on the methodology of "downscaling," a concept originally introduced by von Storch *et al.* (1993, 2000). In the context of the Hamburg–Qingdao collaboration, we have compiled a review of accomplishments achieved in both Qingdao and Geesthacht, which is presented in Chapter 1, as given in our article "The concept of large-scale conditioning of climate model simulations of atmospheric coastal dynamics: Current state and perspectives" (von Storch *et al.*, 2018). The article "Added value of high-resolution regional climate models: Selected cases over the Bohai Sea and the Yellow Sea areas" by Li, 2017 reprinted in Chapter 2 addresses the fundamental question of what can be achieved through downscaling.

We would like to touch on the important issue of uncertainty arising from the use of different global analyses as driving factors, which is outlined in our article "Testing reanalyses in constraining dynamical downscaling" (full reference provided in the reference list), however, space limits us. Instead we would like to present the abstract here and highly recommend that readers search out the full article:

Reanalysis data sets have been widely used in regional climate dynamical downscaling studies. In this study, we test the use of various reanalysis data sets in constraining dynamical downscaling by assessing the reconstruction skill of the Yellow Sea coastal winds using the COSMO model in Climate Mode (CCLM) with 7-km resolution. Four reanalysis forcing data sets are used as lateral boundary conditions and internal large-scale constraints (spectral nudging): the National Centers for Environmental Prediction and National Center for Atmospheric Research (NCEP/NCAR) reanalysis data set (NCEP1) is downscaled to an intermediate domain with 55-km resolution (CCLM_55km), ERA-interim reanalysis data set (ERAint), NCEP climate forecast system reanalysis data set (CFSR), and Japanese 55-year reanalysis data set (JRA55). Several statistical analysis methods are employed to assess the modeled winds through comparison with observed offshore wind data from 2006, and it is found that the downscaled simulations yield good quality wind speed products. However, they all tend to overestimate observed low wind speeds and underestimate observed high wind speeds. Furthermore, the quality of the modeled wind direction is strongly associated with the wind speed intensities, exhibiting a much better reproduction of wind direction at strong wind speeds than at light wind speeds.

The downscaling simulations driven by ERAint, JRA55, and CFSR are consistent with each other in the reproduction of local wind speed and direction; the simulations driven by ERAint and JRA55 are slightly better for strong winds and those driven by CFSR are better for light winds. All three simulations generate local wind estimates that are superior to those of the simulation driven by CCLM_55km. This superiority reflects the better quality of the CFSR, ERAint, and JRA55 reanalyses with regard to assimilated local observations compared with the CCLM_55km hindcast, which exploits only upper-air large scale NCEP1 wind fields.

Chapter 1

The Concept of Large-Scale Conditioning of Climate Model Simulations of Atmospheric Coastal Dynamics: Current State and Perspectives*

H. von Storch[†], L. Cavicchia[‡], F. Feser[†], and D. Li[§]

[†]*Institute of Coastal Research, Helmholtz Center Geesthacht, Germany*
[‡]*School of Earth Sciences, University of Melbourne, Australia*
[§]*Key Laboratory of Ocean Circulation and Waves, Institute of Oceanology,
Chinese Academy of Sciences, Qingdao, China*

We review the state of dynamical downscaling with scale-constrained regional and global models. The methodology, in particular spectral nudging, has become a routine and well-researched tool for hindcasting climatologies of sub-synoptic atmospheric disturbances in coastal regions. At present, the spectrum of applications is expanding to other phenomena, but also to ocean dynamics and to extended forecasting.

Additionally, new diagnostic challenges are appearing such as spatial characteristics of small-scale phenomena such as Low Level Jets.

1. Introduction

Polar Lows, Medicanes, Coastal Low Levels Jets, and other meso-scale storm phenomena represent significant cases of dangerous weather phenomena. Large-scale atmospheric dynamical state shapes the occurrences and features of such events. They are particularly prevalent and energetic above coastal waters, so that for planning coastal infrastructure as well as offshore activity, knowledge about the statistics, including extreme value statistics, is needed. Because of the large-scale conditioning, such statistics vary in time, reflecting not only decadal variability but also long-term climate change.

Deriving such statistics represents a challenge not only because of the size of such disturbances, but also because of the changing quality and density of the observational basis. Satellite products provide a good source of data (e.g., Blechschmidt, 2008) but working with them is labor-intensive, and statistics based on such data may be compromised by subjective choices and the limited lifetimes of different satellites.

An alternative approach of processing is dynamical downscaling, which extrapolates in the state space from large to smaller scales. A climatology of Polar Lows over the North Atlantic was obtained by implementing this method (Zahn and von Storch, 2008), and later of Polar Lows over the North Pacific (Chen and von Storch, 2013),[1] of Medicanes over the Mediterranean Sea (Cavicchia *et al.*, 2014a), of Tropical Cyclones over the Northwestern Pacific (Barcikowska *et al.*, 2017), and of Coastal Low Level Jets in the Bohai/Yellow sea region (Li *et al.*, 2018).[2] The methodology has matured now and is routinely used in many applications — mainly for deriving atmospheric states in the past decades.

We review past developments, address the present state and prospects of such approaches, with simulating dynamical properties of small-scale cyclones like Polar Lows (e.g., Zahn *et al.*, 2008), and

---

[1]Reprinted in this collection as Chapter 7.
[2]See Chapter 10.

the possibility of implementing such methodology in global models (Yoshimura and Kanamitsu, 2008).

2. Simulating Small Synoptic Features Conditioned by the Large-Scale State Constraining by Spectral Nudging

The idea to constrain regional dynamical models to follow the large-scale trajectory provided by weather analysis scheme, was introduced into atmospheric sciences in the 1990s. Earlier methods involved nudging spatial means (Kida *et al.*, 1991; McGregor and Katzfey, 1998; Sasaki *et al.*, 1995). Waldron *et al.* (1996) introduced the term "spectral nudging" into weather forecasting contexts, and von Storch *et al.* (2000), independently of the earlier work, suggested and tested the usage in climate simulations, extended to seasonal and decadal timescales. Later contributions and tests were provided by Miguez-Macho *et al.* (2004).

The basic idea is related to the downscaling concept, according to which the large-sale state together with smaller scale physiographic details, such as mountain ranges or coastal configurations, would condition the smaller scale dynamical state. Originally, the concept was introduced as an empirical variant (e.g., von Storch, 1995), while simulations with limited area models were later recognized as representing a dynamical variant of downscaling (e.g., Giorgi *et al.*, 2001). In a strict sense, however, limited area modelling employing the conventional boundary forcing does not induce consistent smaller scales by processing large-scale states. Instead, the states along the narrow strips along the lateral boundaries of the considered region are processed in the spirit of boundary value problems, even if it is known that in this case this mathematical concept is not well posed (Davies, 1976; Laprise, 2008; Oliger, 1978; Staniforth, 1997).

This changed when a constraining of the larger scales was introduced. Trigonometric expansions, cosine expansions (Denis *et al.*, 2002), and also spherical harmonics expansion allow the needed separation of scales. "Nudging terms" were added to the equations, which penalize deviations from a given state on large scales, but leave small scales unconstrained.

Constraining the large scales in a model simulation represents a kind of data assimilation (e.g., Robinson *et al.*, 1998), but in a different manner than conventionally done. In data assimilation, two equations are formulated, "the state space equation" describing the forward development of the systems trajectory, and "the observation equations", which relates observables to state variables in the state space equation. Both equations represent a kind of knowledge, namely theoretical knowledge, about the system (the dynamics in the state space) and empirical knowledge (the observed quantities); both types of knowledge are incomplete, and the two equations are integrated forward in tandem for combining the two types of knowledge efficiently. First, the state space equation is used to estimate the future state; this suggested state is then transformed into an observable using the observation equation.

Eventually, the estimated future state is corrected by a term proportional to the difference of the actual observable and the estimated observable — and so forth. In most cases, the observables are local observations, but in the case of downscaling it is the large-scale state, e.g., states in a coordinate system spanned by orthogonal functions sorted according to scale. The assumption is that the driving coarse resolution data is of satisfactory quality for large scales; thus the dynamical model (featured in the state space equation) is supposed to follow these states closely, while it is asked to provide additional details for those smaller scales, where the coarse resolution data are considered to need improvement.

Figure 1 illustrates the success in a spectrally constrained three-months simulation. The curves describe the similarity of the state of the driving reanalysis (here: National Centers for Environmental Prediction/National Center for Atmospheric Research (NCEP/NCAR) reanalysis; there are of course various other reanalyses, but this one has the advantage of being built for many more decades than more recent reanalyses, and is based on four-dimensional data throughout) and the constrained limited area model simulations (here: REMO regional model). A similarity value of 1 indicates full representation of the reanalysis by the limited area model. The two upper curves show the similarity of the large scales in a constrained run and another free run. Obviously, the constrained simulation follows the trajectory of the reanalysis closely, whereas the free run intermittently simulates significant deviations (for details, refer to von Storch *et al.* (2000)). The lower two curves show the same similarity measure but for medium scales (medium scales are here: all non-large scales, excluding the smallest scales affected by the truncation of the

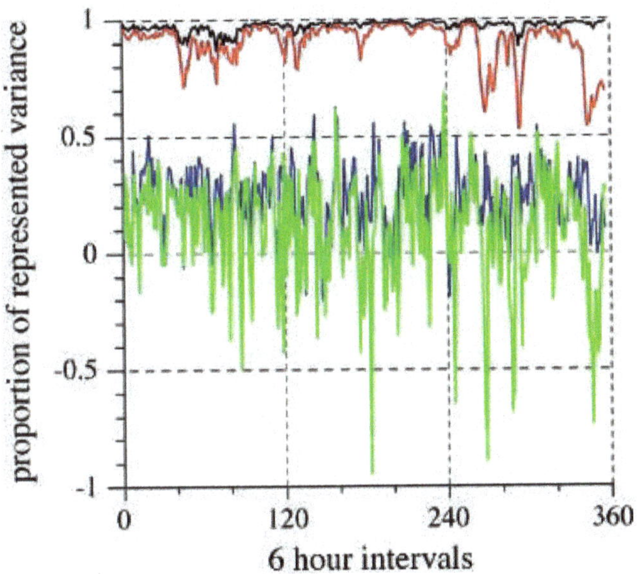

Fig. 1.   Similarity of zonal wind at 850 hPa between simulations and (driving) NCEP/NCAR reanalyses in constrained (black and blue) and unconstrained (red and green) simulations with a limited area model. Top: Large spatial scales. Bottom: Regional spatial scales (von Storch *et al.*, 2000), © Copyright 2000 AMS.

finite grids). In both, the constrained and unconstrained simulation the deviation from the global re-analyses is large, which indicates that the limited area model is generating additional details on such scales. That this additional detail is indeed realistic, and thus represents an added value over the re-analysis was shown by Feser *et al.* (2011).

When the simulation region is large (such as all of contiguous US), the improvements can be dramatic (Rockel *et al.*, 2008), but when the region is small then the effect becomes insignificant (Schaaf *et al.*, 2017). Obviously, such developments are more likely when the region is farther away from the boundaries because of the large constraints they involve.

Sometimes, the argument is brought forward that such a correction would violate principles of conservation, e.g., mass or momentum. This is indeed true, but accepted in all reanalyses schemes; also the violations are not large, when employed regularly, and the systems are not closed, at least in terms of energy and momentum.

Obviously, a number of choices have to be made when implementing the spectral nudging method in particular referring to the variables, the intensity of the nudging, conditional upon wave length and height, frequency of spectral nudging; in the classical paper by von Storch *et al.* (2000), the variables were the two components of the wind above 850 hPa (so that surface details are permitted to influence the lower part of the atmosphere, with less and less influence higher up in the troposphere and stratosphere). Several sensitivity experiments have been conducted since then, for instance by Alexandru *et al.* (2009), Omrani *et al.* (2012), Kang *et al.* (2005), Miguez-Macho *et al.* (2004, 2005), Park and Hwang (2017), Radu *et al.* (2008), Ramzan *et al.* (2017), and Tang *et al.* (2010). Schubert-Frisius *et al.* (2017) extensively tested different settings for employing spectral nudging in a global model (see Section 4).

In most applications, the method is used to downscale reanalysis — in particular the NCEP/NCAR reanalysis, which has covered the globe since 1948 and has the advantage of providing multi-decadal histories of weather. Its grid resolution is about 210 km. It certainly suffers from inhomogeneities, in particular the advent of globally covering satellites which represented a major change outside of some well-observed regions mostly in Europe and North America. Another popular data set is ERA-Interim from the European Centre for Medium-range Weather Forecasts, which has been available since 1979, with a considerably higher grid resolution (of about 80 km). In recent years, further reanalyses such as the NCEP Coupled Forecast System Reanalysis CFSR and the National Aeronautics and Space Administration Modern Era Retrospective-analysis for Research and Applications MERRA2 entered the "market", usually covering the satellite era (post-1979), with resolutions of about 0.5° (in detail, c.f. https://climatedataguide.ucar.edu/).

Generally, the global reanalyses are not homogeneous, since the observational data, which are processed in the (frozen) analysis-schemes, are non-stationary in terms of coverage, density, and quality. However, the development of the large-scale features of the atmospheric dynamics, are plausibly homogeneous, because less data are needed to determine their state. Thus, the spectral nudging method employs a qualitatively stationary part of the reanalysis, and determines a consistent and homogeneous addition to the reanalyses.

Besides spectral nudging, there is also another nudging technique, which forces a simulation to follow a given trajectory in phase space, namely "grid nudging" or "analysis nudging". Obviously such a constraint is limiting the development of the state much more than that of selective nudging since not only the large scales, which are supposed to be less affected by the finite grid resolution, but all scales, including those heavily affected by the grid truncation and the interrupted energy cascade, are affected. A number of studies have considered the two techniques (Liu *et al.*, 2012; Ma *et al.*, 2016; Spero *et al.*, 2018).

The latter, the grid-point ansatz, is useful when sub-grid-scale processes are examined and fitted, while for hindcasting spectral nudging is to be preferred, as it allows improved small-scale dynamics and better descriptions of physiographic detail.

3.   Issue: Divergence in Phase Space[3]

From predictability studies, it has long been known that global models starting from the same modified initial conditions and subject to the same forcing end up in very different trajectories, albeit within a corridor given by climatology, if a minuscule change is introduced somewhere in the process (e.g., Chervin *et al.*, 1974). Interestingly, it was not recognized for another 20 years that limited area models show similar behavior, namely that the state in the interior of the area is not determined by the lateral boundary conditions, but that different trajectories may form for limited time periods. When initiated with very slightly different initial conditions, significant variability emerges in the model region at a possibly much later time, as demonstrated by Ji and Vernekar (1997), Rinke and Dethloff (2000), Weisse *et al.* (2000), Weisse and Feser (2003), and Alexandru *et al.* (2007). After a while, the system returns to a preferred trajectory, but may again generate a bundle of very different trajectories for a limited time at any later time.[4]

The examples demonstrate that the different trajectories of the simulated systems are for extended times similar, but that every now

---

[3]This issue is taken up again in Part II of this book.

[4]Later, its has been demonstrated that the same divergence-in-phase space can be initiated when running the same code, with identical initial and boundary values, but on different computing platforms.

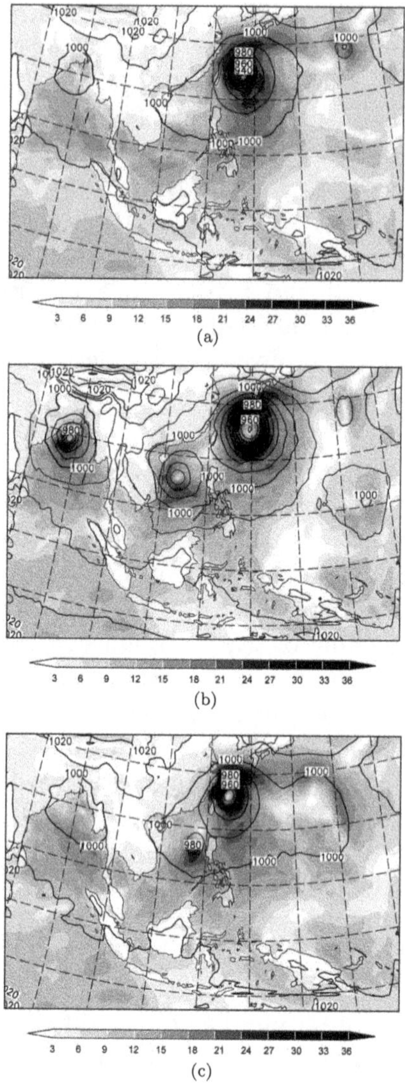

Fig. 2.   Simulated sea level pressure (hPa; isobars) and near-surface wind speed fields (m/s^{-1}; shaded) of Typhoon Winnie representing 0000 UTC 17 August 1997. (a) With constraining; (b, c) two simulations without constraining. The simulations were initialized with (b) conditions of 0000 UTC 3 August 1997 and (c) conditions of 0000 UTC 7 August 1997. (Model: COSMO-CLM; forcing: NCEP 1; 50 km grid resolution. Feser and von Storch (2008a), © American Meteorological Society. Used with permission.

and then a major split takes place, which results in significant differences for a limited time. It means, regional models generate intermittently different developments ("divergence in phase space") — which is not indicating "falseness" of a model, but is rightly reflecting the stochastic character of "weather formation".

Examples of such episodes are given by Weisse *et al.* (2000), Weisse and Feser (2003), Zahn *et al.* (2008), and Feser and von Storch (2008a). As an example, Figure 2(a) shows an example from seasonal simulations for a time, when in reality a typhoon formed in East Asia (Feser and von Storch, 2008a). In one simulation, not one but three typhoons formed (Figure 2(b)), and in the second, one major storm plus a secondary storm formed (Figure 2(c)).

When the large scales in a simulation are constrained the effect is largely suppressed, and the differences are mostly insignificant. Figure 2(a) shows the result of a constrained simulation, and only one typhoon emerges, at the right location and time (albeit with a too shallow depth).

4. Simulating Small Synoptic Features Conditioned by the Large-Scale State in Regional Models[5]

Spectrally nudged regional climate models have been used to construct climatologies, and scenarios of possible future changes, of small-scale synoptic phenomena in various regions, among them Europe, the South Atlantic, East Asia, the Mediterranean Sea, the Yellow Sea, and the North Pacific. Of course, other meteorological phenomena were also described and studied, such as precipitation connected with the East Asian summer monsoon (Lee *et al.*, 2004).

The first cases were Polar Lows over the North Atlantic (Zahn and von Storch, 2008; Zahn *et al.*, 2008) and over the North Pacific (Chen and von Storch, 2013; Chen *et al.*, 2012). These relatively small energetic storms form over the subarctic sea, mostly in cold air outbreaks, and it is has only been possible to accurately identify them since the advents of satellites. Using NCEP/NCAR reanalysis as driving data, a history of the formation of such storms could be constructed beginning in 1948. The annual cyclogenesis frequency was found to

---

[5]More on this, see Part III of this book.

be mostly stationary, with no remarkable trends towards more or less storms.[6] The used grid resolution of about 50 km was seemingly not sufficient to allow for realistic deepening of the storms. Future scenarios were also constructed[7]; in the regions of the North Atlantic and North Pacific, the frequency of storm occurrence decreased due to more stable atmospheric conditions as the higher atmosphere warmed faster than the surface; an assessment of the change in intensity was not made (Chen *et al.*, 2012; Zahn and von Storch, 2010).

Medicanes are also small, intense storms over the Mediterranean Sea, with features similar to those found in tropical storms, such as vertical symmetry, a thermal profile characterized by a warm core, and a spiral shape in the cloud cover with a central cloud-free eye. The horizontal resolution of gridded models or reanalysis data are particularly sensitive issues for the case of Medicanes, with the smaller among such storms showing a radius as small as 70 km. Using a double-nested regional model, with a grid resolution of 10 km in the inner region, and NCEP/NCAR reanalyses as forcing data, Cavicchia and von Storch (2012) demonstrated that realistic Medicanes were formed at the right time and location. Figure 3 shows the features of a Medicane which occurred in January 1995 as simulated in a downscaling simulation; a Medicane with realistic features is formed, whereas in the forcing data the storm is not visible due to the coarse resolution. No noteworthy trends for the time since 1948 were detected by Cavicchia *et al.* (2014a) in the main formation regions. For future climate scenarios, a decrease in storm occurrence was found caused by increased atmospheric stability in analogy to the Polar Lows case. On the other hand, a slight intensification of the simulated future storms was found (Cavicchia *et al.*, 2014b). As an example, Figure 4 shows the spatial densities of Medicanes (given as number of tracks passing through a grid box).

Low Level Jets and other phenomena in the Bohai and Yellow Sea region: Coastal areas are featured with plenty types of meso- or small-scale atmospheric processes. A simulation with a resolution of about 7 km was conducted over the Bohai and Yellow Sea region in East Asia during 1979–2013, forced by the ERA-Interim reanalysis dataset (Li *et al.*, 2016). Li (2017) revealed that the simulation

---

[6]See Chapter 7.
[7]See Chapter 8.

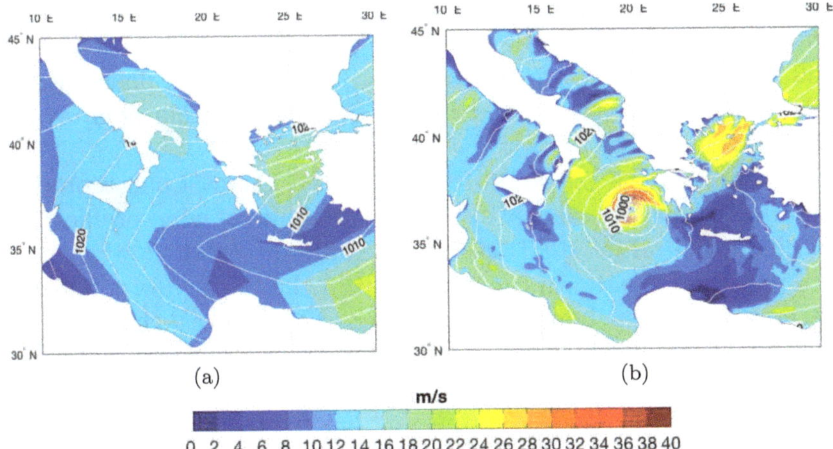

Fig. 3.   Snapshots of January 1995 Medicane on 15 January at 18 UTC. (a) Sea level pressure (2 hPa contours) and wind field (m/s, color shaded) from the forcing NCEP reanalysis. (b) Same as (a) but from COSMO-CLM simulation at 10 km resolution.

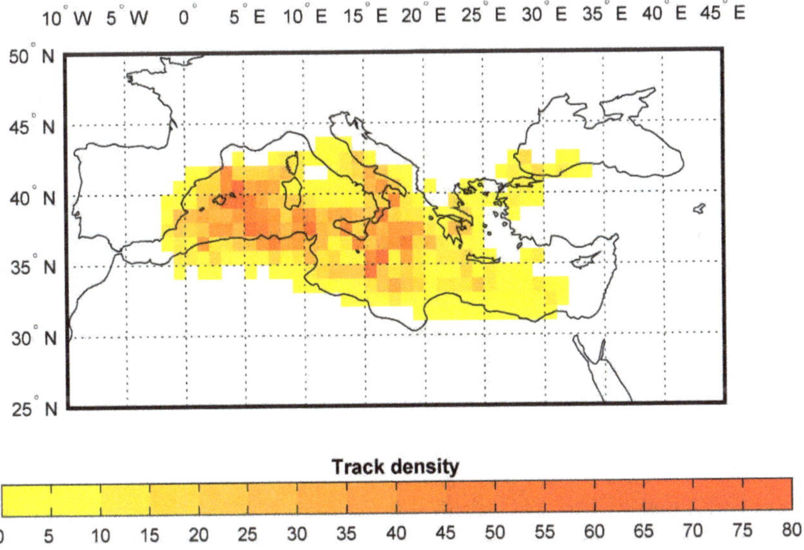

Fig. 4.   Track density of simulated Medicanes in the 1948–2010 period (COSMO-CLM, 10 km grid resolution, forcing: NCEP; (Cavicchia *et al.*, 2014a).

outperforms ERA-Interim in capturing the detailed spatial and temporal structures of cases of meso-scale phenomena such as a typhoon and a cold surge. A vortex street, an orography-related phenomenon, can be realistically generated by the simulation rather than by the ERA-Interim.[8]

Furthermore, Li *et al.* (2018) proved the hindcast can reproduce the climatology, the diurnal cycle, the variability of wind profiles, and specific low level jet (LLJ) cases, which are mesoscale-flow phenomena with horizontal wind maxima within the lowest few kilometers of the troposphere. Long-term statistics of LLJs reveal that they feature a strong diurnal cycle, intra-annual, and interannual variability, but weak decadal variability. LLJs are more frequent in April, May, and June (defined as LLJ season) and less frequent in winter season. Figure 5 shows that LLJs mostly occur at a height of 200–400 m, with intensities generally less than 16 m/s. The dominant wind directions are southwesterly and southerly. Li *et al.* (2018) also identified a low-frequency link between anomalies of LLJ occurrence over the Bohai and Yellow Sea and regional large-scale barotropic circulation over the East Asia-northwest Pacific region.[9]

Tropical cyclones affect large parts of tropical coastal areas worldwide and their inhabitants. Because of their vast damage potential, it is essential to analyze their possible future changes and past variability. A commonly used approach to generate homogeneous tropical cyclone (TC) statistics relies on general circulation and regional climate models. They have the advantage of using a non-changing model system over time which is important to derive long-term statistics. The spectral nudging technique has been applied successfully in a number of studies in the Western Pacific to gain realistic TC distributions of the past (Feser and von Storch, 2008a, 2008b; Feser and Barcikowska, 2014). Spectral nudging reduces the number of TCs simulated by the regional climate model just forced at the lateral boundaries and not using spectral nudging in the model domain's interior by about half, but generally retains the ones which were observed (Feser and Barcikowska, 2012). The TC generation and development is very dependent on the quality of the forcing data set and storm intensity is sensitive to the model's resolution. A TC

---

[8]See also Chapter 2.
[9]See also Chapter 10.

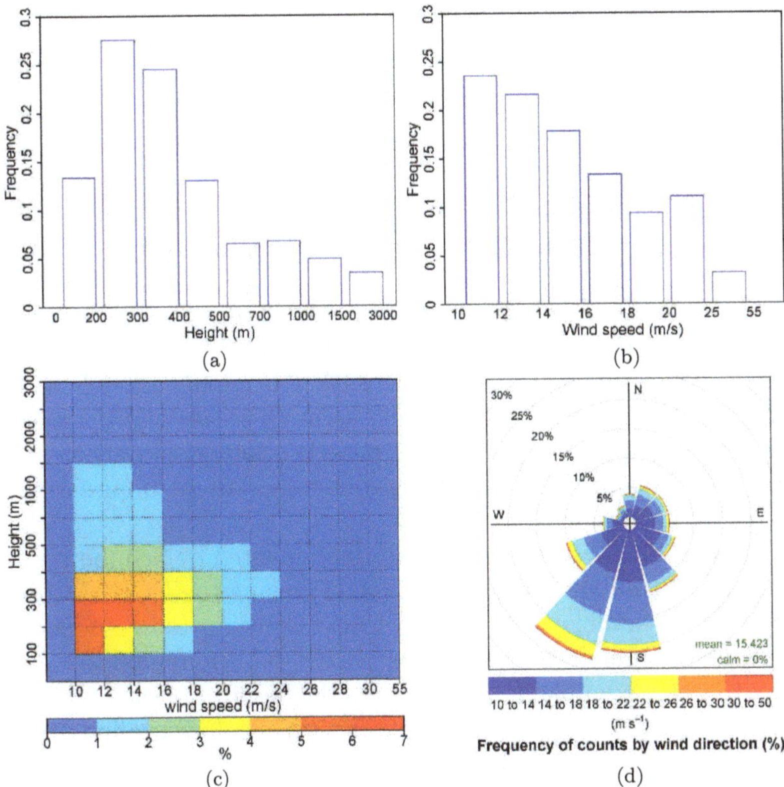

Fig. 5. Low Level Jets statistics over the Bohai Sea and Yellow Sea during April, May, and June (1979–2013): (a) Jet height histogram (%), (b) jet wind speed histogram (%), (c) jet height-wind speed distribution, and (d) jet wind rose (Li *et al.*, 2018), © American Geophysical Union. Used with permission.

climatology for the last decades (Barcikowska *et al.*, 2017) shows a large similarity to observation-derived best track data (Barcikowska *et al.*, 2012) for the most recent times which feature the best measurement quality and availability. Both modelled- and observation-based data sets show an increase in TC activity, in terms of annually accumulated TC days for the last three decades. For earlier decades, statistics differ between the model and observations. An upward shift in TC intensities in the regional model is apparent around the end of the 1970s which is presumably based on the introduction of satellite measurements in the forcing reanalysis at that time. This shows the large dependency of the regional climate model on its global forcing data and that for some cases spectral nudging cannot completely cure data inhomogeneities, which were presumably inherent in the forcing data.

5.   Simulating Small Synoptic Features Conditioned by the Large-Scale State Constraining in Global Models

It was Yoshimura and Kanamitsu (2008) who noticed that the idea of spectral nudging may also be implemented into global models. Indeed, conceptually the implementation in global models is more attractive than in regional models, as in this case the downscaling concept is applied in a purer sense, namely a forcing of only the larger scales and not a hybrid of forcing along the lateral boundaries and of large-scale components.

The concept was also tested and implemented by Schubert-Frisius *et al.* (2017) and von Storch *et al.* (2017), using the ECHAM6 global atmospheric models with a T255 grid resolution (corresponding to about 60 km), and the large-scale NCEP/NCAR reanalysis as constraint. As an example, Figure 6 shows the description of a typhoon in the NCEP/NCAR reanalysis together with the downscaling achieved by the constrained global simulation.

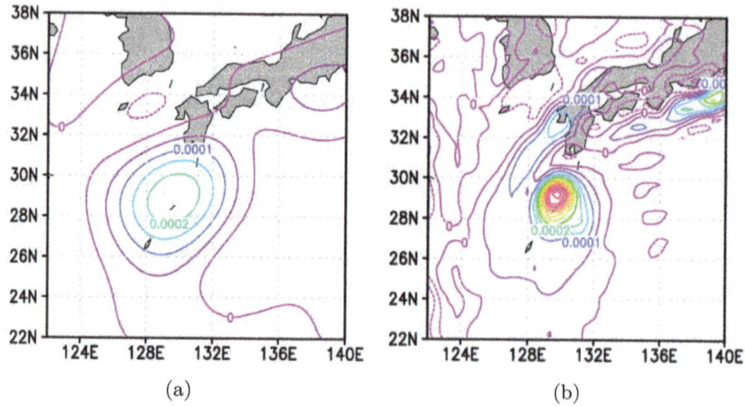

(a)   (b)

Fig. 6.   Vorticity fields of typhoon "Tokage" at the time of its maximum strength 1800 UTC 19 October 2004, (a) in the NCEP/NCAR re-analysis and (b) in the constrained global model (Schubert-Frisius *et al.*, 2017), © Copyright 2017 AMS.

A variety of parameters, such as the vertical profile of the nudging parameters, were tested (Schubert-Frisius *et al.*, 2017). In addition, the performance of the global simulation with respect to regional simulations was confirmed (von Storch *et al.*, 2017). It turned out that if the global and the regional models had a comparable grid resolution, the hindcasts were also similar, as demonstrated by Figure 7, which compares the global simulation with two regional models in terms of the Brier Skill Score using the satellite product QuikSCAT as a reference. For the wind in regions marked in red, the regional model performs better than the global model, whereas blue regions mark a superiority of the global model. The global model is a bit better than the regional model when a comparable grid resolution is employed, but the reduction of the grid resolution to 7 km in the regional model leads to a significant improvement. It remains to be seen what the situation is if the global model is run with a resolution which is close to being convection-permitting (<4 km according to Prein *et al.* (2015)).

Thus, instead of running many regional simulations in different domains, introducing further potential sources of inhomogeneity

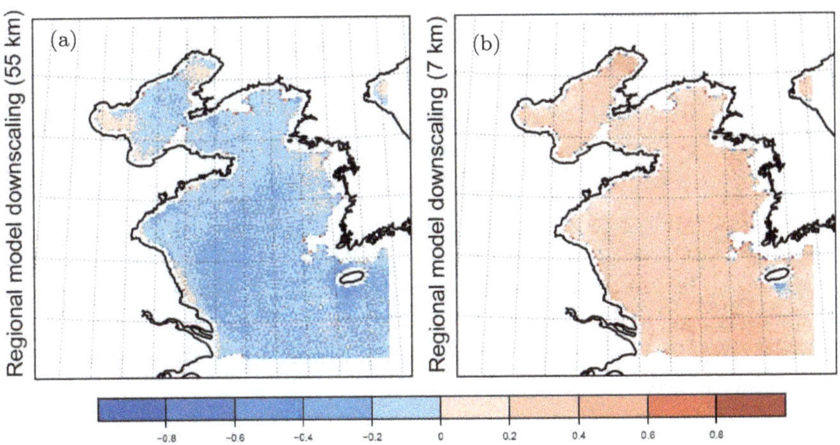

Fig. 7.   Brier Skill Scores (red: improvements; blue: no improvement) of a regional hindcast using: (a) 55 km grid resolution; (b) 7 km grid resolution with the constrained global model over the period of December 1999–November 2009 for marine surface wind speeds of 3–25 m/s relative to observational reference data ("truth") QuikSCAT (von Storch *et al.*, 2017), © American Geophysical Union. Used with permission.

due to boundary conditions, it may be sufficient to run the global simulations for all regions at the same time. In conclusion: running a global model with enforced global (large-scale) circulations allows the simulations of all regional climates at the same time.

6.   Concluding Outlook: Purposes and Challenges

In general, spectral nudging has several advantages as outlined in this chapter. However, it may also gloss over certain model deficiencies; indeed, both failures and successes with spectrally nudged models draw their skill from both the model formulation and the large-scale evidence provided by reanalyses. If one of the two sources of information is failing, it may be possible that the other is strong enough to overcome the insufficiency in the employed model or the employed reanalysis.

Constraining large scales in simulations, but letting small scales develop freely, conditional upon the state of the large scales, may be used for a variety of purposes. The most common application is the construction of regional climatologies for different parts of the world, namely Europe (Geyer, 2014) including the Mediterranean region (Cavicchia *et al.*, 2014a), East Asia and the Northwestern Pacific (Barcikowska *et al.*, 2017; Feser and von Storch, 2008a, 2008b; Li *et al.*, 2016; Platonov *et al.*, 2017), the South Atlantic Tim *et al.* (2015), Central Siberia Klehmet *et al.* (2013); other applications are studying meteorological processes (Kolstad *et al.*, 2016) and regional details in forecasting (Zhao *et al.*, 2016). These climatologies have been used to study changing weather-related phenomena, such as storms, ocean waves, storm surges, atmospheric deposition and transport of chemical elements, marine biota modelling, carbon cycle studies, plant productivity analyses, but also for economic applications such as oil spill simulations or ship routing and design.

The usage of a scale-dependent constraining, with the better observed large scales limiting the space of possible developments of the smaller scales, may also be applied in other dynamical systems, which include a downscaling hierarchy. Consistently, recent approaches for using this method in ocean models are implemented and tested (Katavouta and Thompson, 2013, 2016; Wright *et al.*, 2006).

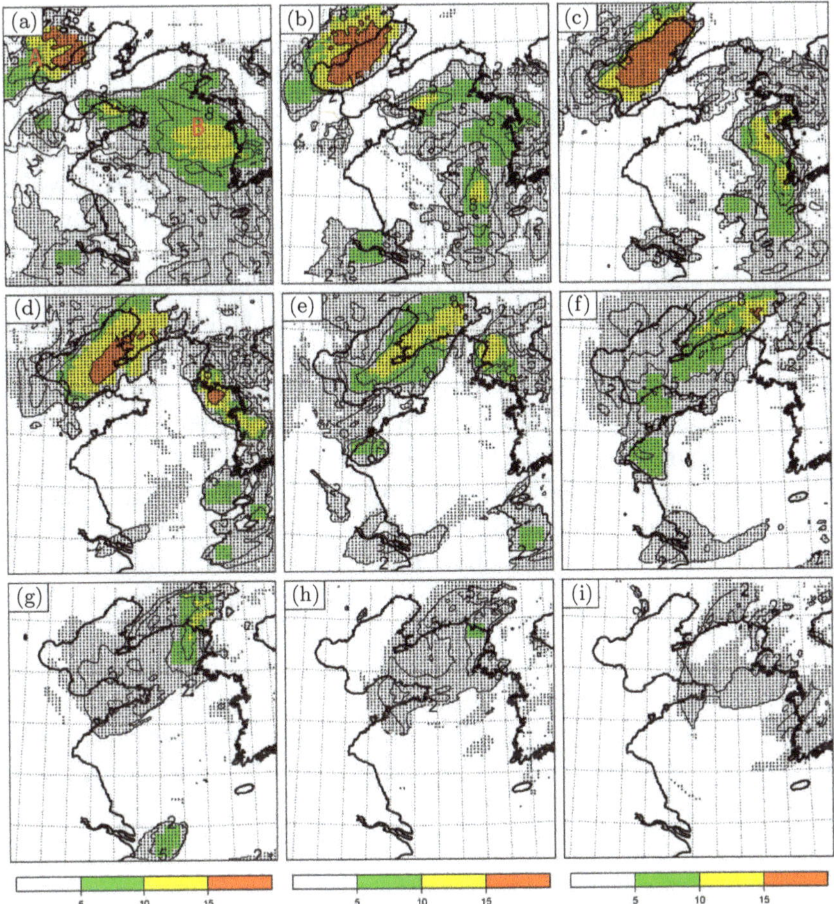

Fig. 8.   (a–i) Two coastal Low Level Jets (LLJ) A and B simulated in a constrained COSMO-CLM simulation with a grid resolution of 7 km, forced by ERA-Interim with three-hourly increments starting from 1 April 2006. The dotted area features maximum wind speed larger than 10 m/s at height below ∼3 km, the coloured area stands for Low Level Jets, with differences between the wind maximum and minimum larger than or the wind speed at ∼3 km greater than 5 m/s. LLJ A: Starting from weak and small LLJ, then growing and becoming strong LLJ, then decaying and disappearing. LLJ B: A well-developed LLJ at the beginning, then disintegrates into LLJ pieces, decaying and disappearing.

Modelling of regional weather streams across several decades permits the construction of climatologies of certain phenomena. We have discussed Polar Lows, Medicanes, and typhoons above, and briefly touched Low Level Jets. Identifying and characterizing the members of such climatological ensembles of phenomena may pose new challenges in describing such phenomena (say, counting, determining scales, intensities, and tracks). This challenge grows when dealing with the output of convection-permitting resolutions, simply because of the more detailed output. There are certainly some aspects which need further consideration — in particular, the question of how well different configurations of the spectral nudging (variables, vertical profile, frequency, for instance) function. In addition, the implementation of a large-scale constraint using orthogonal functions, which may be associated with spatial scales (such as trigonometric or spherical harmonics), requires a uniform gridding; however, the implementation of a large-scale constraint in models with different grid resolutions will pose an extra challenge.

An example are Low Level Jets, which have been studied mostly as a local phenomena, namely connected to certain vertical profiles. After having gridded data, Low Level Jets should be described by their full three- or four-dimensional structures. Figure 8 shows the developments of two Low Level Jets in the Bohai/Yellow Sea region to help illustrate the challenge. The LLJs do not represent a trajectory with a growing, a mature, and a decaying phase, but a disintegrating object. To deal with LLJs as a spatial pattern, pattern recognition methods need to be implemented for defining patterns, sizes, and dynamic properties.

Funding

The work was partly supported through the Cluster of Excellence 'CliSAP' (EXC177), University of Hamburg, funded through the German Research Foundation (DFG). It is a contribution to the Helmholtz Climate Initiative REKLIM (Regional Climate Change), a joint research project of the Helmholtz Association of German Research Centres (HGF). The work was also supported by the National Key Research and Development Program of China (2017YFA0604100) and the National Natural Science Foundation of China (41706019). The research was partially supported through funding from the Earth System and Climate Change Hub of the Australian Government's National Environmental Science Programme (NESP).

Acknowledgements

Thanks to Jens Hesselbjerg-Christensen and Rene Laprise for advice. The German Climate Computing Center (DKRZ) provided the computer hardware for the climate model simulations.

References

Alexandru, A., de Elia, R. and Laprise, R. (2007). Internal variability in regional climate downscaling at the seasonal scale, *Monthly Weather Review* **135**, 9, pp. 3221–3238, doi:10.1175/MWR3456.1.

Alexandru, A., de Elia, R., Laprise, R., Separovic, L. and Biner, S. (2009). Sensitivity study of regional climate model simulations to large-scale nudging parameters, *Monthly Weather Review* **137**, 5, pp. 1666–1686, doi:10.1175/2008MWR2620.1.

Barcikowska, M., Feser, F. and von Storch, H. (2012). Usability of best track data in climate statistics in the Western North Pacific, *Monthly Weather Review* **140**, 9, pp. 2818–2830, doi:10.1175/MWR-D-11-00175.1.

Barcikowska, M., Feser, F., Zhang, W. and Mei, W. (2017). Changes in intense tropical cyclone activity for the Western North Pacific during the last decades derived from a regional climate model simulation, *Climate Dynamics* **49**, 9, pp. 2931–2949, doi:10.1007/s00382-016-3420-0.

Blechschmidt, A.M. (2008). A 2-year climatology of polar low events over the Nordic Seas from satellite remote sensing, *Geophysical Research Letters* **35**, 9, doi:10.1029/2008GL033706.

Cavicchia, L. and von Storch, H. (2012). The simulation of medicanes in a high-resolution regional climate model, *Climate Dynamics* **39**, 9, pp. 2273–2290, doi:10.1007/s00382-011-1220-0.

Cavicchia, L., von Storch, H. and Gualdi, S. (2014a). A long-term climatology of medicanes, *Climate Dynamics* **43**, 5, pp. 1183–1195, doi:10.1007/s00382-013-1893-7.

Cavicchia, L., von Storch, H. and Gualdi, S. (2014b). Mediterranean tropical-like cyclones in present and future climate, *Journal of Climate* **27**, 19, pp. 7493–7501, doi:10.1175/JCLI-D-14-00339.1.

Chen, F. and von Storch, H. (2013). Trends and variability of north pacific polar lows, *Advances in Meteorology* **2013**, pp. 1687–9309, doi:10.1155/2013/170387.

Chen, F., Geyer, B., Zahn, M., and von Storch, H. (2012). Towards a multi-decadal climatology of North Pacific Polar Lows employing dynamical downscaling. *Terrestrial, Atmospheric and Oceanic Sciences* **23**, 291.

Chen, F., von Storch, H., Zeng, L. and Du, Y. (2014). Polar low genesis over the North Pacific under different global warming scenarios, *Climate Dynamics* **43**, 12, pp. 3449–3456, doi:10.1007/s00382-014-2117-5.

Chervin, R., Gates, W. and Schneider, S. (1974). The effect of the time averaging on the noise level of climatological statistics generated by atmospheric general circulation models, *Journal of the Atmospheric Sciences* **31**, pp. 2216–2219.

Davies, H. C. (1976). A lateral boundary formulation for multi-level prediction models, *Quarterly Journal of the Royal Meteorological Society* **102**, 432, pp. 405–418, doi:10.1002/qj.49710243210.

Denis, B., Côté, J. and Laprise, R. (2002). Spectral decomposition of two-dimensional atmospheric fields on limited-area domains using the discrete cosine transform (dct), *Monthly Weather Review* **130**, 7, pp. 1812–1829, doi:10.1175/1520-0493(2002)130<1812:SDOTDA> 2.0.CO;2.

Feser, F. and Barcikowska, M. (2012). The influence of spectral nudging on typhoon formation in regional climate models, *Environmental Research Letters* **7**, 1, p. 014024, doi:10.1088/1748-9326/7/1/014024.

Feser, F. and Barcikowska, M. (2014). Changes in typhoons over the last decades as given in observations and climate model simulations, in *In Natural Disasters — Typhoons and Landslides — Risk Prediction, Crisis Management and Environmental Impacts*, pp. 1–20, Nova Science Publishers, ISBN 978-1-63463-309-3.

Feser, F. and von Storch, H. (2008a). A dynamical downscaling case study for typhoons in Southeast Asia using a regional climate model, *Monthly Weather Review* **136**, 5, pp. 1806–1815, doi:10.1175/ 2007MWR2207.1.

Feser, F. and von Storch, H. (2008b). Regional modelling of the Western Pacific typhoon season 2004, *Meteorologische Zeitschrift* **17**, 4, pp. 519–528, doi:10.1127/0941-2948/2008/0282.

Feser, F., Rockel, B., von Storch, H., Winterfeldt, J. and Zahn, M. (2011). Regional climate models add value to global model data: A review and selected examples, *Bulletin of the American Meteorological Society* **92**, 9, pp. 1181–1192, doi:10.1175/2011BAMS3061.1.

Geyer, B. (2014). High resolution atmospheric reconstruction for Europe 1948–2012: CoastDat2. *Earth System Science Data* **6**, 147–164.

Giorgi, F., Christensen, J., Hulme, M., von Storch, H., Whetton, P., Jones, R., Mearns, L. and Fu, C. (2001). Regional climate information-evaluation and projections, in J.T. Houghton *et al.*, ed., *Climate Change 2001: The Scientific Basis. Contribution of Working Group to the Third Assessment Report of the Intergovernmental Panel on Climate Change*, pp. 583–638, Cambridge University Press.

Ji, Y. and Vernekar, A. D. (1997). Simulation of the Asian summer monsoons of 1987 and 1988 with a regional model nested in a global gcm, *Journal of Climate* **10**, pp. 1965–1979.

Kang, H.-S., Cha, D.-H. and Lee, D.-K. (2005). Evaluation of the mesoscale model/land surface model (mm5/lsm) coupled model for East Asian summer monsoon simulations, *Journal of Geophysical Research: Atmospheres* **110**, D10, doi:10.1029/2004JD005266.

Katavouta, A. and Thompson, K. (2013). Downscaling ocean conditions: Experiments with a quasi-geostrophic model, *Ocean Modelling* **72**, pp. 231–241, doi:10.1016/j.ocemod.2013.10.001.

Katavouta, A. and Thompson, K. R. (2016). Downscaling ocean conditions with application to the gulf of maine, scotian shelf and adjacent deep ocean, *Ocean Modelling* **104**, pp. 54–72, doi:10.1016/j.ocemod.2016.05.007.

Kida, H., Koide, T., Sasaki, H. and Chiba, M. (1991). A new approach for coupling a limited area model to a gcm for regional climate simulations, *Journal of the Meteorological Society of Japan* **69**, pp. 723–728.

Klehmet, K., Geyer, B. and Rockel, B. (2013). A regional climate model hindcast for Siberia: Analysis of snow water equivalent, *The Cryosphere* **7**, 4, pp. 1017–1034, doi:10.5194/tc-7-1017-2013.

Kolstad, E. W., Bracegirdle, T. J. and Zahn, M. (2016). Re-examining the roles of surface heat flux and latent heat release in a "hurricane-like" polar low over the Barents Sea, *Journal of Geophysical Research: Atmospheres* **121**, 13, pp. 7853–7867, doi:10.1002/2015JD024633.

Laprise, R. (2008). Regional climate modelling, *Journal of Computational Physics* **227**, 7, pp. 3641–3666, doi:10.1016/j.jcp.2006.10.024.

Lee, D.-K., Ch, D.-H. and K, H.-S. (2004). Regional climate simulation of the 1998 summer flood over East Asia, *Journal of the Meteorological Society of Japan Series II* **82**, 6, pp. 1735–1753, doi:10.2151/jmsj.82.1735.

Li, D. (2017). Added value of high-resolution regional climate model: Selected cases over the Bohai Sea and the Yellow Sea areas, *International Journal of Climatology* **37**, 1, pp. 169–179, doi:10.1002/joc.4695.

Li, D., von Storch, H. and Geyer, B. (2016). High-resolution wind hindcast over the Bohai Sea and the Yellow Sea in East Asia: Evaluation and wind climatology analysis, *Journal of Geophysical Research: Atmospheres* **121**, 1, pp. 111–129, doi:10.1002/2015JD024177.

Li, D., von Storch, H., Yin, B., Xu, Z., Qi, J., Wei, W. and Guo, D. (2018). Low-level jets over the Bohai Sea and Yellow Sea: Climatology, variability, and the relationship with regional atmospheric circulations,

Journal of Geophysical Research: Atmospheres **123**, 10, pp. 5240–5260, doi:10.1029/2017JD027949.

Liu, P., Tsimpidi, A. P., Hu, Y., Stone, B., Russell, A. G. and Nenes, A. (2012). Differences between downscaling with spectral and grid nudging using wrf, *Atmospheric Chemistry and Physics* **12**, 8, pp. 3601–3610, doi:10.5194/acp-12-3601-2012.

Ma, Y., Yang, Y., Mai, X., Qiu, C., Long, X. and Wang, C. (2016). Comparison of analysis and spectral nudging techniques for dynamical downscaling with the wrf model over China, *Advances in Meteorology* **2016**, p. 4761513, doi:10.1155/2016/4761513.

McGregor, J. L. and Katzfey, K. C., J. J. and Nguyen (1998). Fine resolution simulations of climate change for Southeast Asia: Final report for a research project commissioned by Southeast Asian regional committee for start (sarcs), Tech. rep., CSIRO Atmospheric Research.

Miguez-Macho, G., Stenchikov, G. L. and Robock, A. (2004). Spectral nudging to eliminate the effects of domain position and geometry in regional climate model simulations, *Journal of Geophysical Research: Atmospheres* **109**, D13, doi:10.1029/2003JD004495.

Miguez-Macho, G., Stenchikov, G. L. and Robock, A. (2005). Regional climate simulations over North America: Interaction of local processes with improved large-scale flow, *Journal of Climate* **18**, 8, pp. 1227–1246, doi:https://doi.org/10.1175/JCLI3369.1.

Oliger, J. and Sundström, A. (1978). Theoretical and practical aspects of some initial boundary value problems in fluid dynamics. *SIAM Journal on Applied Mathematics* **35**, pp. 419–446.

Omrani, H., Drobinski, P. and Dubos, T. (2012). Spectral nudging in regional climate modelling: How strongly should we nudge? *Quarterly Journal of the Royal Meteorological Society* **138**, 668, pp. 1808–1813, doi:10.1002/qj.1894.

Park, J. and Hwang, S.-O. (2017). Impacts of spectral nudging on the simulated surface air temperature in summer compared with the selection of shortwave radiation and land surface model physics parameterization in a high-resolution regional atmospheric model, *Journal of Atmospheric and Solar-Terrestrial Physics* **164**, pp. 259–267, doi: 10.1016/j.jastp.2017.09.001.

Platonov, P., Kislov, A., Rivin, G., Varentsov, M., Rozinkina, I., Nikitin, M. and Chumakov, M. (2017). Mesoscale atmospheric modelling technology as a tool for creating a long-term meteorological dataset, in *Proceedings of the International Conference on Computational Information Technologies for Environmental Sciences (CITES-2017)*, pp. 22–30.

Prein, A. F., Langhans, W., Fosser, G., Ferrone, A., Ban, N., Goergen, K., Keller, M., Tölle, M., Gutjahr, O., Feser, F., Brisson, E., Kollet, S., Schmidli, J., van Lipzig, N. P. M. and Leung, R. (2015). A review on regional convection-permitting climate modeling: Demonstrations, prospects, and challenges, *Reviews of Geophysics* **53**, 2, pp. 323–361, doi:10.1002/2014RG000475.

Radu, R., Déqué, M. and Somot, S. (2008). Spectral nudging in a spectral regional climate model, *Tellus A: Dynamic Meteorology and Oceanography* **60**, 5, pp. 898–910, doi:10.1111/j.1600-0870.2008.00341.x.

Ramzan, M., Ham, S., Amjad, M., Chang, E.-C. and Yoshimura, K. (2017). Sensitivity evaluation of spectral nudging schemes in historical dynamical downscaling for South Asia, *Advances in Meteorology* **2017**, p. 7560818, doi:10.1155/2017/7560818.

Rinke, A. and Dethloff, K. (2000). On the sensitivity of a regional arctic climate model to initial and boundary conditions. *Climate Research* **14**, pp. 101–113.

Robinson, A., Lermusiaux, P. and N. Q. Sloan, I. (1998). Data assimilation, in K. Brink and A. Robinson. eds., *The Sea: The Global Coastal Ocean I, Processes and Methods*, Vol. 10, pp. 541–594, John Wiley and Sons.

Rockel, B., Castro, C. L., Pielke Sr., R. A., von Storch, H. and Leoncini, G. (2008). Dynamical downscaling: Assessment of model system dependent retained and added variability for two different regional climate models, *Journal of Geophysical Research: Atmospheres* **113**, D21, doi:10.1029/2007JD009461.

Sasaki, H., Kida, H., Koide, T. and Chiba, M. (1995). The performance of long-term integrations of a limited area model with the spectral boundary coupling method, *Journal of the Meteorological Society of Japan* **73**, pp. 165–181.

Schaaf, B., von Storch, H. and Feser, F. (2017). Does spectral nudging have an effect on dynamical downscaling applied in small regional model domains? *Monthly Weather Review* **145**, 10, pp. 4303–4311, doi:10.1175/MWR-D-17-0087.1.

Schubert-Frisius, M., Feser, F., von Storch, H. and Rast, S. (2017). Optimal spectral nudging for global dynamic downscaling, *Monthly Weather Review* **145**, 3, pp. 909–927, doi:https://doi.org/10.1175/MWR-D-16-0036.1.

Spero, T. L., Nolte, C. G., Mallard, M. S. and Bowden, J. H. (2018). A maieutic exploration of nudging strategies for regional climate applications using the WRF model, *Journal of Applied Meteorology and Climatology* **57**, 8, pp. 1883–1906, doi:https://doi.org/10.1175/JAMC-D-17-0360.1.

Staniforth, A. (1997). Regional modeling: A theoretical discussion, *Meteorology and Atmospheric Physics* **63**, 1, pp. 15–29, doi:10.1007/BF01025361.

Tang, J., Song, S. and Wu, J. (2010). Impacts of the spectral nudging technique on simulation of the East Asian summer monsoon, *Theoretical and Applied Climatology* **101**, 1, pp. 41–51, doi:10.1007/s00704-009-0202-1.

Tim, N., Zorita, E. and Hünicke, B. (2015). Decadal variability and trends of the Benguela upwelling system as simulated in a high-resolution ocean simulation, *Ocean Science* **11**, 3, pp. 483–502, doi:10.5194/os-11-483-2015.

von Storch, H. (1995). Inconsistencies at the interface of climate impact studies and global climate research, *Meteorologische Zeitschrift* **4**, 2, pp. 72–80, doi:10.1127/metz/4/1992/72.

von Storch, H., Langenberg, H. and Feser, F. (2000). A spectral nudging technique for dynamical downscaling purposes, *Monthly Weather Review* **128**, 10, pp. 3664–3673, doi:10.1175/1520-0493(2000)128<3664:ASNTFD>2.0.CO;2.

von Storch, H., Feser, F., Geyer, B., Klehmet, K., Li, D., Rockel, B., Schubert-Frisius, M., Tim, N. and Zorita, E. (2017). Regional reanalysis without local data: Exploiting the downscaling paradigm, *Journal of Geophysical Research: Atmospheres* **122**, 16, pp. 8631–8649, doi:10.1002/2016JD026332.

Waldron, K. M., Paegle, J. and Horel, J. D. (1996). Sensitivity of a spectrally filtered and nudged limited-area model to outer model options, *Monthly Weather Review* **124**, 3, pp. 529–547, doi:10.1175/1520-0493(1996)124<0529:SOASFA>2.0.CO;2.

Weisse, R. and Feser, F. (2003). Evaluation of a method to reduce uncertainty in wind hindcasts performed with regional atmosphere models, *Coastal Engineering* **48**, 4, pp. 211–225, doi:10.1016/S0378-3839(03)00027-9.

Weisse, R., Heyen, H. and von Storch, H. (2000). Sensitivity of a regional atmospheric model to a sea state dependent roughness and the need of ensemble calculations, *Monthly Weather Review* **128**, 10, pp. 3631–3642.

Wright, D. G., Thompson, K. R. and Lu, Y. (2006). Assimilating long-term hydrographic information into an eddy-permitting model of the North Atlantic, *Journal of Geophysical Research: Oceans* **111**, C9, doi:10.1029/2005JC003200.

Yoshimura, K. and Kanamitsu, M. (2008). Dynamical global downscaling of global reanalysis, *Monthly Weather Review* **136**, 8, pp. 2983–2998, doi:10.1175/2008MWR2281.1.

Zahn, M. and von Storch, H. (2008). A long-term climatology of North Atlantic polar lows, *Geophysical Research Letters* **35**, 22, doi:10.1029/2008GL035769.

Zahn, M. and von Storch, H. (2010). Decreased frequency of North Atlantic polar lows associated with future climate warming, *Nature* **467**, 7313, pp. 309–312, doi:10.1038/nature09388.

Zahn, M., von Storch, H. and Bakan, S. (2008). Climate mode simulation of North Atlantic polar lows in a limited area model, *Tellus A* **60**, 4, pp. 620–631, doi:10.1111/j.1600-0870.2008.00330.x.

Zhao, Y., Wang, D., Liang, Z. and Xu, J. (2016). Improving numerical experiments on persistent severe rainfall events in Southern China using spectral nudging and filtering schemes, *Quarterly Journal of the Royal Meteorological Society* **142**, 701, pp. 3115–3127, doi:10.1002/qj.2892.

Chapter 2

Added Value of High-Resolution Regional Climate Model: Selected Cases Over the Bohai Sea and the Yellow Sea Areas[*]

D. Li

*System Analysis and Modelling, Helmholtz-Zentrum Geesthacht
Centre for Materials and Coastal Research,
Geesthacht, Germany*

Added value (AV) from dynamical downscaling has long been a crucial and debatable issue in regional climate studies. To assess AV generated by COSMO Climate Local Model (CCLM), a model-reconstructed hindcast with 7-km grid resolution was compared against the forcing data set ERA-Interim (ERA-I) over the Bohai Sea and the Yellow Sea, with satellite and in situ observation as reference. Both quantitative metrics and qualitative assessments have been used in the investigation of AV by CCLM. Land surface winds, extreme winds, a typhoon, a cold surge and a vortex street have been selected in the assessment process. Statistical analysis on surface winds reveals that high-resolution CCLM hindcast can add value to ERA-I in reproducing wind intensities and direction,

---

[*]This chapter was originally published in *International Journal of Climatology*, 37(1), 169–179, doi:10.1002/joc.4695. Copyright (2017), with permission from John Wiley and Sons.

29

probability distribution and extreme winds mainly at mountain areas, which may be related to the highly resolved and accurate description of terrain roughness length and orographic barriers in CCLM. With respect to atmospheric processes, CCLM outperforms ERA-I in resolving detailed temporal and spatial structures for phenomena of a typhoon and of a cold surge; CCLM generates some orography-related phenomena such as a vortex street, which is not captured by ERA-I. These AVs demonstrate the utility of the 7-km resolution CCLM for regional and local climate studies and applications.

1.   Introduction

Regional climate models (RCMs) have been widely used in the climate research and applications, including but not limited to the process studies of various atmospheric phenomena, weather forecasting, the dynamics of extreme atmospheric features as well as applications such as the adaptation planning, the investigation of possible damages and opportunities related to climate variability and change (Rummukainen, 2009; Tapiador *et al.*, 2007). With a high spatial resolution and better representation of the surface boundary conditions (orography, coastline, vegetation and soil characteristics), RCMs are expected to have advantage in describing regional or local processes in more detail than global atmospheric models.

In recent years, the added value (AV) issue received increased attention and plenty of studies have been dedicated to this issue. Feser *et al.* (2011) reviewed the AV from dynamical downscaling and concluded that RCM may add value to general circulation model (GCM) but improvements essentially depend on the specific application, the experiment configurations, the investigated atmospheric parameters and the area interested; polar lows and tropical cyclones have been shown to be described with AV by RCMs, as well as coastal wind speeds (Feser and von Storch, 2008; Winterfeldt *et al.*, 2011; Zahn and von Storch, 2008). Aspects such as precipitation especially over complex orography are typically found improved, with more realistic fine-scale spatial and temporal features, relative to the forcing GCM (Di Luca *et al.*, 2012; Lee and Hong, 2013; Torma *et al.*, 2015).

Over the past years, the typical grid resolution of RCMs has increased from 100 km to 20–50 km. With the increasing of computing power, more models have been set-up with high grid resolution of 5–10 km, some even down to 1 km (Rockel, 2015; Rummukainen, 2009). To determine the AV of those high-resolution RCM data, observation grid data sets with comparable resolution are generally needed, which are however only available in certain cases (Di Luzio *et al.*, 2008; Rauthe *et al.*, 2013).

When it comes to East Asian areas, there are increasing numbers of studies on atmospheric processes, extreme events, climate variability and change by using high-resolution RCMs (Lee and Hong, 2013; Li *et al.*, 2016a, 2016b; Liu *et al.*, 2010; Yu, 2012). Liu *et al.* (2010) examined the impacts of horizontal resolutions of RCM on resolving extreme rainfall over Yangtze River Basin in the summer of 1998 and revealed that higher resolution simulation outperforms lower resolution simulation in reproducing rainfall intensity and rain-belt distribution. Lee and Hong (2013) evaluated the AV of precipitation and temperature by increasing spatial resolution used in an RCM over Korea and concluded that increased resolution contributes to capturing interannual and daily variabilities, and extreme weather conditions. Li *et al.* (2016b) assessed the AV in sea surface wind speed in a high-resolution (7 km) simulation with the COSMO Climate Local Model (CCLM) in the Bohai Sea (BS) and the Yellow Sea (YS) by comparing with QuikSCAT grid data and site observations and concluded that CCLM outperforms its forcing reanalysis data ERA-Interim (ERA-I, Dee *et al.*, 2011) in reproducing coastal wind speeds.

In this paper, we consider land surface winds and several atmospheric processes generated in the above-mentioned multiyear 7-km grid CCLM simulation (Li *et al.*, 2016b) forced and large-scale constrained by reanalysis data ERA-I. We aim to investigate the AV by CCLM to ERA-I over and around the BS and the YS (BYS, Figure 1).

To assess the AV from dynamical downscaling, the meanings of AV in the paper should be defined firstly. By now, neither general formalisms nor systematic approaches have been universally accepted for defining the meaning of AV in the RCM community. A list of categories is regarded as AV including observational AV, conjectural AV, potential AV and AV from user's perspective (Di Luca *et al.*, 2015). Observational AV is a straightforward way, aiming to assess the improvement in some quantitative metric scores by RCM relative to its forcing data set, with observation as reference. This method is widely accepted and used. However, reliable and consistent observations are not always available for doing so. The conjectural AV and potential AV could be used when observations are unavailable. The former one is either based on theoretical knowledge or on previous studies under similar conditions to infer on the possibility of AV. Clearly, it is more subjective than that of observational AV. For example, fine-scale atmospheric processes such as hail or tornado represented in a physically sophisticated way in convective-permitting simulations can be regarded as conjectural AV. Potential AV is defined as the presence of fine-scale information in a given climate statistics or atmospheric process by RCMs. The presence of potential AV is not a sufficient condition but a prerequisite for AV (Di Luca *et al.*, 2012). For example, a climate statistic such as temporal variability or 99th percentile of precipitation in complex orographic areas can greatly attribute from fine scales by a RCM, which is referred as potential RCMs, while the actual AV may be null or negative when the RCM simulates the climate statistic very poorly. These two AVs are applied to characterize the possibility of AV rather than the actual AV and can be used to assess the AV of RCMs in future climate simulations.

In applications of RCM assessment (e.g. Lucas-Picher *et al.* 2012), the authors generally do not limit the concept of AV to one of these categories. For a comprehensive assessment of AV, the combination of observational AV and potential AV is used in this paper, with a focus on surface wind fields and several atmospheric processes in BYS areas.

2. Data and Method

The model data are derived from a hindcast over the entire BYS during 1979–2012 with 0.0625 (about 7 km) grid resolution (Li *et al.*, 2016b) produced by the non-hydrostatic RCM CCLM (Rockel *et al.*, 2008; http://www.clm-community.eu/). The temporal interval of model output is hourly for surface variables such as wind speed, 2-m temperature, surface pressure and total precipitation.

The model was forced with the reanalysis data ERA-I (6-hourly frequency and approximately 80 km spatial resolution) at the lateral and lower boundaries, and in the interior of the large-scale wind field above 850 hPa height, using the spectral nudging technique (von Storch *et al.*, 2000). This large-scale constraining was implemented to keep the model solution close to the forcing at large scales while permitting the development of the local-scale processes on its own.

The forcing data set ERA-I and observational data have also been used in the evaluation process. For doing so, ERA-I data have been spatially and temporally interpolated to the CCLM grid.

A 3-hourly observation wind data set comprising records of 67 land stations (Figure 1) during 1979–2010 were obtained from HadISD — global sub-daily station data set with quality control (Dunn *et al.*, 2012; http://www.metoffice.gov.uk/hadobs/hadisd/).

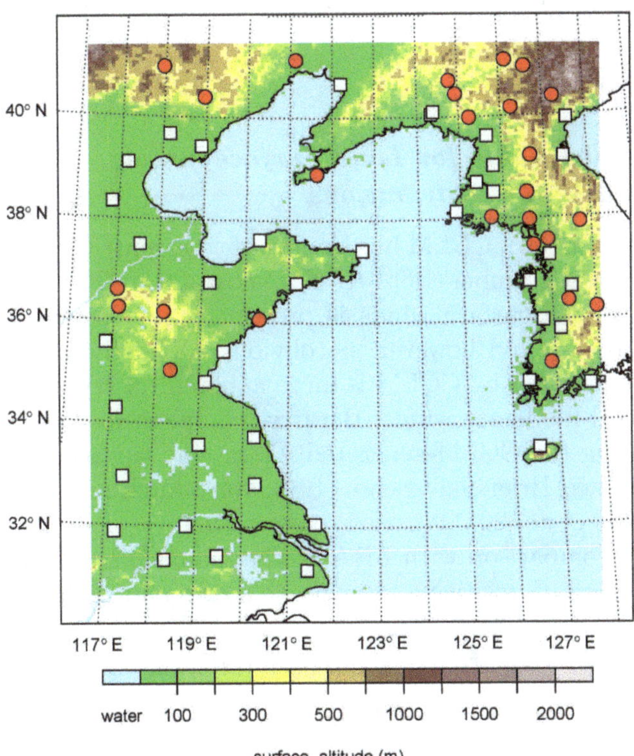

Fig. 1.   Orography of the simulation domain: Circles indicate stations with height larger than 70 m; squares represent stations with height less than 70 m. The white frame represents the sponge zone.

Wind speed and direction are included in the data set; U and V wind components were calculated accordingly. Based on the orography type and height, we found that the stations can be divided into two groups: stations in the mountain areas with height larger than 70 m and stations in the plain areas with height less than 70 m. When compared with the land observation, both CCLM and ERA-I data were interpolated to the stations' locations using the nearest-neighbour method and were masked temporally by land observation data.

In the assessment of AV by CCLM in reproducing atmospheric phenomena, observational data such as tropical cyclone best track data (BTD) from Japan Meteorological Agency (JMA) and Moderate Resolution Imaging Spectroradiometer (MODIS) were used in perspective of quantitative and qualitative assessments, respectively.

3. Winds

3.1. *Added Value for Land Surface Winds and Wind Distributions*

In Li *et al.* (2016b), CCLM has been verified against QuikSCAT wind data during December 1999 to November 2009. The results show that AV of sea surface winds is present in the coastal areas with complex orography; however, no obvious AV in the offshore areas. The AV generated by CCLM is more apparent in strong winds than in light and moderate winds. Here, we complement this analysis by investigating the CCLM-generated AV in land surface winds.

A modified Brier Skill Score (BSS, von Storch and Zwiers, 1999; Winterfeldt *et al.*, 2011) is used to assess the AV of surface winds by CCLM in comparison with ERA-I. The modified BSS simplifies the comparability of positive and negative scores and is shown as

$$\text{BSS} = \begin{cases} 1 - \sigma^2(x_C, x_O)/\sigma^2(x_E, x_O), & \text{if } \sigma^2(x_C, x_O) \leq \sigma^2(x_E, x_O) \\ \sigma^2(x_E, x_O)/\sigma^2(x_C, x_O) - 1, & \text{if } \sigma^2(x_C, x_O) > \sigma^2(x_E, x_O) \end{cases}$$

$$(1)$$

where $\sigma^2(x_C, x_O) = \epsilon((x_C - x_O)^2)$ and $\sigma^2(x_E, x_O) = \epsilon((x_E - x_O)^2)$ indicate the error variances of CCLM wind speed (x_C) and ERA-I wind speed (x_E), respectively, x_O refers to observed wind speed.

Based on the definition, BSS varies between −1 and 1, while negative BSS implying ERA-I being more in agreement with observations, and positive BSS indicates CCLM has generated AV over the ERA-I forcing data set.

Figure 2 shows the BSS distributions at land stations for wind speed (WS), and U and V components. As we can see, positive BSS of WS indicating AV emerge in complex orography areas of the Korean Peninsula, while negative values in Chinese plain areas. When it comes to BSS of U and V components, we see similar distribution patterns over the Korean Peninsula; however, there are more stations with positive BSS in Chinese areas than those of WS. To classify the relationship between AV and orography conditions and the variables considered, the summaries of BSS values for WS, U and V components are given in Table 1 for stations in mountain (>70 m)

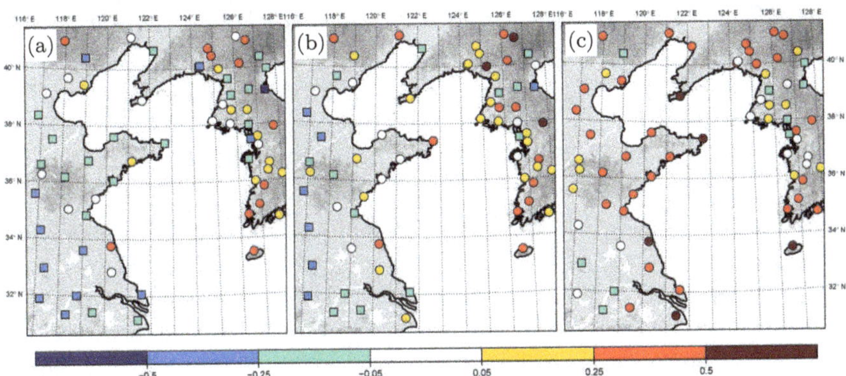

Fig. 2. Brier Skill Scores for (a) wind speed, (b) U component and (c) V component. The grey background indicates orography condition. Locations, where positive AV is detected, are marked as filled circles. BSS < −0.05 are drawn as filled squares to enable differentiation from BSS > −0.05.

Table 1. Counts of positive and negative BSS for wind speed (WS), U and V components over all land stations (all), mountain stations (>70 m) and plain stations (<70 m).

BSS	WS		U		V	
	>0.05	<−0.05	>0.05	<−0.05	>0.05	<−0.05
>70 m (26)	12	8	18	5	21	3
<70 m (41)	10	24	19	15	26	6
All (67)	22	32	37	20	47	9

and plain areas (<70 m). The percentages of positive BSS at stations in mountain areas are much larger than those in plain areas. More stations have positive BSS for U and V than in WS both in mountain and plain areas. Apparently, the AV for winds is highly related to orography features, with more improvement in mountain areas than in plain areas. Additionally, CCLM tends to generate more improvements in wind vectors than in wind intensities.

Another metric for assessing AV in land winds is the two-sample Kolmogorov–Smirnov (KS) distance D(F,G), which measures the maximum distance between two cumulative distribution functions (CDFs), in this case, of modelled and observed wind speeds (cf., Torma *et al.*, 2015):

$$D(F, G) = \sup|F(x) - G(x)| \tag{2}$$

where $F(x)$ and $G(x)$ are the two CDFs and *sup* indicates the supremum function (Chakravarti *et al.*, 1967). $D(F, G)$ varies between 0 and 1, of which $D(F, G) = 0$ indicates identity of the distributions between $F(x)$ and $G(x)$, and $D(F, G) = 1$ represents no overlap.

The distribution of K–S distances between CDFs of modelled and observed wind speed for all land stations (Figure 3(a)) does not reveal much difference between ERA-I and CCLM. Mostly D(F,G)$<$0.4, indicating the distributions of ERA-I and CCLM wind speed have some agreement with those derived from the observation. Figure 3(b) and (c) represents the distributions of K–S distances for stations in mountain and plain areas separately. At mountain stations, CCLM has more stations with small K–S distance than ERA-I, but at plain-area stations CCLM fares worse than ERA-I, which implies that the AV in wind distributions tends to be generated by CCLM in mountain areas rather than over plain areas.

3.2.   *Added Value for Extreme Winds*

Besides the AV shown above, we further investigate the improvement in annual extreme wind speed (AEWS, annual-99th percentile winds) by CCLM. Figure 4 shows the climatology mean of AEWS of ERA-I, CCLM and land observations (points) during 1979–2010 over the BYS areas. Some spatial distribution features in ERA-I and in the CCLM simulation are similar: the AEWS over ocean areas generally decreases from offshore to coastal areas and is much stronger

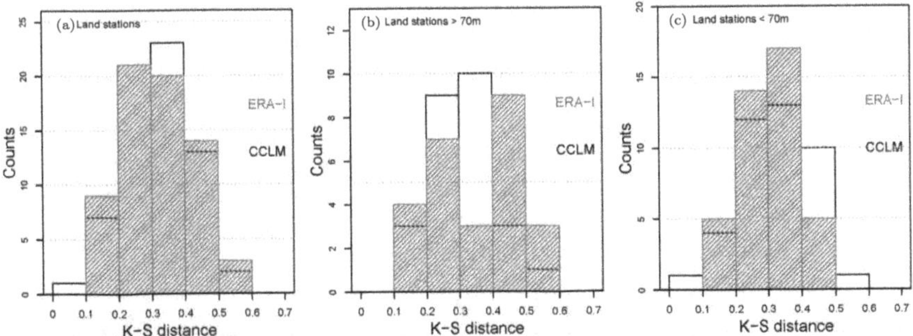

Fig. 3. Distribution of Kolmogorov–Smirnov (KS) distances for the driving ERA-I (gray) and CCLM (black) wind speed CDFs versus the observed CDFs: (a) all land stations, (b) stations with height larger than 70 m and (c) stations with height less than 70 m.

than over land areas, of which plain land areas are characterized with stronger AEWS than most mountain areas. We see more fine-scale features and spatial variability resolved by CCLM over land areas. Significant variations can be detected over distances of tens of kilometres for CCLM winds.

The reliability of resolved fine-scale structures by CCLM can be verified by comparing against observations, as at many stations both over mountain and plain areas AEWS of CCLM tend to be closer to AEWS of observations than ERA-I. The difference map shown in Figure 4(c) reveals that AEWS of CCLM is larger than the one of ERA-I mainly over the BS, the coastal areas of YS and plain land areas, while AEWS of ERA-I is larger over mountain areas such as Korean Peninsula, the southwest and northwest part of the model domain and part of Shandong Peninsula. Over most water areas of YS, there is no much difference between two. Based on the distribution pattern, we can see that the AEWS difference in Figure 4(c) is negatively related with the differences of surface roughness length in CCLM and in ERA-I, both over land and coastal areas (Figure 4(d)), except in part of northern Korean Peninsula and Liaodong Peninsula. The present result is consistent with the illustration by Kunz *et al.* (2010) that the near-surface wind filed is mainly modified by the terrain's roughness length by changing the vertical transfer of horizontal momentum (Wiernga, 1993).

The presence of orographic barriers in high-resolution CCLM is another factor that contributes to the changing of AEWS

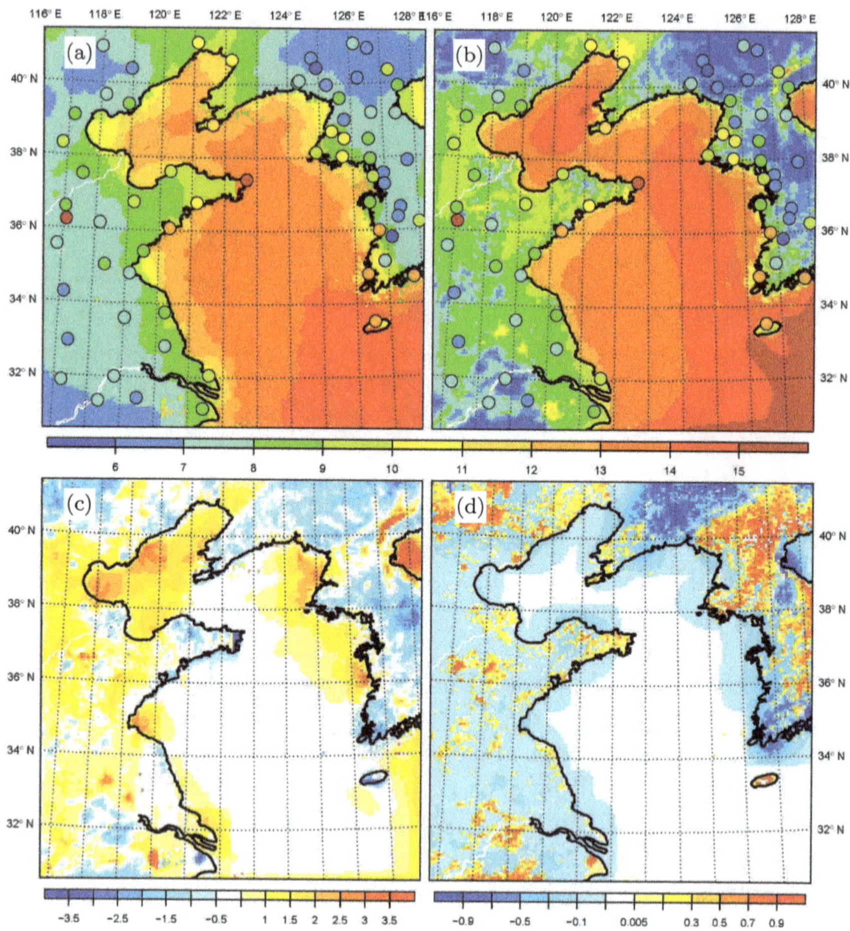

Fig. 4.   Climatology mean of annual extreme wind speed (ms^{-1}) during 1979–2010 for (a) ERA-I and (b) CCLM, points indicate observation results; (c) difference (ms^{-1}) between (b) and (a); (d) the surface roughness length difference (m) between CCLM and ERA-I.

by deflecting and blocking airflows (Barry, 2008; Kunz *et al.*, 2010), which has been verified by investigating the climatology difference between CCLM and ERA-I in wind vectors at different pressure levels (surface, 925 and 850 hPa, figures not shown here). Our results reveal larger difference in the mountain areas than in plain areas, and the modification on wind vectors decreases with increasing height. This also implies that the potential AV by CCLM highly depends on the lower boundary conditions and tends to be generated in the surface field rather than in the upper levels. For the coastal areas, one example of orographic barriers effect is the generation of strong coastal gap winds, which are found being characterized with finer-scale information and stronger wind speed in CCLM than in ERA-I based on case studies.

3.3. *Scale-Dependent Added Value*

RCMs are expected to add value at the scales that they are constructed for (Feser *et al.*, 2011). Therefore, we used the spatial digital filter (Feser and von Storch, 2005) to measure scale-dependent AV. Ideally, the scale-dependent AV should be assessed using observed grid data as reference, which however is unavailable for the study domain. Therefore, the definition of potential AV is used here to indicate a possibility of actual AV. The low-pass filter only left the large-scale wind features larger than 300 km (which are supposedly well resolved by ERA-I); while the meso-scale features between 65 and 150 km are retained during the band-pass filter process. The surface wind speeds of ERA-I and CCLM during the same period have been filtered to large-scale and meso-scale using these filters. We use the relative change of standard deviation (RCSD) to reveal the change of scale-dependent variability: $RCSD = SD_{CCLM}/SD_{ERA-I} - 1$. If RCSD is zero, the variability is unchanged, if it is positive, CCLM has added variability, when it is negative ERA-I varies stronger than CCLM.

In most water areas, CCLM adds little wind variability in large scales which are well-resolved by ERA-I, but it tends to underestimate large-scale wind variability over some complex land or island areas and overestimate the one in some plain or coastal areas (Figure 5(a)). Obviously, the high-resolution orography has an effect on the surface large-scale flow in RCM as indicated in Becker *et al.* (2015).

CCLM adds large variability to ERA-I in meso-scale features (Figure 5(b)). Over the water areas, RCSD ranges from 0.1 to 0.5 in the offshore areas and in coastal and island areas it is generally larger than 0.5. Over complex mountain areas, the variability is much stronger with RCSD up to more than 2.0. We see that the modification of variability by the RCM is mainly in medium scales rather than on large-scale features especially over complex orographic areas,

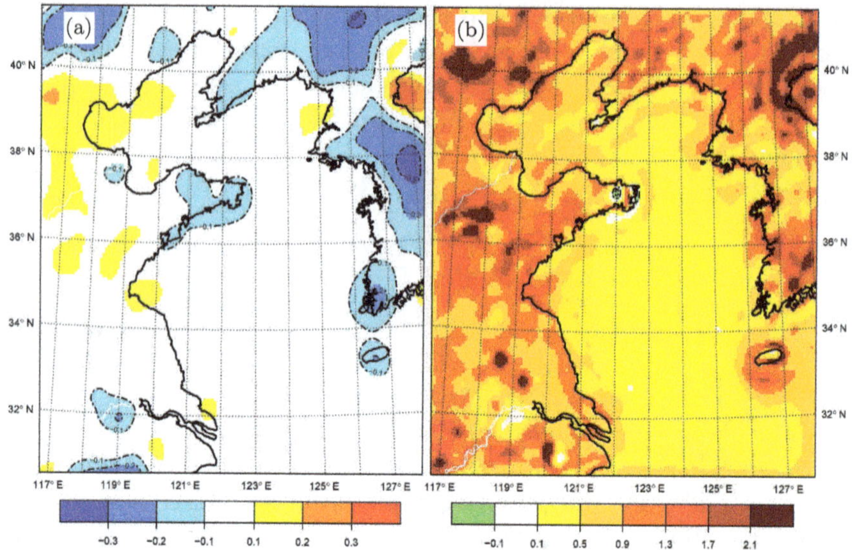

Fig. 5. Relative change of standard deviation of wind speeds by CCLM to ERA-I for (a) large scale and (b) medium scale. Negative values (<-0.1) are also drawn as dot-dashed contour lines to enable differentiation from positive values (>-0.1).

which is generally consistent with the distribution areas of positive AV of surface winds (Section 3.1 and Li *et al.*, 2016b), implying that the AV of surface winds is most likely from medium-scale features captured by CCLM. However, we should be aware that the added variability in medium scale by CCLM, that is, an indicator of potential AV, is only a prerequisite but not a sufficient condition for actual AV. The actual AV may be null or negative even that the potential AV is very large.

Above all, we see that CCLM can add value to ERA-I in land surface winds including U and V wind components, wind intensities, cumulative distributions and extreme winds, especially over mountain areas. The AV of AEWS by CCLM is strongly related to the well-resolved boundary physiography parameters by CCLM such as roughness length and orography detail. The high possibility of real AV in wind variability is mainly in medium scales rather than in large scales especially in complex orography areas, where the orographic-related meso-scale phenomena usually occur.

3.4. *Added Value in Atmospheric Phenomena*

In the previous section, statistical quantitative methods were used to assess the AV in surface winds by CCLM. However, this approach has its limits. As demonstrated in Di Luca *et al.* (2015), a statistical score may be improved by a RCM relative to its forcing data set, while at the same time some underlying physical processes may be deteriorated (and vice versa). Therefore, in this section, some qualitative assessments will be made by investigating the description of the complexity of several atmospheric phenomena simulated by CCLM. This will shed some light on better resolved dynamical processes, which may contribute to the AV described by the statistical scores given in the previous section and in Li *et al.* (2016b).

3.5. *Typhoon*

CCLM has shown a reliable ability in reproducing tropical cyclones in the northwestern Pacific Ocean (when run with 50 or 18 km grid resolution; Feser and von Storch, 2008). The BYS is highly affected by typhoons, even if not as strongly and frequently as, say the South China Sea. Here, we investigate the improvement by CCLM hindcast of describing a typhoon, which passed through the BYS.

In Figure 6, the mean sea level pressure (MSLP) and 10 m wind speed from ERA-I and CCLM are shown at 00:00 UTC 8 August 2011, when the typhoon MUIFA passed through the BYS. The general patterns of typhoon MUIFA are captured by both ERA-I and CCLM, while there are differences in fine-scale structures. Typhoon MUIFA, as described by CCLM, is stronger, with larger wind speed and lower MSLP and spatially more variable than that described by ERA-I. In CCLM fields, some sea-level pressure isolines and strong wind speed zones are distorted in the mountain areas of Korean Peninsula, which is related to the more complex orography employed by CCLM. Furthermore, the tracking of typhoon MUIFA has been conducted in both the ERA-I and CCLM data during 12:00 UTC 6 August 2011 to 12:00 UTC 8 August 2011 based on a tracking algorithm developed by Feser and von Storch (2008). The tracks in CCLM are generally consistent with those in ERA-I (Figure 6) and are similar to the BTD analyses of JMA. In terms of core pressure

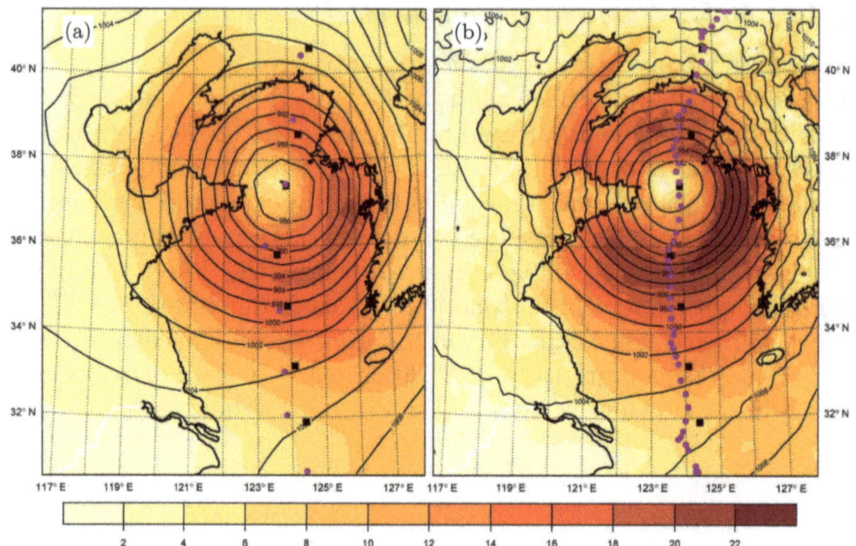

Fig. 6.　Mean sea-level pressure (isolines; hPa) and 10 m wind speed (shaded; ms^{-1}) at 00:00 UTC 8 August 2011 for ERA-I (a) and CCLM (b); the black squares represent locations of typhoon core of Best Track Data, while the dark-magenta dots for those of ERA-I and CCLM.

and maximum wind speed, the output of CCLM is closer to the BTD than those of ERA-I (not shown here).

Another case of typhoon in July 2001 named Kai-Tak revealed a similar improvement by CCLM. We conclude that high-resolution CCLM has the ability to improve the description of typhoons relative to the reanalysis data ERA-I over the BYS, both in terms of spatial details and temporal development.

3.6.　*Cold Surge*

Cold surge is a hazardous weather phenomenon in East Asia and is generally characterized by strong northerly winds, a sharp temperature drop and a steep rise of surface pressure (Chen *et al.*, 2002). A very cold Siberian air outbreak occurred during 24–26 January 2000 (Dorman *et al.*, 2004). The cold air flowed southward and southeastward (wind vectors in Figures 7(b) and (c)) and highly affected the BYS. The temperature dropped sharply about 15 and wind speed rose quickly from 22 to 24 January 2000 at station Dalian, and then

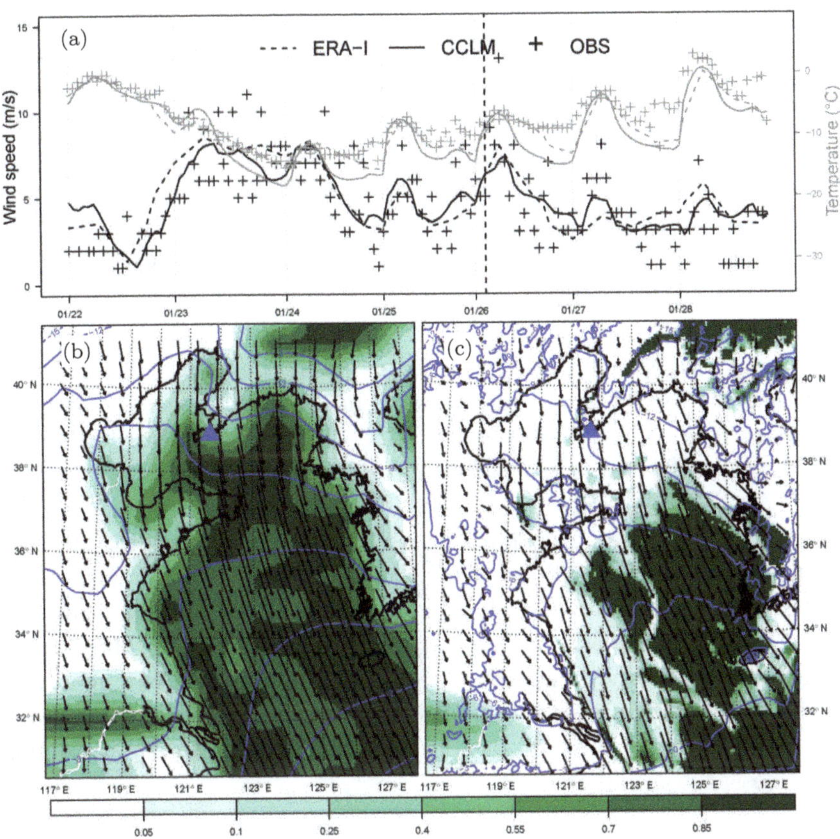

Fig. 7. (a) Time series of air temperature (gray) and surface wind speed (black) during 00:00 UTC 22–00:00 UTC 28 January 2000 at station Dalian (121.633°E 38.9°N) for ERA-I (dashed line), CCLM (solid line) and observation (cross). The vertical black dashed line represents the time at 03:00 UTC 26 January 2000, when the distributions of total cloud cover (shaded), wind vectors and 2 m air temperature (contours, unit is °C) are shown for (b) ERA-I and (c) CCLM.

the cold surge weakened slowly and expired around 28 January 2000 (Figure 7(a)). The daily and multi-day variability of observed temperature and wind speed can be captured by both ERA-I and CCLM. Based on statistical metrics (not shown), the wind speeds during cold surge can be better reproduced by CCLM than by ERA-I at station Dalian, while it is not the case for surface temperature: the temperature of CCLM is much colder than ERA-I and observation from late afternoon to early morning the next day (UTC+8).

A geostationary satellite image of the YS for 02:32 UTC 26 January 2000 (Figure 12(b) in Dorman *et al.* (2004)) showed sheet clouds filling over the BYS, which is caused by roll vortices and generally forms when very cold continental air flows across the sea in the marine boundary layer (Dorman *et al.*, 2004; Houze, 2014). Clear zones are located over the BS and coastal areas of the northern YS. These patterns of clouds are well captured by CCLM but less so by ERA-I (Figure 7(b) and (c)). However, the structure details of sheet clouds cannot be resolved by CCLM.

When it comes to wind vectors, CCLM differs little from ERA-I over ocean areas, while there are some distortions or reduction over land especially for mountain areas. The air temperature is largely modified in the CCLM simulation. ERA-I has regular mostly uniform temperature distribution with increasing temperature southward, while CCLM tends to have more fine-scale structure and spatial variability specially over land areas with complex orography. The temperature isotherm lines of CCLM over water areas have shifted southward relative to ERA-I, which indicates that surface temperature of CCLM is relatively colder than the one of ERA-I. The aerosol treatment, some non-radiative processes including latent heat and surface sensible fluxes or dynamical processes in the atmosphere, may contribute to the cold bias in CCLM. Further investigation may be necessary on this issue.

3.7. *Atmospheric Vortex Street*

Figure 8 shows a series of swirl-pattern clouds to the south of Jeju Island at 02:45 UTC 22 October 2004 obtained from the MODIS image. It is known as atmospheric vortex street (AVT), which generally forms when strong northerly air streams flow across the island in cold seasons, and can extend a distance of 800-km south of Jeju Island (Chung and Kim, 2008).

The wind speed fields of CCLM and ERA-I at 03:00 UTC 22 October 2004 (Figure 9(d) and (e)) show that no AVT is present in ERA-I, while an strong wake with relative low wind speeds is detected in the downwind path of the Jeju Island in the CCLM simulation, even though details of the vortex structure are not captured.

Fig. 8. MODIS image at 02:45 UTC 22 October 2004 over East China Sea. Atmospheric vortex street is shown within the rectangular area.

Figure 9(a)–(d) reveals the temporal evolution of CCLM wind speed between 09:00 UTC 21 and 03:00 UTC 22 October 2004. Figure 9(a) shows an oscillatory like wave downwind of the island and obvious weak wind region; at 15:00 UTC, some vortices break away from the island and flow southeastward, and the vortex street turns to be notable; at 21:00 UTC, the vortices flow southward with the changing of wind direction, two weak wind regions as vortices are located more than 200 km southeastward of Jeju Island. After 03:00 UTC, the vortex street weakens. The whole processes including generation, evolution and dissipation last for almost 2 days and are described by CCLM. However, the details of its fine structure cannot be reproduced by the present CCLM with 7 km resolution; a higher resolution modelling may provide further details of the AVT.

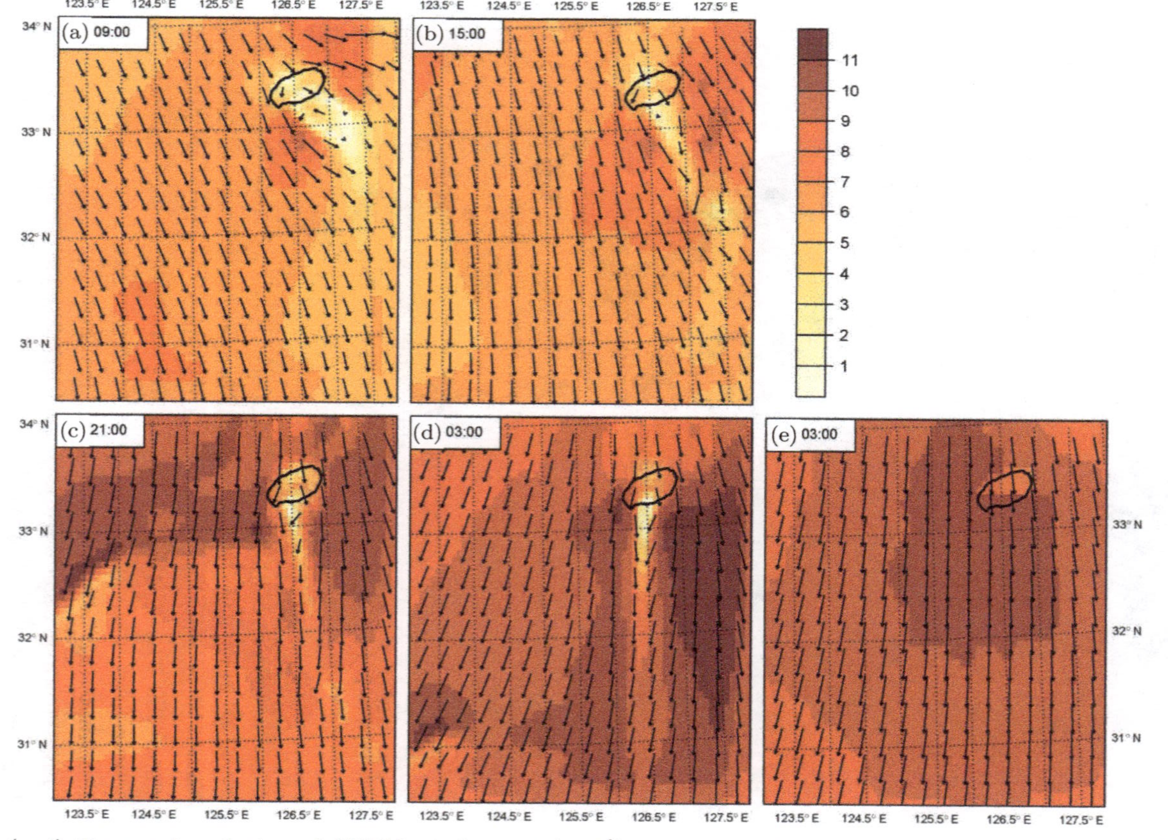

Fig. 9. (a–d) Temporal evolution of CCLM wind speed (ms^{-1}) and vectors between 09:00 UTC 21 and 03:00 UTC 22 October 2004, (e) ERA-I wind speed (ms^{-1}) and vectors at 03:00 UTC 22 October 2004 around Jeju Island.

4. Concluding Remarks

Based on a long-term (1979–2012) atmospheric hindcast over the BS and the YS with 7-km grid resolution, we assessed the AV from dynamical downscaling by analysing surface winds, typhoon, cold surge, AVT, with reanalysis ERA-I, satellite and in situ observation as reference. Both quantitative and qualitative approaches are used.

The results demonstrate the robust ability of CCLM in reproducing land surface winds and some meso-scale phenomena and reveal some AV by dynamical downscaling. The AV of land surface winds by CCLM is assessed using statistical quantitative approach by comparing with observation data and we found AV by CCLM is mostly over mountain areas in terms of wind intensities, wind vectors, wind distributions as well as extreme wind events; while over plain areas, AV in wind vectors and extreme winds can be generated as well. The difference in extreme winds between ERA-I and CCLM is highly affected by the terrain's roughness length and orographic barriers effects, which are also applicable to difference of mean wind speeds between ERA-I and CCLM (not shown).

The AV in simulating atmospheric processes is assessed mainly using a qualitative approach. Phenomena such as typhoons and cold surges can be better simulated with better resolved fine-scale features on spatial and temporal scales, while some phenomena can be captured by a high-resolution RCM when no obvious signal exists in reanalysis data set such as AVTs. Except those mentioned phenomena in the present study, some other phenomena such as atmospheric fronts and coastal mountain gap winds were also added with detailed spatial and temporal structures by CCLM relative to ERA-I.

With the adoption of high-resolution RCM, the local physiographic conditions can be better captured, which is a pre-condition for AV by dynamical downscaling especially over complex orography areas. Whether there is AV or not also depends on the model setups, sub-grid scale parametrization, forcing data sets, variables or processes investigated.

Most papers mainly focus on AV in terms of precipitation or of temperature. Winds are rarely investigated. The present study demonstrates the AV by CCLM in reproducing surface winds and several atmospheric processes, which supplies supports for further investigation in climatology and variability features of winds and particular atmospheric process.

However, we need to notice the fact that the definition of potential AV is used when observation data are not available. We should keep in mind that the potential AV such as added variability in mesoscale wind speed field in Section 3.3 only indicates a possibility not a determination of real AV. Additionally, the simulation was constrained with adoption of spectral nudging method. Similarly, the results will be different when simulations are considered, which are not constrained by spectral nudging. Furthermore, finer-scale processes such as the detailed structure of cloud cover and of vortex streets cannot be resolved by present 7-km grid resolution. Similarly, an even finer resolution of 1–2 km grids may perform markedly better in this respect and is indispensable in the future.

Funding

The work has been funded by China Scholarship Council (CSC201206330070) and the REKLIM project (LK14401KSG0101).

Acknowledgements

The author would like to thank Hans von Storch, Beate Geyer, Benjamin Schaaf and Markus Schultze for valuable comments and discussions on this work. The following institutions are acknowledged for providing the data: the reanalysis data set ERA-I from the European Centre for Medium-Range Weather Forecasts (ECMWF); MODIS image obtained from the NASA/GSFC MODIS Service website; the observation data at Dalian station from the National Climate Data Center (NCDC).

References

Barry, R. G. (2008). *Mountain Weather and Climate*, 3rd edn. Cambridge, UK: Cambridge University Press.

Becker, N., Ulbrich, U. and Klein, R. (2015). Systematic large-scale secondary circulations in a regional climate model, *Geophysical Research Letters* **42**, 10, pp. 4142–4149, doi:10.1002/2015GL063955.

Chakravarti, I. M., Laha, R. G., and Roy, J. (1967). *Handbook of Methods of Applied Statistics*, Volume I, pp. 392–394, John Wiley: Hoboken, NJ.

Chen, T. C., Yen, M. C., Huang, W. R. and Gallus, W. A. (2002). An East Asian cold surge: Case study, *Monthly Weather Review* **130**, 9, pp. 2271–2290, doi:10.1175/1520-0493(2002)130<2271:AEACSC>2.0.CO;2.

Chung, Y. S. and Kim, H. S. (2008). Mountain-generated vortex streets over the korea south sea, *International Journal of Remote Sensing* **29**, 3, pp. 867–877, doi:10.1080/01431160701281080.

Dee, D. P., Uppala, S. M., Simmons, A. J., Berrisford, P., Poli, P., Kobayashi, S., Andrae, U., Balmaseda, M. A., Balsamo, G., Bauer, P., Bechtold, P., Beljaars, A. C. M., van de Berg, L., Bidlot, J., Bormann, N., Delsol, C., Dragani, R., Fuentes, M., Geer, A. J., Haimberger, L., Healy, S. B., Hersbach, H., Holm, E. V., Isaksen, L., Kallberg, P., Koehler, M., Matricardi, M., McNally, A. P., Monge-Sanz, B. M., Morcrette, J. J., Park, B. K., Peubey, C., de Rosnay, P., Tavolato, C., Thepaut, J. N. and Vitart, F. (2011). The ERA-interim reanalysis: Configuration and performance of the data assimilation system, *Quarterly Journal of the Royal Meteorological Society* **137**, 656, pp. 553–597, doi:10.1002/qj.828.

Di Luca, A., de Elia, R. and Laprise, R. (2012). Potential for added value in precipitation simulated by high-resolution nested Regional Climate Models and observations, *Climate Dynamics* **38**, 5–6, pp. 1229–1247, doi:10.1007/s00382-011-1068-3.

Di Luca, A., de Elia, R. and Laprise, R. (2015). Challenges in the quest for added value of regional climate dynamical downscaling, *Current Climate Change Reports* **1**, 1, pp. 10–21, doi:10.1007/s40641-015-0003-9.

Di Luzio, M., Johnson, G. L., Daly, C., Eischeid, J. K. and Arnold, J. G. (2008). Constructing retrospective gridded daily precipitation and temperature datasets for the conterminous United States, *Journal of Applied Meteorology and Climatology* **47**, 2, pp. 475–497, doi: 10.1175/2007JAMC1356.1.

Dorman, C. E., Beardsley, R. C., Dashko, N. A., Friehe, C. A., Kheilf, D., Cho, K., Limeburner, R. and Varlamov, S. M. (2004). Winter marine atmospheric conditions over the Japan Sea, *Journal of Geophysical Research-Oceans* **109**, C12, p. C12011, doi:10.1029/2001JC001197.

Dunn, R. J. H., Willett, K. M., Thorne, P. W., Woolley, E. V., Durre, I., Dai, A., Parker, D. E. and Vose, R. S. (2012). HadISD: A quality-controlled global synoptic report database for selected variables at long-term stations from 1973–2011, *Climate of the Past* **8**, 5, pp. 1649–1679, doi:10.5194/cp-8-1649-2012.

Feser, F. and von Storch, H. (2005). A spatial two-dimensional discrete filter for limited-area-model evaluation purposes, *Monthly Weather Review* **133**, 6, pp. 1774–1786, doi:10.1175/MWR2939.1.

Feser, F. and von Storch, H. (2008). Regional modelling of the Western Pacific typhoon season 2004, *Meteorologische Zeitschrift* **17**, 4, pp. 519–528, doi:10.1127/0941-2948/2008/0282.

Feser, F., Rockel, B., von Storch, H., Winterfeldt, J. and Zahn, M. (2011). Regional climate models add value to global model data: A review and selected examples, *Bulletin of the American Meteorological Society* **92**, 9, pp. 1181–1192, doi:10.1175/2011BAMS3061.1.

Houze, R. J. (2014). *Cloud Dynamics*, Academic Press.

Kunz, M., Mohr, S., Rauthe, M., Lux, R. and Kottmeier, C. (2010). Assessment of extreme wind speeds from Regional Climate Models — Part 1: Estimation of return values and their evaluation, *Natural Hazards and Earth System Sciences* **10**, 4, pp. 907–922, doi:10.5194/nhess-10-907-2010.

Lee, J.-W. and Hong, S.-Y. (2013). Potential for added value to downscaled climate extremes over Korea by increased resolution of a regional climate model, *Theoretical and Applied Climatology* **117**, 3, pp. 667–677, doi:10.1007/s00704-013-1034-6.

Li, D., von Storch, H. and Geyer, B. (2016a). Testing reanalyses in constraining dynamical downscaling, *Journal of the Meteorological Society of Japan* **94A**, pp. 47–68, doi:10.2151/jmsj.2015-044.

Li, D., von Storch, H. and Geyer, B. (2016b). High-resolution wind hindcast over the Bohai Sea and the Yellow Sea in East Asia: Evaluation and wind climatology analysis, *Journal of Geophysical Research: Atmospheres* **121**, 1, pp. 111–129, doi:10.1002/2015JD024177.

Liu, H., Zhang, D.-L. and Wang, B. (2010). Impact of horizontal resolution on the regional climate simulations of the summer 1998 extreme rainfall along the Yangtze River Basin, *Journal of Geophysical Research-Atmospheres* **115**, p. D12115, doi:10.1029/2009JD012746.

Lucas-Picher, P., Wulff-Nielsen, M., Christensen, J. H., Adalgeirsdottir, G., Mottram, R. and Simonsen, S. B. (2012). Very high resolution regional climate model simulations over Greenland: Identifying added value, *Journal of Geophysical Research-Atmospheres* **117**, p. D02108, doi:10.1029/2011JD016267.

Rauthe, M., Steiner, H., Riediger, U., Mazurkiewicz, A. and Gratzki, A. (2013). A Central European precipitation climatology — Part I: Generation and validation of a high-resolution gridded daily data set (HYRAS), *Meteorologische Zeitschrift* **22**, 3, pp. 235–256, doi:10.1127/0941-2948/2013/0436.

Rockel, B. (2015). The regional downscaling approach: A brief history and recent advances, *Current Climate Change Reports* **1**, 1, pp. 22–29, doi:10.1007/s40641-014-0001-3.

Rockel, B., Will, A. and Hense, A. (2008). The regional climate model COSMO-CLM(CCLM), *Meteorologische Zeitschrift* **17**, 4, pp. 347–348, doi:10.1127/0941-2948/2008/0309.

Rummukainen, M. (2009). State-of-the-art with regional climate models, *Wiley Interdisciplinary Reviews — Climate Change* **1**, 1, pp. 82–96, doi:10.1002/wcc.8.

Tapiador, F. J., Sanchez, E. and Gaertner, M. A. (2007). Regional changes in precipitation in Europe under an increased greenhouse emissions scenario, *Geophysical Research Letters* **34**, 6, p. L06701, doi:10.1029/2006GL029035.

Torma, C., Giorgi, F. and Coppola, E. (2015). Added value of regional climate modeling over areas characterized by complex terrain-Precipitation over the Alps, *Journal of Geophysical Research-Atmospheres* **120**, 9, pp. 3957–3972, doi:10.1002/2014JD022781.

von Storch, H. and Zwiers, F. W. (1999). *Statistical Analysis in Climate Research*, Cambridge University Press.

von Storch, H., Langenberg, H. and Feser, F. (2000). A spectral nudging technique for dynamical downscaling purposes, *Monthly Weather Review* **128**, 10, pp. 3664–3673, doi:10.1175/1520-0493(2000)128<3664:ASNTFD>2.0.CO;2.

Wiernga, J. (1993). Representative roughness parameters for homogeneous terrain, *Boundary-Layer Meteorology* **63**, 4, pp. 323–363, doi:10.1007/BF00705357.

Winterfeldt, J., Geyer, B. and Weisse, R. (2011). Using QuikSCAT in the added value assessment of dynamically downscaled wind speed, *International Journal of Climatology* **31**, 7, pp. 1028–1039, doi:10.1002/joc.2105.

Yu, E. (2012). High-resolution seasonal snowfall simulation over Northeast China, *Chinese Science Bulletin* **58**, 12, pp. 1412–1419, doi:10.1007/s11434-012-5561-9.

Zahn, M. and von Storch, H. (2008). A long-term climatology of North Atlantic polar lows, *Geophysical Research Letters* **35**, 22, doi:10.1029/2008GL035769.

https://doi.org/10.1142/9781800615816_0005

Part II

Internal Variability of Marginal Oceans

The scientific community has been deeply intrigued by the inquiry into whether the widespread fluctuations in atmospheric and oceanic dynamics stem from internal stochastic processes or reactions to particular external influences ever since modern physical thought became a part of atmospheric and oceanic sciences. It became clear that certain variations were almost indistinguishable from random fluctuations, such as the emergence of a thunderstorm tomorrow. While conditions could forecast the likelihood of a thunderstorm forming in the region, pinpointing the exact time and location proved challenging. Therefore, we could regard the statistics of thunderstorms as being influenced or "conditioned" by the surrounding environment but not firmly predetermined by it. Victor Starr, renown meteorologist from the mid 20th century, summarized this by the following observation 1942:

> *The General Nature of Weather Forecasting. The general problem of forecasting weather Conditions may be subdivided conveniently into two parts. In the first place, it is necessary to predict the state of motion of the atmosphere in the future; and, secondly, it is necessary to interpret this expected state of motion in terms of the actual weather which it will produce at various localities. The first of these problems is essentially of a dynamic nature, in as much as it concerns itself with the mechanics of the motion of a fluid. The second problem involves a large number of details because, under exactly similar conditions of motion, different weather types may occur, depending upon the temperature of the air involved, the moisture content of the air, and a host of, local influences.*

This observation demonstrates the ubiquitous randomness in atmospheric dynamics, but is this truly random or simply our incapacity to determine all aspects of the state, rendering us unable to identify a deterministic driver? In such an instance, stochasticity would simply be a convenient mathematical model to deal with what is otherwise insurmountable (von Storch, 2023). Practically, it does not matter; instead we have to adopt the view, or construction, of random variations, and we are left with the challenge to separate deterministic, thus predictable, changes from random, thus unpredictable variations.

The presence of developments, which we cannot conceptualize other than (conditional) random, or noise, has significant implications for the short term but also for the determination of the effects in numerical experimentation and simulation. Here, simulation refers to the construction of past, or possible future atmospheric or oceanic states, which is affected by some external factors, such as elevated greenhouse gas concentrations (e.g., Yuan *et al.*, 2023). Numerical experimentation refers to running a dynamical model repeatedly, with controlled modifications. In this context, a suitable example is the case of the role of tidal activity in a hydrodynamical model of the Yellow Sea (Chapter 6). Also, all downscaling efforts (see Part IV) are subject to this uncertainty.

The emergence of random variations in extended atmospheric model simulations was first described by Chervin *et al.* (1974), and after that statistical tests became routine practice when evaluating global numerical experiments in atmospheric sciences. That this happened first with atmospheric models is likely due to the fact that macroturbulence (in particular storms) were resolved in such global models in the 1970s, while this took another one or two decades or so in ocean models. That this was also a property of limited area models of atmospheric dynamics was recognized only in the mid-1990s (Ji and Vernekar, 1997; Rinke and Dethloff, 2000; Weisse *et al.*, 2000); in regional models of oceanic dynamics, this took place even later and is still not common knowledge (e.g., Waldman *et al.*, 2017; Penduff *et al.*, 2018). This part of the anthology documents how the Qingdao–Hamburg cooperation contributed to the disclosure of the hydrodynamical noise in marginal seas.

The study of Zhang *et al.* (2019) "Temporal and spatial statistics of travelling eddy variability in the South China Sea" was originally based on the assumption that meso-scale eddy activity, such as number of eddies or intensity, could be "downscaled" from large-scale currents — in this case in the South China Sea (Chapter 3). All efforts to do so failed, and it turned out that these statistics were mostly independent of the variations of the large-scale currents. Variations of the meso-scale eddy activity appeared as random.

The emergence of noise (Chapter 4) in the South China Sea was then studied in ensembles of simulations with different spatial resolution by Tang *et al.* (2019). A key measure is the signal-to-noise ratio, which compares the time standard deviation of the deterministic (ensemble mean) variations with the time mean of standard deviations at each time of the unforced variations (the deviations of the ensemble mean). It reflects the relative importance of the atmospheric driver and the internal variability — when the former dominates, the signal-to-noise ratio is larger, while a smaller ratio indicates a dominance of the noise. Tang *et al.* (2019) introduced this measure and found that the overall signal-to-noise ratio grows in simulations with increasing grid resolution. In a next step (Chapter 5), Tang *et al.* (2020) found the internal variability mostly on medium and small scales.

The last reprinted paper (Chapter 6) in that series by Lin *et al.* (2022), repeated the study for the shallower Bohai/Yellow Sea system and found the same patterns as Tang *et al.* (2019) had found in the deep South China Sea before. An additional experiment, when tides were turned off and on, demonstrated the sensitivity of the internal variability to the presence of tides in marginal sea — when tides are on, the internal variability is strongly reduced. A deepening analysis of how to understand on the system's level the sensitivity of the noise generation to the presence of tides and the seasonal cycle was based on the concept of the Stochastic Climate Model developed by Klaus Hasselmann, using the memory as key analytical tool, which is outlined in our article entitled "The Stochastic Climate Model helps reveal the role of memory in internal variability in the Bohai and Yellow Sea" (Lin *et al.*, 2023). It was accepted for publication late in the process of assembling this anthology, therefore it must suffice to reproduce the abstract:

Hasselmann's theory elucidates how short-term random noise leads to longer-term unprovoked variations, i.e., red spectra. Here, we study ensembles of numerical model simulations of the hydrodynamics of the Bohai and Yellow Sea concerning internal variability formation. Short(/long) term variations are associated with small(/large) spatial scales, and the internal variability of long-term temporal and large-scale variations is markedly enhanced, even without external forcing on these scales, when the tides are turned off. This pattern is well explained by Hasselmann's theory. A critical element in this theory is the concept of memory, which in our ensembles exhibits a scale dependence that aligns with the scale-dependent nature of redness. Additionally, this framework clarifies why there is a significant reduction of long-term fluctuations during winter and when tides are active: the system's memory is notably diminished under these conditions.

Chapter 3

Temporal and Spatial Statistics of Travelling Eddy Variability in the South China Sea*

M. Zhang[†], H. von Storch[†,‡], X. Chen[‡], D. Wang[§], and D. Li[¶]

†Institute for Coastal Research, Helmholtz-Zentrum Geesthacht Centre for Materials and Coastal Research, Geesthacht, Germany
‡College of Oceanic and Atmospheric Sciences, Ocean University of China, Qingdao, China
§State Key Laboratory of Tropical Oceanography, South China Sea Institute of Oceanology, Chinese Academy of Sciences, Guangzhou, China
¶Key Laboratory of Ocean Circulation and Waves, Institute of Oceanology, Chinese Academy of Sciences, Qingdao, China

The variability across decades of years of migrating eddy activities in the South China Sea (SCS) has not yet been documented. We employ a daily global eddy-resolving (0.1°) model product called STORM that covers a period of 1950–2010 to fill this gap. The frequency and pattern of eddy

---

*This chapter was originally published in *Ocean Dynamics*, 69, 879–898, doi:10.1007/s10236-019-01282-2. This chapter is licensed under the terms and conditions of the Creative Commons Attribution (CC BY) license (https://creativecommons.org/licenses/by/4.0/), which permits unrestricted use, distribution, and reproduction in any medium.

occurrence in the simulation is broadly consistent with satellite-based (AVISO) data.

On average, annually, 28 anticyclonic travelling eddy (AE) tracks and 54 cyclonic travelling eddy (CE) tracks with long travel lengths were derived from the discrete sea surface height anomaly fields of STORM. Eddy centers most frequently pass by the Luzon Strait and along the continental slope in the northern SCS to the Vietnam coast. The lifespans range from 6 to 240 days for AEs and to 293 days for CEs, and the longest travel lengths are 1941 km and 1988 km, respectively.

EOFs of the spatial fields of eddy diameter (ED), eddy intensity (EI), and eddy number (EN) show almost white eigenvalue spectra, when calculated on the model's 0.1-degree grid, but when the data are coarsened to grids with 1-degree and 2-degree grid spacing, meaningful structures emerge.

EI and ED are highly correlated on both seasonal and interannual time scales. In general, CEs are much more active than AEs, but the AEs with high intensities or large diameters are more frequent than similar CEs.

The monthly ED, EI, and EN exhibit annual cycles, which are, however, not very stable. The variabilities of annual means of ED, EI, and EN are large at interannual time scales, are little at interdecadal sales, and exhibit hardly a trend. The sizes and intensities of eddies in the SCS are hardly connected to the ENSO-variability in the tropical Pacific.

The EOFs, the weakness of the annual cycle stability and the absence of a correlation with ENSO, point to a massive presence of internal variability (as opposed to variability provoked by large-scale drivers).

1.   Introduction

The South China Sea (SCS) is the largest marginal sea in the Northwest Pacific and has an average depth of more than 2000 m. Every year, a large number of oceanic eddies occur in the SCS. Because of their vital roles in ocean circulation and in the marine ecosystem, oceanic eddies in the SCS have drawn a large amount of attention from oceanographers and ecologists. Ocean eddies regulate the distribution of ocean properties by transporting oceanic energy, mass, and materials in both the horizontal and vertical directions (Chen *et al.*, 2012; Zhang *et al.*, 2014). The rich nutrients that are trapped by eddy-induced upwelling contribute extensively to marine productivity (Lee *et al.*, 1991; Mahadevan, 2014).

The upper layer ocean circulation in the SCS is mainly dominated by the monsoon and has a strong western boundary current and a large gyre. The boundary current, along with the gyre, reverses its direction during the summer monsoon and winter monsoon. In the northern SCS, one branch of the Kuroshio intrudes into the SCS through the Luzon Strait and results in exchanges of heat, salt, and moment (Nan *et al.*, 2015; Wang *et al.*, 2006). Wang *et al.* (2003) summarized several generation mechanisms of eddies in different regions of the SCS by analyzing 86 mesoscale eddies derived from altimeter data from 1993 to 2001. The frontal instability southwest of Taiwan, related to the Kuroshio, is thought to be a factor in shedding eddies. In addition, the strong eastward current jet during the southwesterly monsoon has been found to generate eddies offshore of Vietnam, particularly a stationary pair of an anticyclonic eddy (AE) and a cyclonic eddy (CE). The vorticity from the Kuroshio front, the wind stress curl, and the interaction of strong currents with the topography also influence the eddy generation mechanisms.

Many case studies of eddies in the SCS have demonstrated and revealed the details of these generation mechanisms. Most of the case studies focused on the entire lifetime of the eddies, including their formation, evolution, propagation, dissipation, and three-dimensional (3D) structure (Chu *et al.*, 2014, 2017; Geng *et al.*, 2016, 2017; Li *et al.*, 2015; Wang *et al.*, 2015; Yuan *et al.*, 2007; Zu *et al.*, 2013). Wang *et al.* (2008) found that the relaxation of Ekman transport anomalies may have contributed to the shedding of two AEs in the northeastern SCS. Zhang *et al.* (2016) captured the full-depth 3D structure of an AE and CE eddy pair near the Luzon Strait in 2013/2014 based on a multi-month "SCS Mesoscale Eddy Experiment" and suggested that the dominant dissipation mechanism of the eddy pair may have been the generation of submesoscale motions. In the summer, an eddy dipole, with one strong AE and one weak CE is always located near 11°N. The formation of this eddy pair is related to the vorticity transport from the western boundary current, which is driven by the wind stress curl (Chu *et al.*, 2017).

In our paper, we examine the multidecadal climatology of travelling SCS eddies, specifically the annual cycle, interannual variability, interdecadal variability, and long-term trend in the number, intensity, and spatial size. For doing so, we identified and tracked eddies, and determine their intensities and sizes, in a multidecadal simulation found to properly reproduce the main features of the SCS circulation (Zhang and von Storch, 2017).

Recently, several statistical analyses of the multiyear variability in SCS eddies have been performed (Chen *et al.*, 2011, 2012; Feng *et al.*, 2017; Li *et al.*, 2011; Nan *et al.*, 2011; Sun *et al.*, 2016; Wang *et al.*, 2003; Xiu *et al.*, 2010). Xiu *et al.* (2010) investigated the eddy activity in water deeper than 1000 m from 1993 to 2007, using both 7-day interval AVISO altimeter data and the output from a Regional Ocean Model System (ROMS). The authors identified approximately 32.8 eddies each year from the AVISO data (of which 52% were cyclonic eddies) and found that the interannual variabilities in eddy number and the area occupied by eddies are not correlated with El Niño events. Chen *et al.* (2011) focused on mesoscale eddies and identified 827 mesoscale eddies (approximately 48.6 per year) in the SCS during 1993–2009 from 7-day interval AVISO satellite data with different detection parameters. Their study found that more AEs occurred than CEs. However, using the same data set, Nan *et al.* (2011) detected many more CEs (41) than AEs (27) in southwest of Taiwan. Feng *et al.* (2017) used daily AVISO data to identify mesoscale eddies in the SCS from 1993 to 2007 and found more CEs than AEs. The number of detected eddies is sensitive to the region and the identification parameters. In the analysis of Chen *et al.* (2011), the eddy intensities over 17 years exhibited a weak negative correlation with the sea surface temperature anomalies in the central Pacific (given by the Nino3 index), but no correlation was found between the eddy number and El Niño activity. However, Chu *et al.* (2017) found in a composite analysis that the El Nino–Southern Oscillation (ENSO) influences the stationary eddy pair off the Vietnam coast. This eddy pair is one of the most important phenomena off the Vietnam coast during the summer monsoon periods, but it was absent during the ENSO transition years. They suggested that the ENSO transition events led to changes in the southwesterly monsoon and then resulted in the eastward current jet turning northward and the eddy pair disappearing. Based on the output of an eddy-resolving ocean simulation for the Earth Simulator (OFES), Sun *et al.* (2016) investigated the interannual variability in the eddy kinetic energy (1980–2014) in the northeastern SCS and determined that the Luzon Strait transport exhibited a modulating effect, but their study did not consider the eddy activity.

An issue, which has not been completely resolved, is to what extent the statistics of eddy formation and lifecycle may be seen as being forced by large-scale conditions, say the seasonal mean barotropic or baroclinic state (or the atmospheric conditions, which, however, likely would not act directly but indirectly via changing currents). The experiment by Tang *et al.* (2019) demonstrated that at least a significant part, if not the dominant part of eddy activity, is reflecting internal dynamic processes, unprovoked by external causes.

The previous analyses of eddy activity, such as the eddy number distribution and the seasonal and interannual variabilities in the SCS, have suffered from several limitations. The limited accuracy of the satellite data used for the analyses has made it necessary to limit the analysis to mesoscale eddies with high intensity. In addition, only time scales up to a few years could be considered because of the relatively short time span of no more than 30 years, starting from 1993.

The data studied in our work have some advantages, such as no observational errors, and uninterrupted, homogeneous coverage across six decades. However, the data are the output of a model and may be affected by unknown shortcomings, related to the limited spatial and temporal resolutions, the incompleteness of the physical processes, and the absence of an interactive exchange of properties between the ocean and atmosphere. However, we have no indications that the STORM simulation is inconsistent with the limited observational data in the South China Sea; a systematic comparison of some large-scale features as derived from satellite data and the simulation revealed no such inconsistencies (Zhang and von Storch, 2017).

Our paper is intended to improve our understanding of the multidecadal statistical characteristics and the variability of eddy activity by examining a 61-year time model hindcast. This simulation called STORM is part of the German consortium project STORM (J. von Storch *et al.*, 2012; https://swprojects.dkrz.de/redmine/projects/storm/wiki/STORM_list_of_experiments).

Our work investigates the statistical feature and the variability on different time scales of the travelling eddies in the SCS. It does not directly contribute to an improvement of our knowledge about their dynamics. However, the provision of the statistical characteristics and spatiotemporal variabilities of long-term oceanic eddy activities are needed as a prerequisite for assessing the roles and impacts of eddies, as well as projecting the future activities. In particular, such knowledge will allow building empirical downscaling models, which allow estimating future eddy statistics, given a change in the large scales.

This chapter is organized as follows. The "Data sets" section describes the details of the eddy-permitting and multidecadal STORM simulation, as well as the AVISO altimeter data set for validating STORM data. To avoid problems from the calculation of differential and integral operators on discrete fields (Chelton *et al.*, 2011), we developed an eddy detection and tracking method that was exclusively based on the discrete sea surface height anomaly (SSHA) fields. It is explained in the "Methods" section. In the "Simulation validation" section, the STORM simulation was assessed. The "Results and discussion" section focuses on the statistical characteristics and the variabilities on different time scales. In addition, the relationship with El Niño is briefly discussed. The "Summary and discussion" section provides a summary of our work and discusses some related aspects.

2.   Data Sets

2.1.   *Satellite Observations*

We use the "AVISO" data set (Archiving, Validation, and Interpretation of Satellite Data in Oceanography (AVISO) [AVISO 1996, 2015; https://www.aviso.altimetry.fr/en/home.html]) for validating our simulation with respect to the eddy activity. The "Delayed Time" and "All-Sat merged" AVISO has merged data from all altimeter missions available (up to four at a given time) (Faghmous *et al.*, 2015).

We use the gridded AVISO data set, with a grid resolution of $0.25°$ and time resolution of 1 day for the time period 1993–2010. An issue is the accuracy of the data set. The information provided by the AVISO handbook (Taburet *et al.*, 2018) is not really conclusive, but we may expect the root mean square error to be larger than 1 cm, possibly much larger. While the grid resolution is about 25 km, the effective resolution in describing phenomena is also expected to be larger, namely 65 km and possibly much more (Dufau *et al.*, 2016).

The study of Amores *et al.* (2018) suggested the gridded AVISO data set underestimated the number of eddies due to its limited spatial resolution.

2.2. *STORM Simulation*

The quasi-realistic modelling of the South China Sea, resolving the specifics of the circulation in that regional ocean, was begun by Pohlmann (1987), who was the first to find an upwelling regime along the Vietnam coast. Today, doing so has become a standard routine with various global and regional models.

Our analyses are based on the daily global ocean model product from the German consortium Project STORM, covering a period of 1950–2010 (J. von Storch *et al.*, 2012). This simulation employs a state-of-the-art ocean model — the Max Planck Institute Ocean Model (MPI-OM). For ensuring an isotropic horizontal resolution, the bipolar grid of the model has been replaced by a tripolar grid. In total, this simulation has 3600×2392 horizontal grid points and 80 uneven vertical levels. The grid resolution in our domain (Figure 1) is approximately $0.1°$ so that eddies are resolved (Hallberg, 2013). Tidal forcing was not activated in this simulation (J. von Storch *et al.*, 2012). The model was forced by the 6-hourly National Centers for Environmental Prediction (NCEP)/National Center for Atmospheric Research (NCAR) reanalysis (Kalnay *et al.*, 1996), after a 25-year spin-up phase using the German Ocean Model Intercomparison Project (OMIP) forcing. The kinetic energy reached a quasi-steady state in the deep ocean after the spin-up. For the following analyses, we refer to the simulation as STORM.

STORM performed well in the research on meso- or small-scale oceanic phenomena, including eddy-related heat and salt fluxes (von Storch *et al.*, 2016), upwelling systems (Tim *et al.*, 2015; Yi *et al.*, 2017), and sea level along the Ghana coast (Evadzi, 2017).

The skill of STORM in representing oceanic dynamics in the SCS has been validated by Zhang and von Storch (2017). STORM agrees well with AVISO and C-GLORS (Storto *et al.*, 2016) in the seasonal means and interannual variabilities of SSHA in the SCS. The finer resolved STORM generates stronger currents and presents more details of the strong upwelling offshore the Vietnam coast during summertime. The assessment demonstrated the

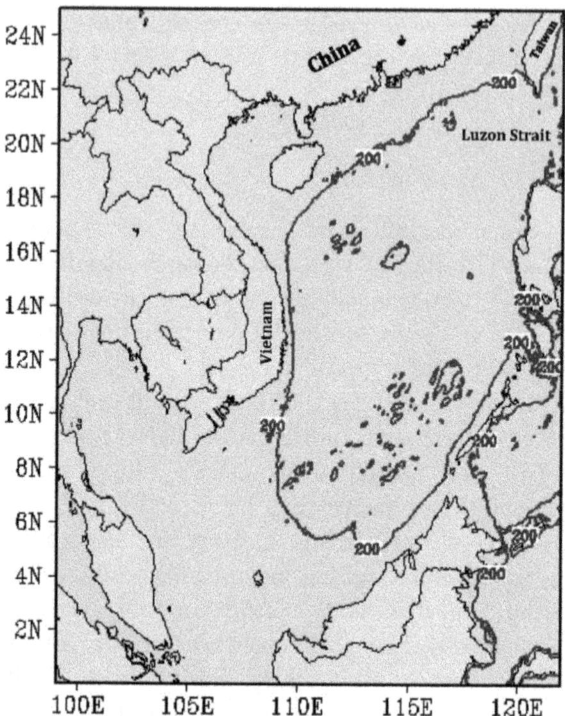

Fig. 1.   The domain in this work. The contour shows the 200-m isobath in the SCS.

ability of STORM to capture the main features of the SCS oceanic hydrodynamics.

3.   Methods

3.1.   *Eddy Detection and Tracking Scheme*

Our study focuses on travelling eddies irrespective of the size. Therefore, in this paper, an eddy is defined to be an extremum that moves in space.

The eddy-detection scheme begins with searching for extrema (minima and maxima) in the SSHA field. A center point in a box will be defined as an extremum if the corresponding SSHA is greater (or smaller) than all other points in the box. The difference between the extremum SSHA and the mean SSHA of all other points in the box defines the relative intensity (RI) of an eddy.

We employ boxes of $5° \times 5°$ grid cells, as suggested by Faghmous *et al.* (2015), who showed that a box of this size is suitable for detecting extrema, while a box with 7×7 cells fails with small eddies, and a box with a 3×3 neighborhood leads to too many extrema.

The next choice is that of a minimum RI. Too strict thresholds may break an eddy track into several pieces. An automated tracking program cannot adequately address such a situation, as it will terminate a track in the middle and derive more than one track. To avoid this problem, a mild threshold is set at the moment. According to sensitivity tests (Zhang *et al.*, 2017), only extrema with RI $\geq$ 3 mm will be considered. In addition, an "eddy core" with at least 5 pixels, with one extremum in the center and one pixel in each of the four directions (north, south, east and west), is requested. Our new method to determine an eddy core will be introduced in detail in the "Eddy size detection methodology" section.

The tracking procedure is used to connect the local extrema at consecutive time steps (days). For each extreme at time T, another extreme at time $T + 1$ will be considered as a member of the same track if it fulfills the following criteria:

- It is the closest extreme to the extreme at time T when all extrema at time $T + 1$ are considered.
- Taking eddy travel speed and the spatial resolution of our data into consideration, the distance between the extrema is less than 25 km.
- The RI properties of the extrema differ by no more than a factor of 1.5.

As we focus on eddies with long track lengths and high intensities, we apply the following filters:

- The accumulated track is required to be at least 100 km.
- In case some eddies move back-and-forth, a restriction on the distance from the initial position to the final position is implemented (at least 50 km).
- In the eddy detection part, all eddy centers with a $|RI| \geq 3$ mm are kept. However, tracks are kept only if the strongest extreme RI surpasses the threshold $RI_{\mathrm{max}} = 6$ mm.

- Many small disturbances occur near the shore (especially in shallow water). Since eddies need vertical space to form and develop, a depth criterion is implemented, namely, a track must travel for over 90% of its lifetime in water deeper than 200 m.

Then, for each eddy track that satisfies the criteria listed above, the track length, the lifetime, and the maximum strength along the track are derived.

3.2. *Eddy Size Detection Methodology*

To properly describe eddies that form in the South China Sea, we need a measure of the sizes of such eddies. Most eddy-detection methods are based on a physical feature, a geometric feature, or a hybrid of them (Nencioli *et al.*, 2010). Two typical methodologies are widely used, the W-based method and the winding angle (WA) method (Chen *et al.*, 2011; Xiu *et al.*, 2010; Zhan *et al.*, 2014).

The W-based method is based on a physical parameter W, which describes the relative importance of rotation compared with the deformation in the flow (Chelton *et al.*, 2011):

$$W = (v_x + u_y)^2 + (u_x - v_y)^2 - (v_x - u_y)^2 \tag{1}$$

u and v represent the eastward and northward velocity components, respectively, and the subscripts x and y indicate partial differentiation. W is used to divide the SSHA field into two parts that are dominated by divergent flow ($W > 0$) or rotational flow ($W < 0$; suspected eddy area). Then, a threshold value of $W(W_0)$ is specified to identify an eddy core (Xiu *et al.*, 2010). The eddy size is determined by the closed contour where $W = W_0$. It is slightly problematic to obtain W through differentiation. In addition, if only SSHA data are available, as is the case for some satellite data, only geostrophic velocity can be derived from the SSHA field. Additionally, Chelton *et al.* (2011) found that the eddy area identified by the closed contour of W does not closely match the closed contour of the SSHA data.

Another popular method is the WA method, which is based on the geometric assumption that the streamline around the eddy core is close to a circle or a spiral. The accumulated angle of each consecutive segment of the streamline is computed. The streamlines with absolute accumulated angles larger than 2π are defined as closed streamlines (circular or spiral curve). An SSHA extremum in the center that is surrounded by a series of closed streamlines constitutes the eddy structure in the WA method (Zhan *et al.*, 2014). The eddy margin is characterized by the outermost closed streamline. The WA method works better than the W-based method in the eastern South Pacific (Chaigneau *et al.*, 2008). However, the WA method also has limitations. The need to approximate the geostrophic velocity field from the SSHA field results in a problem similar to that in the W-based method.

In this paper, we present a geometry-based method that exclusively relies on the discrete SSHA field to determine the size of an eddy. The outermost closed contour of the SSHA around the eddy core is defined as the eddy edge. According to the geostrophic balance theory, water tends to flow along the SSHA contour. Therefore, the flow along the outermost closed contour can also be considered closed. The SSHA distribution within an eddy is characterized by an extremum (minimum or maximum) in the center and a number of closed SSHA contours around the eddy center, whose SSHA displays a monotonic increase (or decrease) towards the center. Because of this monotonicity, we refer to this method as "M-based."

An iterative detection procedure aims to identify the largest (if the center is a minimum; otherwise smallest) closed contour that fulfills the monotonicity condition. The eddy size estimation procedure extends the neighborhoods surrounding the eddy core iteratively until the monotonous condition is violated. The condition is fulfilled unless all neighbors around the eddy core are greater (or less) than one threshold. The initial threshold is set to the center value of one eddy. In the following steps, the threshold is always updated by the value of the most recently detected closed contour. As the detected eddy size grows in this way, the threshold increases (or decreases) step by step. Every time the monotonicity is extended, another larger (smaller) closed contour exists in these neighborhoods, and the contour corresponds to the minimum (or maximum) from these neighborhoods. Then, the eddy size will grow by including one or more points in this new contour, and the value of this contour will be used to update the threshold to check the monotonous condition in next step.

We define the eddy size as $d = 2\sqrt{A/\pi}$, where A indicates the area of the eddy, and d is the diameter of the eddy if it is a circle.

This M-based method is similar to the method of Faghmous *et al.* (2015). However, a key criterion is different. In their method, the condition for an eddy size to stop growing is another extremum included in its interior, which differs from our monotonous condition. Their condition could result in an overestimation of the size. On the other hand, the increment of the threshold in their method is defined by the user in advance. Too coarse a threshold step may fail to obtain an accurate eddy core, but a step that is too fine will require extensive computational time. The appropriate steps vary for different eddies, and it is difficult to select steps that are appropriate for all eddies.

Figure 2 shows the detected results from our M-based method on 2010-01-01 from STORM data. The sizes and shapes of the masked eddies coincide with the contour lines in the SSHA fields, which reveals the ability of the M-based method to capture eddy sizes. In some cases, the eddy size is underestimated due to the interruption of the eddy structure by islands or land. This algorithm also addresses the situation that monotonicity cannot be closed as a result of the islands. The M-based method can detect the smallest eddies with structures containing only five points. In this study, we keep all the small eddies.

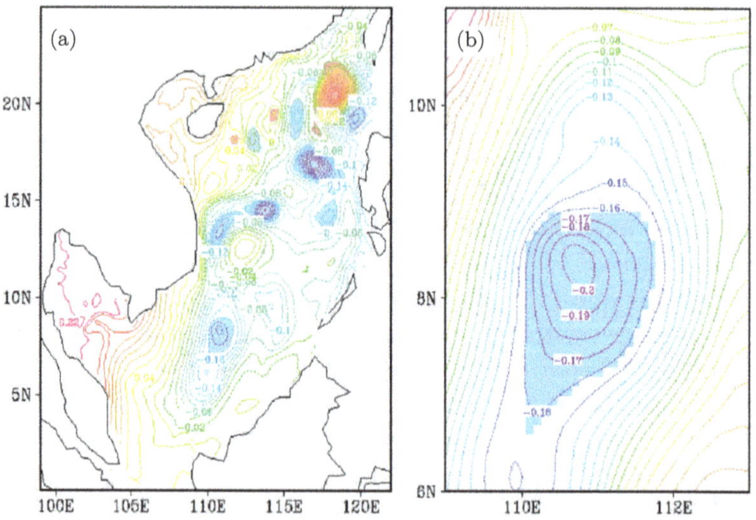

Fig. 2.   (a) The SSHA distribution (contours; units: m) and the eddies detected by the M-based method (blue: CE; red: AE) on 2010-01-01 and (b) the single cyclonic eddies derived from (a).

4. Simulation Validation

In the following, we compare our simulation results with data from the AVISO gridded data set, using the paper by Chen *et al.* (2011) as a reference. Therefore, we use the same criteria for detecting potential eddies as in Chen *et al.*'s work. In our main chapter on the long-term statistics of eddies in the SCS, we use a different set of parameters (see the "Eddy detection and tracking scheme" Section 3.1), because our work is aimed at investigating the statistics of eddies with all sizes, and the absolute intensity limitation will lead the small eddies to be disregarded.

For assessing if STORM simulation could reproduce eddy activity in the SCS, we derived the minima and maxima in the SSHA fields with intensity ≥ 3 cm and diameter $d \leq 35$ km (Chen *et al.*, 2011) from STORM and AVISO during the joint period 1993–2010. These minima, and maxima, may be part of a migrating eddy, but also, in case of AVISO represent observational noise. For fair comparison, the 0.1-degree STORM in this section was interpolated onto the same grids with the 0.25-degree AVISO.

Figure 3(a) and (b) show the occurrence frequency of such minima, or maxima, at all grid points. Both data sets show potential eddy centers occurring most frequently in the Luzon Strait. Starting from the Luzon Strait, the region with high occurrence frequency extends westward along the continental shelf, passing by the southeast of China and reaching Vietnam coast. Apart from these seas, some small areas with high occurrence frequency are scattered in the central and northern part of SCS.

It is notable that much more potential eddies show up in AVISO data than we have in the STORM simulation. However, errors of the magnitude of 1 and more cm prevail in AVISO data set (AVISO, 2015), so that it is plausible that, given the minimum intensity of 3 cm and an RMS error of 1 and more cm, some of the extrema may be artifacts generated in the observational and gridding procedures. We assume that these artifacts are short-lived and do not travel consistently in space. Therefore, we focus the comparison sequences of such minima, or maxima, which form tracks.

It may be worthwhile to examine such cases of short-lived, not-travelling minima, or maxima, in the AVISO data set, but this effort would be beyond the framework of the present study.

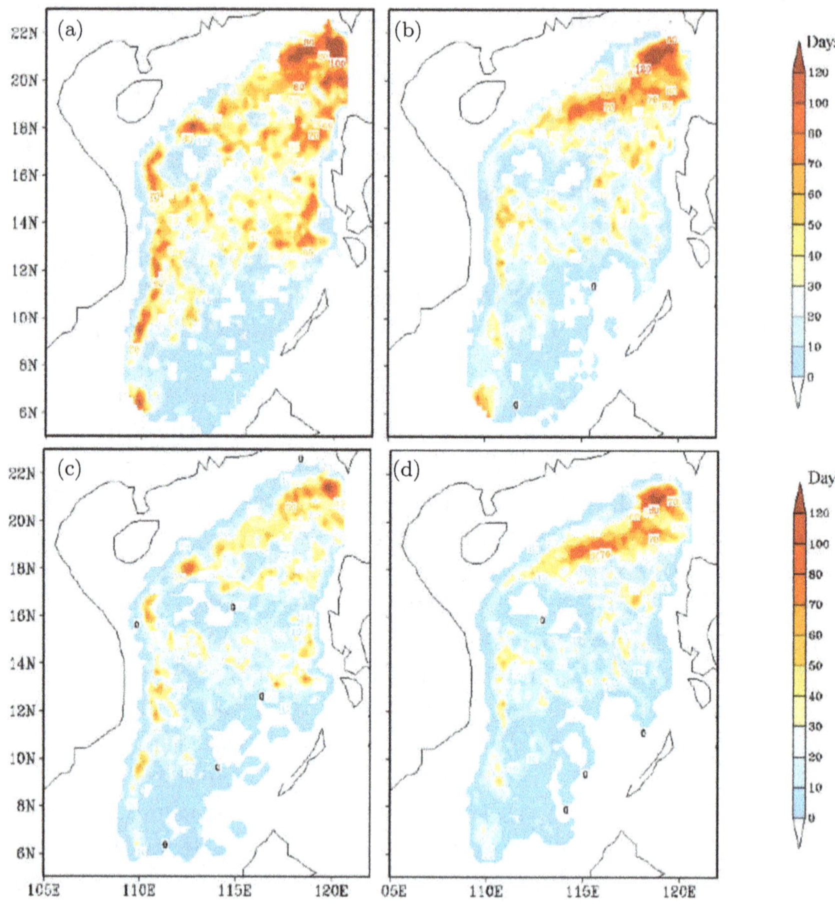

Fig. 3.   Comparison of STORM with AVISO. The frequency of eddy centers in each grid box during 1993–2010 from AVISO (a) and coarsened STORM (b); (c) and (d) are the same with (a) and (b) but after eddy connecting and filtering. Note the difference with Figure 4, where the full resolution of STORM is employed.

Eddy propagation speeds are similar to the phase speed of long baroclinic Rossby waves (Faghmous *et al.*, 2015) and are expected to be less than 30 cm/s. Thus, an eddy cannot travel longer than 30–40 km in 1 day. Given the coarse resolution of AVISO here, we use this maximum daily travel distance, but when deriving in the "Results and discussion" section the statistics of eddy migration for STORM alone, with its higher grid resolution, we use the 25 km. The maximum daily travel distance 40 km is different from 150 km in Chen *et al.*'s work based on the 7-day interval data (about 21.4 km per day) either. So, for the filtering step, there is no need to use the same filtering criteria as in Chen *et al.*'s paper any more. Thus, we use the filters in the "Eddy detection and tracking scheme" section. After this connecting and the filtering, many potential eddies disappear in AVISO. The patterns in AVISO and STORM match better and the difference are weaker (Figure 3(c,d)).

STORM reproduces a too small frequency at 14°N, 119°E and 16°N, 110°E, and a too large one at 19°N, 115°E. These discrepancies are thought to be related to the simulated Kuroshio. In the SCS, ocean models often produce stronger Kuroshio intrusion near the Luzon Strait but a weaker effect from Kuroshio in the middle and southern SCS. We suggest that the too large frequency may be related to the simulated strong Kuroshio intrusion, whereas the small ones may be related to the simulated weak Kuroshio effect in the middle of SCS.

The general pattern (Figure 3(b)) coincides with the eddy probability derived from 7-day interval AVISO shown in Chen *et al.* (2011) as well, but their region with high frequent occurrence is connected, larger, and concentrated. Their work got the eddy probability at each point by figuring out the time when the grid was covered by a vortex, different from our only using the time when grid point is occupied by a minimum or a maximum in SSHA. A vortex can cover many grid points, but an eddy center is only located in one grid point.

In addition, it is worth noting but ignored by the previous researchers that, for altimeter missions, the mesoscale resolution capability is limited (Dufau *et al.*, 2016). Amores *et al.* (2018) analyzed the extent that the gridded altimeter products can characterize ocean eddies. Their results suggested that the gridded altimeter data set underestimated the eddy density and overestimated the amplitudes. The data set can capture less than 16% of the eddies, and the limited spatial resolution of the products is attributed to the underestimation. But no other better observations are available at this time for deriving eddies and assessing the realism of the STORM data.

An additional check of the realism of the detected eddies was made concerning the temperature distribution. Anticyclonic eddies (AEs) are supposed to have a warm core, while cyclonic eddies (CEs) a cold core. We examined the temperature gradient between the core and the outermost point of the eddies at the surface and at 100-m depth at the time of peaking (maximum sea surface height (SSH) deviation of the center from the surrounding). The result of the temperature gradient and the corresponding SSHA gradient is shown in Figure 4 for AEs and CEs in 2010.

For the surface, the signal is weak (top diagrams of Figure 4); for AEs, a relationship of sea surface temperature and SSHA is hardly visible, but obvious albeit not strong for CEs. At 100-m depth (lower diagrams in Figure 4); however, the link between SSHA and temperature becomes clear, in particular for strong eddies ($|SSHA| \geq$ 10 cm). This result is consistent with earlier studies; for instance, Itoh and Yasuda (2010) found most anticyclonic eddies to be warm, but clearly not all (namely 85%).

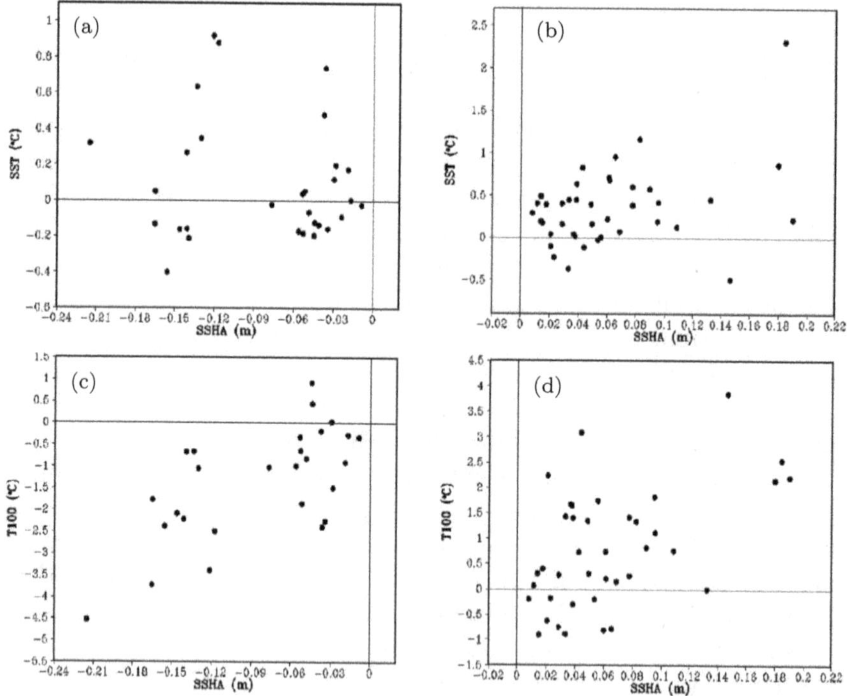

Fig. 4.   Scatter diagram of the SSHA gradients of the peak eddy points in 2010 and the corresponding seawater temperature gradients ((a) and (b) are SST gradients; (c) and (d) are the temperature gradients at a depth of 100 m). (a) and (c) are for anticyclonic eddies, and (b) and (d) are for cyclonic eddies.

5.   Results and Discussion

In this section, we consider parameters such as eddy intensities, diameters, track lengths, and lifespans, and examined climatological statistics, such as frequencies and temporal variability.

5.1.   *Climatological Characteristics*

During the period 1950–2010, a total of 62,317 AE points (maxima along a track) and 115,133 CE points (minima along a track) were detected from the STORM daily data, corresponding to 1709 AE tracks (AEs; 28.0 per year) and 3331 CE tracks (CEs; 54.6 per year), respectively.

The index "IN" is defined to measure the relative portion of CEs and AEs, given by:

$$\text{IN} = \frac{N_C E - N_A E}{N_C E + N_A E} \tag{2}$$

where $N_C E$ and $N_A E$ represent the genesis numbers of CEs and AEs. The IN index can vary between -1 and 1, with 0 indicating that the same genesis numbers of AEs and CEs. IN $= 1$ (-1) indicates that no AEs (CEs) were generated. The monthly IN has an average value of 0.29 and a skewness of -0.67. Negative skewness indicates that the median is larger than the expectation of 0.29; the number of CEs is over 1.8 times greater than the number of AEs in most months.

The combined effect of the enhanced spatial and temporal resolutions of the STORM data, together with the relatively mild sizes and intensity criteria, lead to many more eddies in our results, compared with the work using 7-day interval and 0.25-degree AVISO (Chen *et al.*, 2011; Xiu *et al.*, 2010; see the "Simulation validation" Section 4). The detected AEs are much less than CEs. It is related to the difference in the eddy size distribution between the two kinds of eddies, and will be discussed later, when the eddy diameter (ED) distribution is addressed.

Figure 5 presents the spatial distribution of the frequencies when an eddy center passes by each grid box during 1950–2010. Most eddy centers occur in the northern SCS. Note the difference to Figure 3(d), where the analysis was done with coarsened data and only the years 1993–2010 for allowing a fair comparison with the AVISO data.

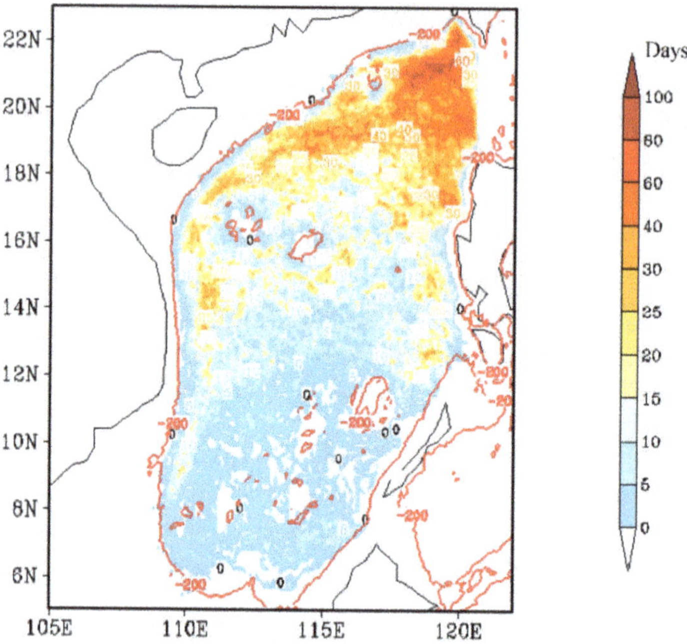

Fig. 5.   The occurrence frequency of eddy centers in each grid box during 1950–2010. Note the difference to Figure 3, where a coarse resolution of STORM is employed for allowing a fair comparison with AVISO.

The regions with the highest frequencies are located near the Luzon Strait and extend southwestward to the Vietnam coast along the continental slope, which correspond to the strong western boundary currents in the northern SCS (e.g., Zhang and von Storch, 2017). Frontal instability due to the Kuroshio intrusion and wind stress curl are the major factors to generate eddies near Luzon Strait and offshore Vietnam coast. On the basis of the westward propagation characteristics of eddies in the northern SCS, the high occurrence frequency along the continental slope might result from the eddy moving from Luzon Strait.

The maximum frequency along the Vietnam coast is not as high as that in Chen *et al.* (2011) from AVISO 7-day interval data. This difference may occur because our algorithm filters stationary eddies, so a stationary pair of an AE and a CE off the Vietnam coast associated with the coastal wind jet (Chu *et al.*, 2017; Lin *et al.*, 2017) is not counted in our analysis.

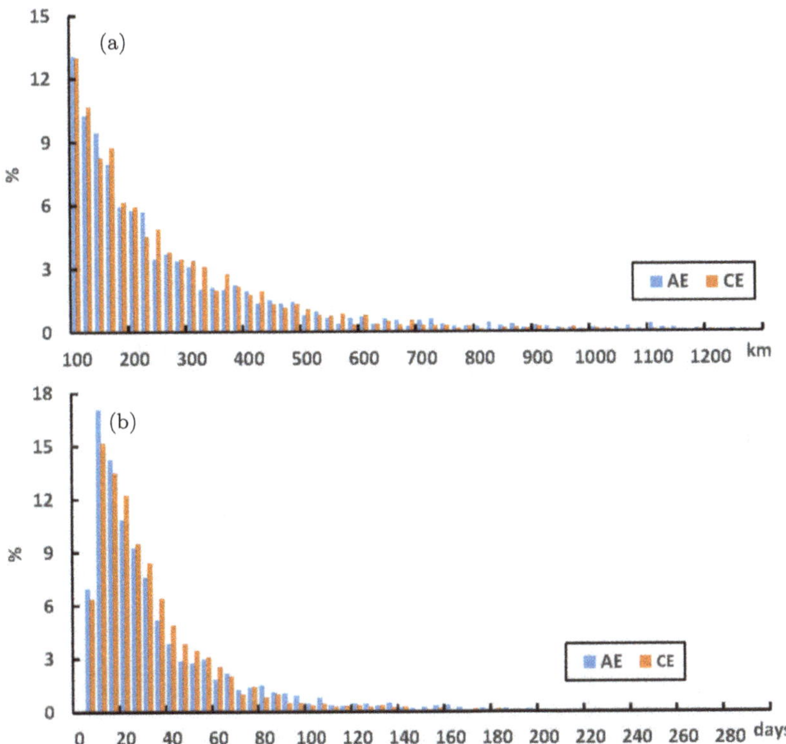

Fig. 6. The probability density function of eddy track length (a) and eddy lifespan distribution (b).

To describe the distributions of eddy track length and eddy lifespan, probability density functions of eddy lengths and lifespans have been calculated (Figure 6). In our time period, AEs traveled no further than 1941 km with an average track length of 292 km, and CEs did not travel further than 1988 km with a mean length of 274 km. The eddy lifespans ranged from 6 to 240 days for AEs and to 293 days for CEs, with mean lifespans of 36.5 days and 34.6 days, respectively. The length and lifespan distributions do not differ much between AEs and CEs, but the ranges of the lengths and lifespans of CEs are wider than those of AEs.

The distributions of the eddy intensities (EIs) and eddy diameters (EDs) of the peak points (those are the eddy points with highest EI along an eddy track) were investigated. The largest intensities range from 0.55 to 37.3 cm, with maxima of 3–4 cm for AEs and 2–3 cm

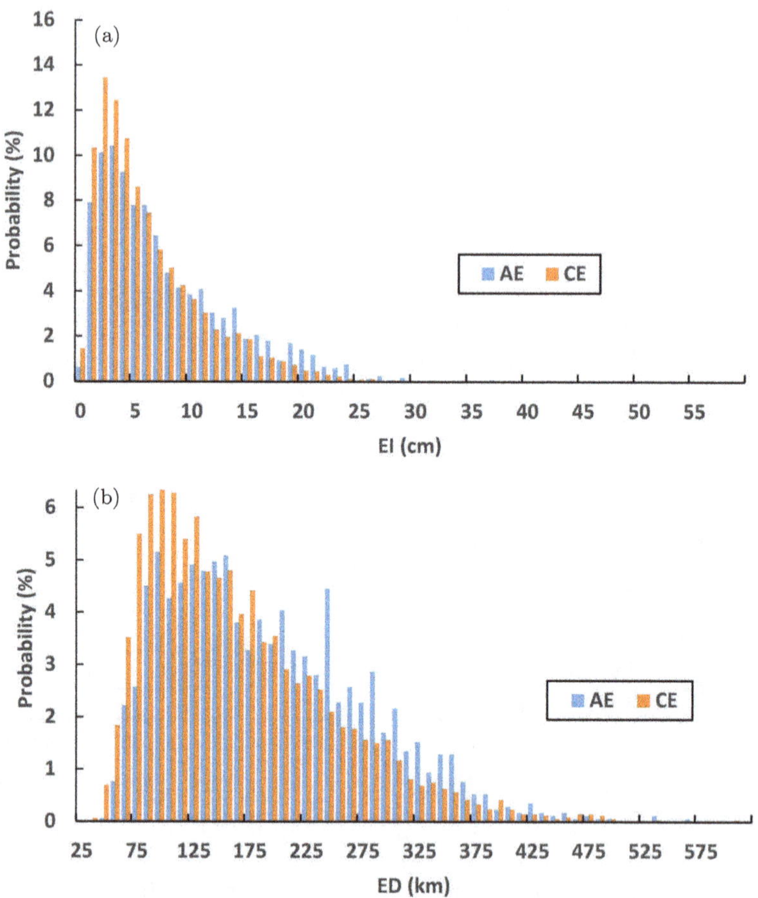

Fig. 7.   The probability density function of eddy intensity of peak eddy points with an interval of 1 cm ((a); start from 0.0 cm) and of eddy diameter of peak eddy points with an interval of 10 km ((b); start from 25 km).

for CEs (Figure 7(a)). Our eddy detection and tracking algorithm removed the eddy tracks that have highest $RI_{max} < 6$ mm. Therefore, many weak eddies with an EI < 1 cm are deleted; thus, the frequency of eddies with intensity 0–1 cm is low. If all eddy points are considered (not shown), the EI values vary from 0.11 to 37.3 cm, with the mean EI of 6.2 cm for AEs and 4.9 cm for CEs. AE points tend to occupy higher percentages of eddies with EI > 6 cm than CE points. More than 40% of the AE points and less than 30% of the CE points have EI > 6 cm.

The corresponding diameters at the strongest eddy points range from 43 to 631 km (Figure 7(b)). The EDs most frequently occur in 95–105 km for the eddy peak points. No eddy peaks associated with a diameter less than 35 km, which is related to the size limitation of 5 pixels. Most of the peak CE points (over 41%) have EDs less than 135 km, which represent more than 29% of the AEs. It is worth noting that the situation reverses with peak AEs, which occupy over 13% more than the peak CEs when the ED increases over 205 km.

If all eddy points are considered (not shown), diameters vary from 30 to 572 km for AEs and to 633 km for CEs. The percentage of large AE eddies is larger than that of CEs: 41.3% of AEs have a diameter over 175 km, whereas the percentage of CEs is only 26.2%. Much more CE cases have small eddy sizes.

We attribute the difference in the detected eddy number of AEs and CEs to the presence of smaller size CEs. The previous work (Chen *et al.*, 2011) was based on the AVISO altimeter data, with a coarse grid space of 0.25°, and detected more AEs than CEs. As suggested by Amores *et al.* (2018), AVISO often underestimates the eddy activity because its coarser spatial resolution for capturing small eddies. However, our work employs the high-resolution STORM simulation, which describes smaller eddies. On the other hand, our improved algorithm allows more smaller eddies to be kept. In those additional small eddies, CEs have a much higher percentage than AEs, finally leading to the number of CEs being higher than that of AEs.

The climatological analysis presents more active CEs, with larger range but a smaller mean value of EI and ED, which means more percentage of AEs have a larger size and higher intensity.

5.2. *Dominant Spatial Patterns*

Empirical orthogonal functions are a convenient and much-used tool to identify dominant spatial patterns of variability. We have used this approach to study the annual number of eddies (EN), the annual mean EIs, and annual mean EDs of the AEs and of the CEs. Thus, our data fields X consist of 61 time "points" and 27,271 spatial "sea-points" in the SCS. To calculate the eigenvalues of $X'X$ is not possible because of the sheer size of the matrix, but the matrix XX' has the same nonzero eigenvalues (von Storch and

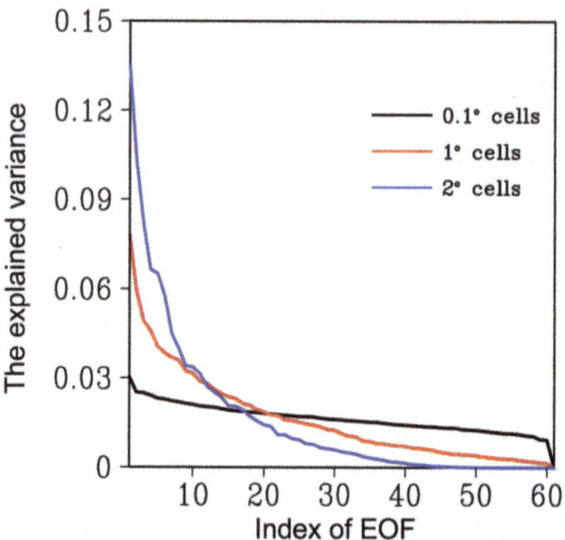

Fig. 8.   Eigenvalue spectra of the covariance matrix of annual anomalies of the mean diameter of annual mean cyclonic eddies in the SCS on different spatial grids — the original 0.1-degree of STORM (black), and after coarsening to 1-degree (red) and 2-degree (blue). Since 61 temporal samples are available (61 years), all eigenvalues beyond number 61 are zero and not plotted.

Hannoschöck, 1984), and the eigenvectors are related through X. Since XX' is much smaller than $X'X$, we solve the eigenproblem by the smaller matrix, without encountering numerical problems.

The result is surprising — when the eigenvalues, for instance for the spatial distribution of cyclonic eddy diameter, are plotted as a spectrum (the black curve in Figure 8), the results is a distribution rather similar to what one would expect from a white noise analysis, using an estimate of the error in estimating eigenvalues following (Lawley, 1956). The same result is obtained for the other two variables, EI and EN (not shown).

This is a rather rare case; in most cases, at least the estimates eigenvalue of the first EOF is well separated from the rest of the estimated eigenvalues, so that only the tail of the eigenvalues, sometimes beginning at the second eigenvalue but in most cases at higher-index eigenvalues becomes consistent with a white eigenvalue spectrum, and with arbitrary eigenvectors.

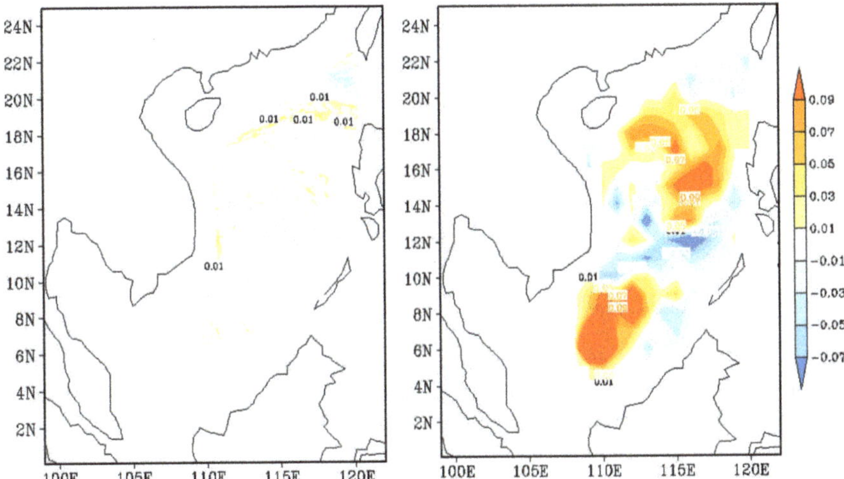

Fig. 9.   The first EOFs of annual man diameters of cyclonic eddies in the South China Sea determined from spatial distribution on the original 0.1-degree grid (left; represents about 3% of overall variance), and on the coarsened 1-degree grid (right; about 8% of overall variance).

The first EOF is shown in Figure 9 (left) — the pattern is indeed quite fragmented, and a pattern can hardly be recognized. The next eigenvectors exhibit a similar texture of randomness.

However, when the spatial field, of originally 0.1-degree resolution, is coarsened, by binning into 1-degree boxes, the situation improves. The spectrum of eigenvalues is displayed in Figure 8, as red curve. This time, the difference between the first two eigenvalues is about the estimated error of the first eigenvalue, so that the first two eigenvalues are about separated, possibly also the second, whereas the tail, beginning with the third number is consistent with a white noise spectrum. Also, the pattern, shown in Figure 9 (right), begins to show some structure. But, the percentages of variance represented by the first two eigenvalues are meager 8% and 6%. Thus, differently from most EOF analyses, the identified patterns do not describe a relevant part of the overall variability.

Indeed, even after coarsening to a 1-degree grid, the variability of the seasonal mean diameters of CEs in the SCS reflects little variations in large-scale conditions, such as currents or wind systems. Similar results are obtained for seasonal EIs and ENs, both CEs and AEs.

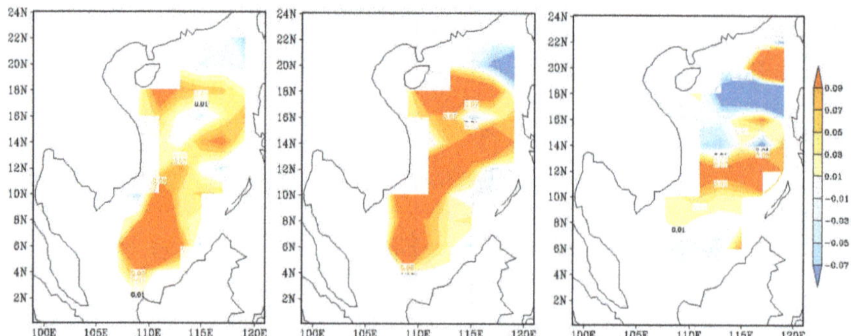

Fig. 10.   First EOFs of annual anomalies of mean diameter (ED; left, 14% of variance) and of mean intensity (EI, middle, 12% of variance), and number of CEs (EN; right; 22% of variance) on a coarsened 2-degree grid.

An even stronger coarsening was done to boxes of size 2°. The spectrum, shown in blue in Figure 8, is getting a little steeper, and the first two EOFs represent about 25% of the overall variance. Maybe, even the third eigenvalue is getting a little more robust. The first eigenvectors for the ED, the EI, and the EN for CEs are shown in Figure 10.

Interestingly, the principal components, i.e., the time coefficients, of the EOFs derived on the three different grids of cyclonic eddy diameters, are positively correlated. For the very noisy EOFs on the 0.1-degree grid, the correlation to that on the 1-degree grid is only 0.44, whereas for the 2-degree grid, it is a meager 0.32. The correlation between the PCs derived in the 1-degree grid and the 2-degree grid is 0.78. Thus, even if the EOFs on the finest grid must be considered mostly useless because of noise contamination, it represents to a minor extent the signal found on the coarsened grids.

The first pattern for the diameter has its center of activity in the southern SCS, with a mostly uniform sign; the spatial variability of the intensity is given by a large area with more intense and a small area near Luzon strait of less intense (and vice versa) CEs; the number of CEs is dominated by a multipole with an overall spatial mean of zero.

Similar results hold for AEs (Figure 11). The AE patterns are limited to the central and northern part of the SCS, reflecting the absence of AEs in the southern SCS. The first EOF of ED describes a general enlarging or shrinking of eddies, while the intensity and

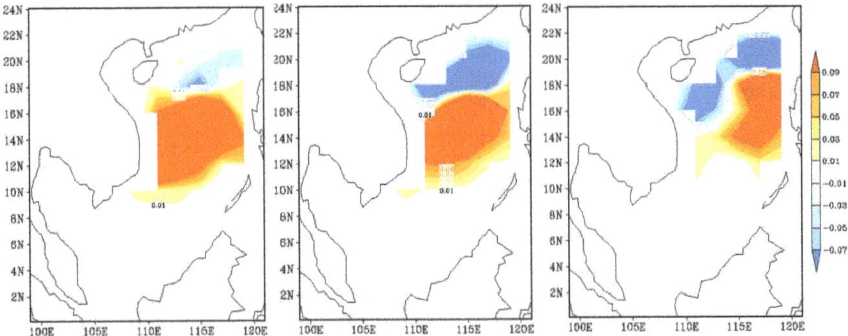

Fig. 11.   First EOFs of annual anomalies of mean diameter (ED; left, 16% of variance) and of mean intensity (EI, middle, 18% of variance), and number of AEs (EN; right; 19% of variance) on a coarsened 2-degree grid.

number are summarized by a dipole, favoring either the southern or the northern part of the Luzon strait throughflow.

Thus, building, in the spirit of empirical downscaling, links between spatially disaggregated distributions of eddy parameters to large-scale patterns of, say, regional wind, barotropic stream function, current shear, or vertical stability, will be a challenge. However, this will be dealt with in a subsequent paper. We hypothesize that the internal variability, unprovoked by changing large-scale conditions — thus "noise" in the spirit of (Tang *et al.*, 2019) — is a significant factor for the variations of eddy activity. Of course, for verifying the hypothesis, more analysis is needed.

5.3.   *Seasonal Variability*

In this section, the seasonal features of accumulated EN, mean EI, and ED are discussed.

More CEs are generated than AEs in all the 12 months (Figure 7(a)). The number of eddies generated in each month exhibits a significant seasonal variation. Most CEs are generated in the winter half year, with a peak of 364 eddies in March. More AEs are formed in the spring than in the autumn. The minimum number of eddies generated is only 105 in September.

For all 1709 AEs and 3331 CEs, the EIs and EDs of the peak eddy points are determined. The monthly means of EI and ED of peak eddy points (Figure 12(b) and 12(c)) exhibit different variations. AEs

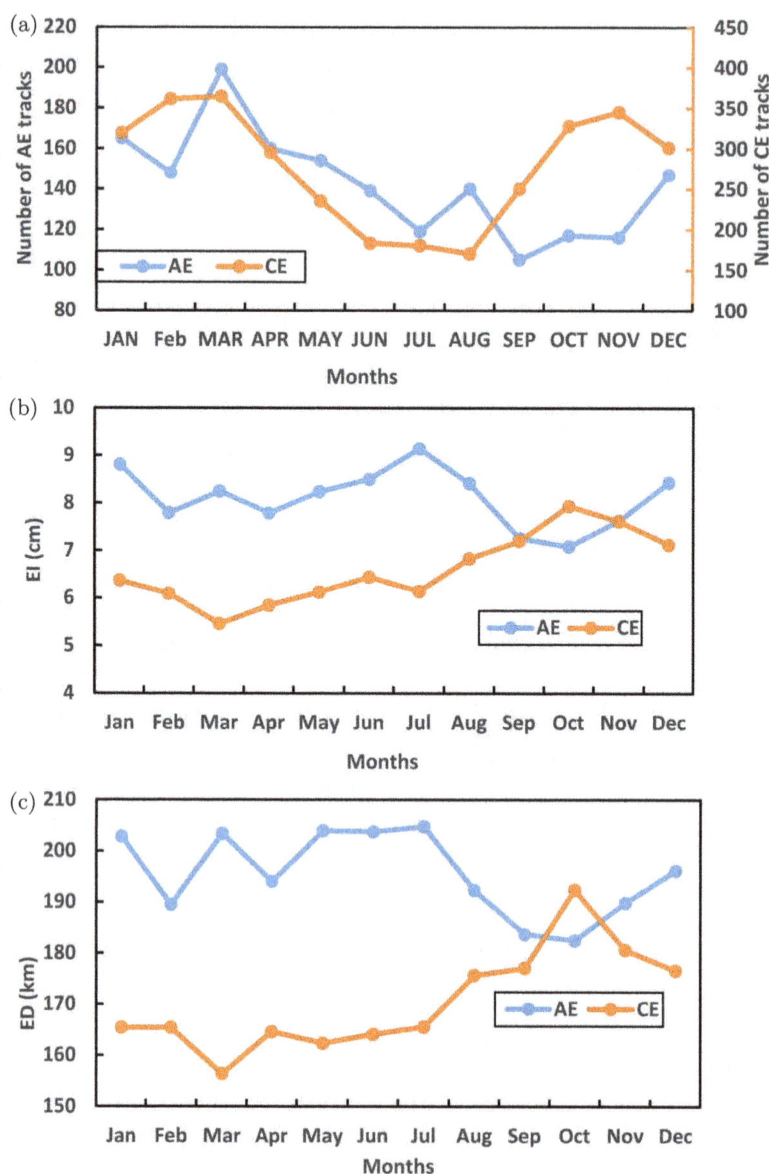

Fig. 12.   The annual cycle of the multiyear sum number of eddy tracks (a) and the means of EI (b) and ED (c) of the peak eddy points.

always have stronger intensities and larger sizes than CEs. In addition, the monthly EI and ED data reveal an annual cycle, peaking in July with EI = 9.14 cm and ED = 205 km for AEs, while these values peak in October for CEs with EI = 7.93 cm and ED = 192 km. From July to October, the ED and EI values of peak AEs decline sharply, from the maxima hitting the minima. On the contrary, rapid growth occurs in both the ED and EI variability of CEs and increase to the maxima during these 4 months. In terms of the monthly means of all 62,317 AE points and 115,133 CE points, an annual cycle is also apparent (not shown), AEs peak in August, with EI = 7.01 cm and ED = 174 km. CEs reach peaks in November when EI = 5.58 cm and ED = 147 km.

The presence of well-defined annual cycles points to the presence of a link between the eddy activity and large-scale conditions, in wind, currents of stability, which is undergoing annual variations (cf. Zhang *et al.*, 2017), in contradiction to the above-formulated hypothesis that the variability, is mostly generated internally.

Therefore, we have examined the distributions of differences between months, when the diameter, the intensity EI, and the number EN have their multiyear maximum and multiyear minimum. This is done for all years, within which at least one eddy is found. The result is presented in Figure 13 as Box-Whisker diagrams, with minimum, lower quartile, median (mean quartile), upper quartile, and maximum, and, sometimes outliers (as dots).

- The annual cycle is not very stable. In up to 25% and more of all years, the normally positive difference between maximum and minimum is reversed.
- The variability for anticyclonic eddies is stronger, in particular for diameter and intensity, so that referring to an annual cycle is of limited usage.
- The annual cycle of cyclonic eddies is more stable, with less than 25% of years showing a reversed sign, in terms of number, diameter, and intensity.

On the other hand, the annual cycle, as being manifested in the large-scale monsoonal changes in winds and large-scale currents, is strong (not shown, but see also Zhang and von Storch, 2017). Therefore, we conclude that also the annual cycle of the eddy properties

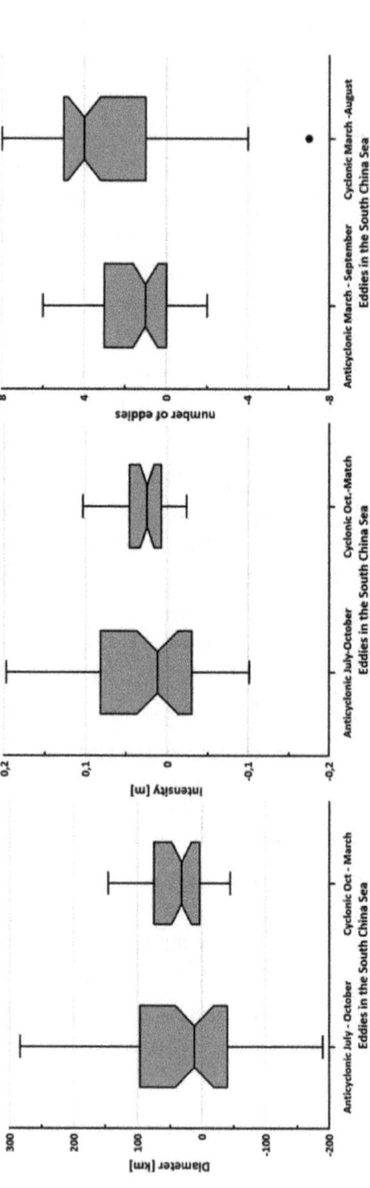

Fig. 13. Box-Whisker diagram of the annual differences of eddy parameters in the month of multiyear maximum and multiyear minimum, for describing the stability of the annual cycles. The gray box represents all cases between the upper and lower quartile, and the contraction the medium. The whiskers mark the minima and maxima, and the lone dot an outlier.

is significantly affected by factors unrelated to the large-scale forcing which represent the annual march of the sun.

5.4. *Interannual Variability*

To assess the long-term variability of the eddy properties, the annual time series of the number of eddies, EIs, and EDs have been plotted (Figure 14). Interannual variability dominates the annual series of eddy number. There is some decadal variability with maxima in the 2000s and 1970s, and minima in the 1980s and 1950s.

Xiu *et al.* (2010) published the annual number of generated eddies derived from the AVISO satellite data and their own modelled data

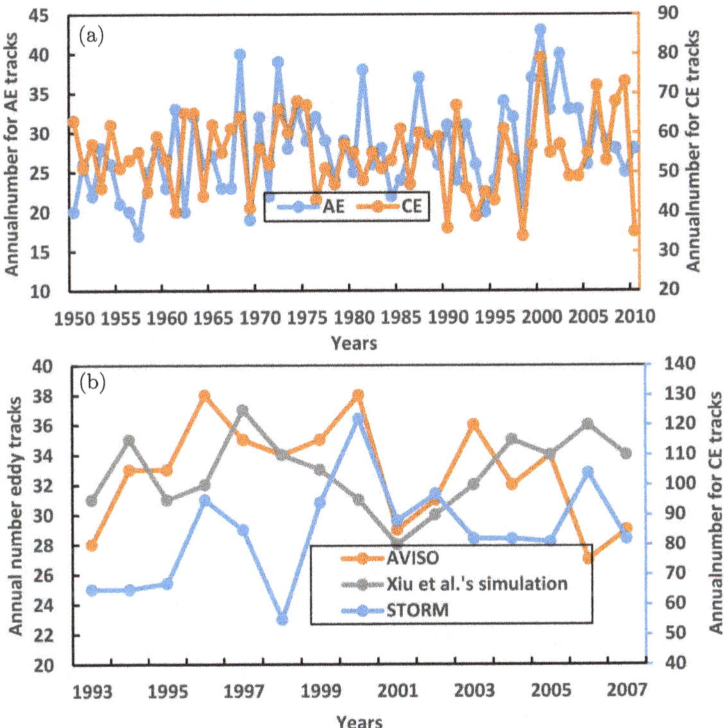

Fig. 14.   The time series of the annual number of eddy tracks. (a) Number found in STORM using the standard set-up of RI = 3 mm and $\mathrm{RI_{max}}$ = 6 mm. The axis on the left (right) is for AEs (CEs). (b) Series according to STORM with a standard set-up, AVISO satellite data (Xiu *et al.*, 2010) and the simulation by Xiu *et al.* (2010). The axis for the STORM result is on the right.

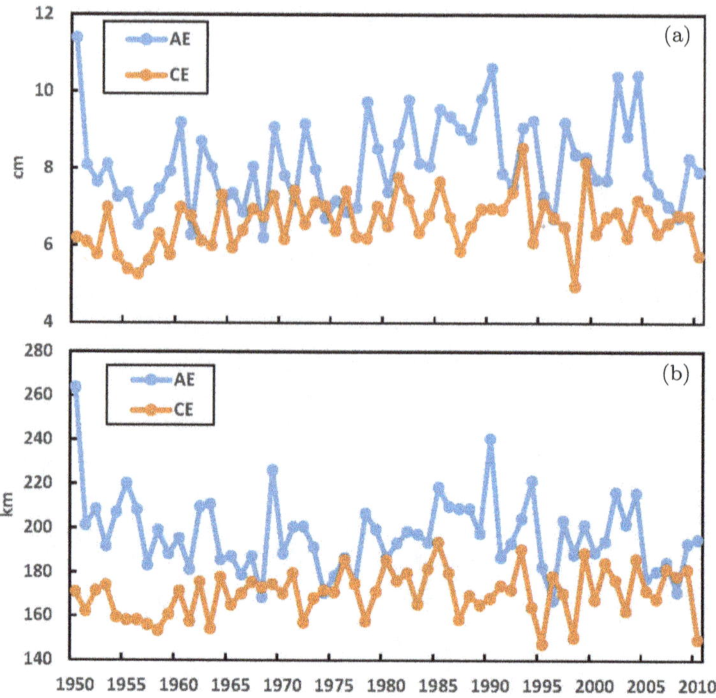

Fig. 15.   The annual time series of EIs (a) and EDs (b) of peak eddy points.

during 1993–2007 by using the W-based eddy detection and tracking method. Figure 14(b) combines our results from STORM and their results to further assess our results. Due to their strict criteria for eddy size, water depth, and so on, the number of eddy tracks identified in their study is less than ours. In terms of variability, STORM outperforms Xiu *et al.*'s simulation when compared with the AVISO satellite data. The correlation coefficient of the annual number of eddies during 1993–2007 between AVISO and STORM is 0.21, which is much higher than the correlation coefficient of 0.03 between AVISO and Xiu *et al.*'s simulation. This comparison suggests that STORM simulation describes more realistically the variability of the travelling eddies in the SCS, which should benefit from the realistic reproduction of the SCS dynamics (Zhang and von Storch, 2017).

With regard to the annual mean EI and ED values of the peak eddy points (Figure 15), the annual series also presents mostly interannual variability with very little decadal variability and no trends.

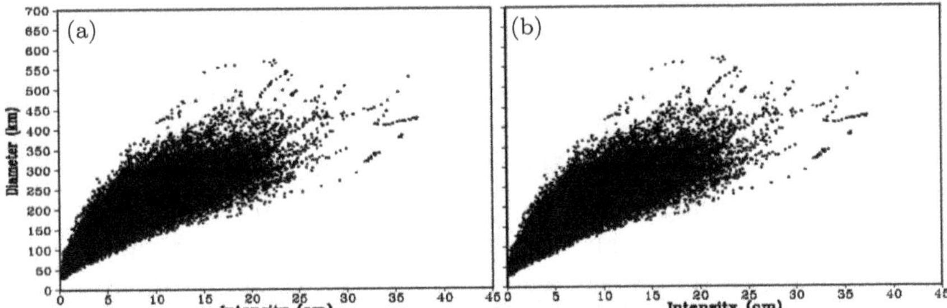

Fig. 16. The scatter diagram between eddy diameter and eddy intensity of AEs (a) and CEs (b) derived for all points along all eddy tracks.

AEs on average have higher EIs and larger EDs than CEs in most years, which is consistent with the climatological results. In the "Climatological characteristics" section, the skewness analysis and the eddy property distribution reveal much more CEs with small size and low intensity than such AEs in the SCS. This feature of the eddies in the SCS may be related to the different generation mechanisms of AEs and CEs.

The AEs vary more strongly than the CEs. The annual mean EDs and EIs are strongly correlated, with correlation coefficients of 0.76 and 0.68 for AEs and CEs, respectively. Thus, strong eddies tend to be large, as is illustrated by the scatter diagrams of the two parameters (Figure 16). The annual mean EDs and EIs of all eddy points (not shown) show similar features, with EI-ED correlation coefficients of 0.88 and 0.78 for AEs and CEs, respectively.

6.   Summary and Discussion

In this work, the long-term variability and statistical features of the travelling eddies in SCS during the period 1950–2010 have been documented by means of the eddy-permitting and multidecadal output from the ocean global model simulation STORM, which was forced by a 6-hourly NCEP atmospheric reanalysis. STORM simulation has a horizontal resolution of 10 km on average, which enables it to resolve eddies in the SCS.

For a limited time, beginning in 1993, the satellite data (AVISO) of the SSHA in the SCS are available — we have used these data to find out if the STORM simulation data generate consistent numbers and characteristics of eddies in the SCS.

Sometimes the claim is made that "Since STORM is running in a realistic hindcast mode, in principle, it should be possible even to reproduce the eddy structures that are observed at a specific day." This claim is wrong, since the model is run in the "climate mode," i.e., driving only through upper and lateral boundary values with no deterministic influence by initial values. What happens in the interior of the South China Sea is in part related to the large-scale forcing of atmospheric states (in particular winds) and to lateral boundary conditions, but in part generated internally ("noise"; see Tang *et al.*, 2019). Thus, we should be able to successfully hindcast the large-scale state of the dynamics in the South China Sea, but not the small scales (even if we have to add the caveat that it is unknown if periods of "intermittent divergence in phase space" take place in such an oceanic set-up; cf. Weisse *et al.*, 2000). This is also the case in the South China Sea. To demonstrate this challenge, we have randomly selected a day, namely 31 December 2010, and have compared the total SSH anomalies as given by AVISO and by STORM, a spatial low-pass components (first 10 EOFs) and a spatial high-pass (difference between full and low-pass filtered fields). The first is quite similar on the large scale, but the difference shows small-scale patterns all over the South China Sea (Figure 17).

Unfortunately, the accuracy of local data in the AVISO gridded data set suffers from significant uncertainties, of the order of 1 cm and more (Taburet *et al.*, 2018); also, the spatial resolution in AVISO is less than in STORM. A large underestimation of eddy activity in the AVISO due to its coarse spatial resolution is found by Amores *et al.* (2018). This contributes to differences between AVISO and STORM, but some of the differences may be due to the insufficiencies of the model simulations. Unfortunately, we cannot quantify the different contributions. But, given the large uncertainty of the AVISO data, we may diagnose a general consistency between AVISO and STORM. Unfortunately, in past studies, which have employed only AVISO data for describing cases and statistics of eddies, these inaccuracies have not always been recognized. Our analysis extends these earlier studies, providing more detail, such as the stability of the annual cycles, the strength of the link of warm/anticyclonic and cold/cyclonic eddies, and the organization of spatial co-variability

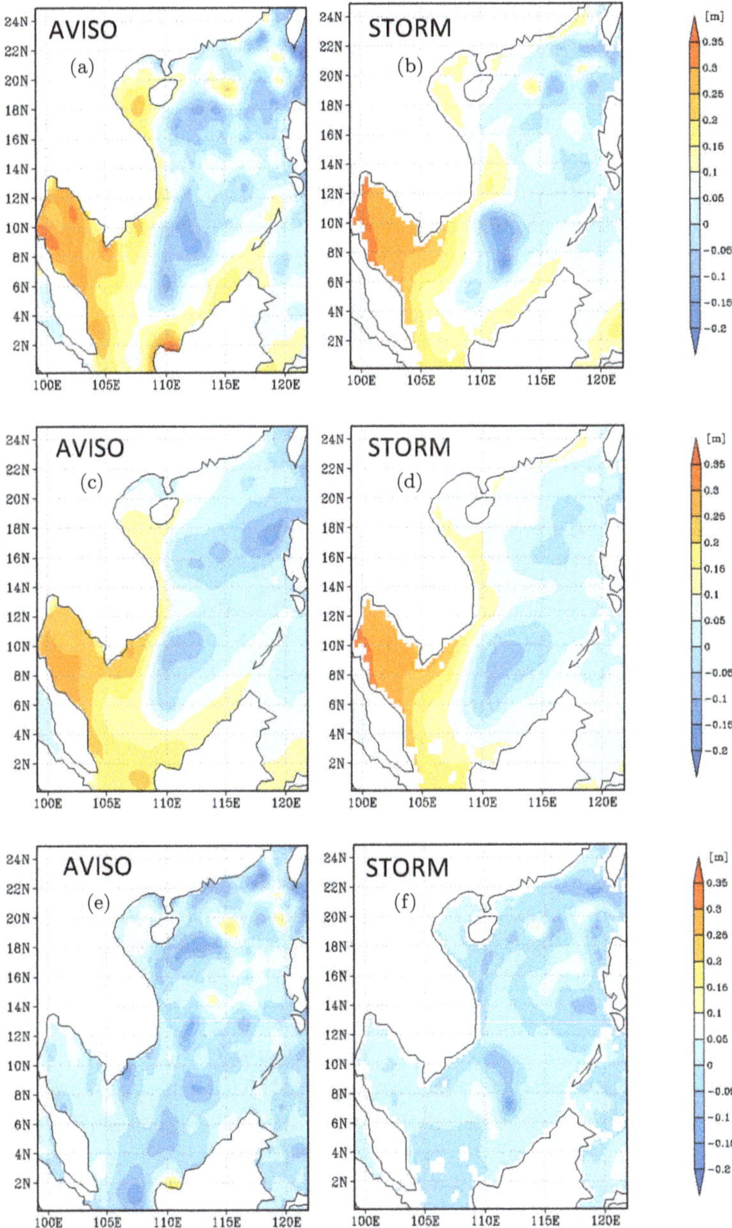

Fig. 17. (a) and (b) are the anomalous fields of the daily detrended SSHA on 2010-12-31 from AVISO and STORM. (c) and (d) are similar to (a) and (b), but the anomalous fields are reconstructed from the first 10 EOFs. (e) and (f) is the difference between the anomalous fields and the reconstructed fields. (Note the same scales).

(EOFs). By having a much longer time series (61 years), we were also able to examine multiyear variability.

For deriving the eddy activity from STORM SSHA data, the M-based eddy detection and tracking method, which makes use of the discrete SSHA fields, has been developed. This algorithm avoids the problems from the calculations of differential or integral operators. In addition, eddies are identified as the moving extremes in space and time. Stationary eddies are not considered in this work.

On average, 28.0 AEs and 54.6 CEs per year are found in the STORM data in the SCS according to the criteria and the parameters in the M-based detection and tracking procedure. Those tracks cover 62,317 points (for AEs) and 115,133 points (for CEs), which means that the average lifespan of AEs is 36.5 days and that of CEs is 34.6 days. More CEs are detected than AEs. The lifespans of eddies in the SCS range from 6 to 293 days and travel up to 1988 km. The spatial distribution results show that the most frequent regions for eddy tracks to travel are the Luzon Strait and towards the continental slope to the Vietnam coast. Those eddy points have an EI range of 0.11 to 37.3 cm and an ED range of 30 to 633 km, while the strongest eddy points along the eddy tracks have an EI range of 0.55 to 37.3 cm and an ED range of 43 to 631 km. For both kinds of eddies, the stronger eddies tend to be larger in size.

CEs are much more active than AEs, but the AEs with high intensities or large diameters are more frequent than similar CEs. Our algorithm and STORM simulation allow much more small eddies to be kept, compared with the previous work (Chen *et al.*, 2011). In addition, AVISO often largely underestimates the eddy activity (Amores *et al.*, 2018), because the spatial resolution is too coarse to capture the small eddies. Then many more small eddies appear in CEs instead of AEs in our work, which could contribute to the large difference in the detected eddy number of AEs and CEs.

When doing an EOF analysis, no significant spatial patterns are found, when the analysis is done on the original 0.1-degree grid. Since the numerical solution is obtained by the eigenanalysis of the smaller (61×61) XX' covariance matrix, instead of the much larger $X'X$ matrix, this strong contamination of the spatial co-variability by internal small-scale dynamics is not an unwanted artifact of solving large numerical problems but real, in consistence with the result of Tang *et al.* (2019). The absence of significant patterns seems inconsistent with the hypothesis that the spatial distribution of eddy activity would be noteworthily be constrained by some large-scale "drivers," such as large-scale wind patterns or current patterns. Instead, the formation of eddies is more an issue of internal variability, largely unprovoked by external drivers. However, more analysis and numerical experimentation are needed to arrive at more robust results.

The STORM data suggest that the variability in the eddy properties in the SCS is dominated by intra-annual variability. Eddy number peaks in March and the fewest number of eddies occurs in September for AE and in August for CEs. For the EI and ED of the peak eddy points, the peaks occur in July for AEs and October for CEs. When comparing the annual cycle across different years, we find that they differ strongly from year to year, also pointing to the above-mentioned internal variability.

The interannual variability is strong, while the decadal variability is weak, and a long-term trend is not found.

One purpose of this analysis was to prepare a data set that was suitable for the statistical downscaling of eddy properties. Our results presented here indicate that a spatially resolving estimation of eddy properties as conditioned by the large scales in the South China Sea cannot be done straightforward. This aspect will be worked out in more detail in a forthcoming paper. However, it may be said already here, that an alternative may be to use spatially aggregated eddy properties, such as the total number of eddies.

As a first attempt to quantify links to large-scale conditions, we have tried to relate the eddy parameters to ENSO conditions. El Niño is one of the most important phenomena in the tropical ocean. For the SCS, El Niño is considered significant because of its impact on the wind stress curl over the SCS and the ocean circulation in the SCS (Fang *et al.*, 2006; Wang *et al.*, 2006). In addition, with regard to the annual time series of eddy number and EI when considering all eddy points, a simultaneous drop is observed during the strong El Niño year of 1998. The study of Tuo *et al.* (2018) found a changing influence of El Niño on the eddy activity in the SCS around 2004. Based on the altimeter data (1993–2013), they found significant and negative correlation between El Niño and the eddy activity showed up before 2004, but disappeared afterward.

Table 1.   The correlation coefficients between El Niño 3.4 index and the eddy properties.

El Niño 3.4 index vs	Genesis number EN	Intensity EI	Diameter ED
AEs	0.09	0.09	0.08
CEs	−0.11	−0.04	0.04

However, we found that the variability of the eddy activity shows weak or no correlation with El Niño. When the El Niño 3.4 index (https://psl.noaa.gov/data/climateindices/list/) is correlated with the number of generated eddies and the eddy parameters of all eddy points on a monthly basis, we find very small correlations, which seem to be irrelevant (see Table 1). To take into account the effect of serial correlations (Zwiers and von Storch, 1995), we assume that a 12-month time difference is needed to have somewhat independent samples. Using this "efficient sample size," we found that none of the links were significant.

Tuo *et al.* (2018) investigated 10-year sliding correlations of eddy number and El Niño 3.4 index and suggested significant negative correlations. However, in their work, after 2005, the negative correlation weakens and reach zero around 2008. After that, weak positive correlation emerged. In different time windows, their correlations are quite different and sometimes even opposite. Our work focuses on their relationship in the whole 61 years, which differs from Tuo *et al.*'s work (2018).

We conclude that a relevant link to the ENSO dynamics on the tropical Pacific does not exist for the formation of eddies in the SCS, again supporting our hypothesis that the main source of variability is internal.

7.   Appendix

To evaluate the influence of the RI threshold set in the eddy-detecting algorithm, we conducted a comparative analysis with increased values of $RI = 6$ mm and $RI_{max} = 10$ mm instead of the standard set of $RI = 3$ mm and $RI_{max} = 6$ mm. The result is shown in Figure 18. The variability of the revised series is similar to the original series

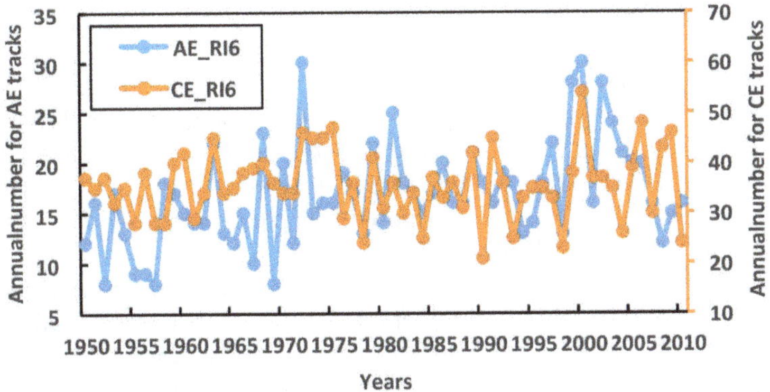

Fig. 18. The time series of the annual number of eddy tracks found in STORM using the set-up of RI = 6 mm and RI_{max} = 10 mm. The axis on the left (right) is for AEs (CEs).

(Figure 14(b)), with correlation coefficients of 0.82 and 0.80 for AEs and CEs, respectively. Additionally, both figures present many more CEs than AEs. It can be concluded that the selected threshold has little effect on the variability of the number of eddy tracks.

Funding

We are grateful to the China Scholarship Council for financially supporting the first author's study abroad (No. 201406330048). This work is financially supported by the National Key Research and Development Plan, 2016YFC1401300 and Taishan Scholars program, and we appreciate that. One of the authors also has been financially supported by the National Key Research and Development Plan (2017YFA0604101).

Acknowledgements

The authors express gratitude to the Max Planck Institute of Meteorology (MPI) for providing the STORM data set and to the German Climate Computing Center (DKRZ) for suggestions, as well as AVISO providing AVISO satellite data, and ICDC of Hamburg University for advice on using AVISO data.

References

Amores, A., Jorda, G., Arsouze, T. and Le Sommer, J. (2018). Up to what extent can we characterize ocean eddies using present-day gridded altimetric products? *Journal of Geophysical Research-Oceans* **123**, 10, pp. 7220–7236, doi:10.1029/2018JC014140.

AVISO. (1996). AVISO user handbook: Merged topex/poseidon products, *Techn. Rep. AVI-NT-02-101-CN* 3.

AVISO. (2015). Ssalto/duacs user handbook:(m) sla and (m) adt near-real time and delayed time products, reference: CLS-DOS-NT-06-034, 74.

Chaigneau, A., Gizolme, A. and Grados, C. (2008). Mesoscale eddies off Peru in altimeter records: Identification algorithms and eddy spatio-temporal patterns, *Progress in Oceanography* **79**, 2–4, pp. 106–119, doi:10.1016/j.pocean.2008.10.013.

Chelton, D. B., Schlax, M. G. and Samelson, R. M. (2011). Global observations of nonlinear mesoscale eddies, *Progress in Oceanography* **91**, 2, pp. 167–216, doi:10.1016/j.pocean.2011.01.002.

Chen, G., Hou, Y. and Chu, X. (2011). Mesoscale eddies in the South China Sea: Mean properties, spatiotemporal variability, and impact on thermohaline structure, *Journal of Geophysical Research-Oceans* **116**, p. C06018, doi:10.1029/2010JC006716.

Chen, G., Gan, J., Xie, Q., Chu, X., Wang, D. and Hou, Y. (2012). Eddy heat and salt transports in the South China Sea and their seasonal modulations, *Journal of Geophysical Research-Oceans* **117**, p. C05021, doi:10.1029/2011JC007724.

Chu, X., Dong, C. and Qi, Y. (2017). The influence of ENSO on an oceanic eddy pair in the South China Sea, *Journal of Geophysical Research-Oceans* **122**, 3, pp. 1643–1652, doi:10.1002/2016JC012642.

Chu, X., Xue, H., Qi, Y., Chen, G., Mao, Q., Wang, D. and Chai, F. (2014). An exceptional anticyclonic eddy in the South China Sea in 2010, *Journal of Geophysical Research-Oceans* **119**, 2, pp. 881–897, doi:10.1002/2013JC009314.

Dufau, C., Orsztynowicz, M., Dibarboure, G., Morrow, R. and Le Traon, P.-Y. (2016). Mesoscale resolution capability of altimetry: Present and future, *Journal of Geophysical Research-Oceans* **121**, 7, pp. 4910–4927, doi:10.1002/2015JC010904.

Evadzi, P. I. K. (2017). *Regional Sea-level at the Retreating Coast of Ghana Under a Changing Climate*, Ph.D. thesis, Staats-und Universitätsbibliothek Hamburg Carl von Ossietzky.

Faghmous, J. H., Frenger, I., Yao, Y., Warmka, R., Lindell, A. and Kumar, V. (2015). A daily global mesoscale ocean eddy dataset from satellite altimetry, *Scientific Data* **2**, p. 150028, doi:10.1038/sdata.2015.28.

Fang, G., Chen, H., Wei, Z., Wang, Y., Wang, X. and Li, C. (2006). Trends and interannual variability of the South China Sea surface winds, surface height, and surface temperature in the recent decade, *Journal of Geophysical Research-Oceans* **111**, C11, p. C11S16, doi:10.1029/2005JC003276.

Feng, B., Liu, H., Lin, P. and Wang, Q. (2017). Meso-scale eddy in the South China Sea simulated by an eddy-resolving ocean model, *Acta Oceanologica Sinica* **36**, 5, pp. 9–25, doi:10.1007/s13131-017-1058-3.

Geng, W., Xie, Q., Chen, G., Zu, T. and Wang, D. (2016). Numerical study on the eddy-mean flow interaction between a cyclonic eddy and Kuroshio, *Journal of Oceanography* **72**, 5, pp. 727–745, doi:10.1007/s10872-016-0366-0.

Geng, W., Xie, Q., Chen, G., Liu, Q. and Wang, D. (2017). A three-dimensional modeling study on eddy-mean flow interaction between a Gaussiantype anticyclonic eddy and Kuroshio, *Journal of Oceanography* pp. 1–15.

Hallberg, R. (2013). Using a resolution function to regulate parameterizations of oceanic mesoscale eddy effects, *Ocean Modelling* **72**, pp. 92–103, doi:10.1016/j.ocemod.2013.08.007.

Itoh, S. and Yasuda, I. (2010). Water mass structure of warm and cold anticyclonic eddies in the western boundary region of the subarctic North Pacific, *Journal of Physical Oceanography* **40**, 12, pp. 2624–2642, doi:10.1175/2010JPO4475.1.

Kalnay, E., Kanamitsu, M., Kistler, R., Collins, W., Deaven, D., Gandin, L., Iredell, M., Saha, S., White, G. and Woollen, J. (1996). The NCEP/NCAR 40-year reanalysis project, *Bulletin of the American Meteorological Society* **77**, 3, pp. 437–472.

Lawley, N. D. (1956). Tests of significance for the latent roots of covariance and correlation matrices, *Biometrika* **43**, 1–2, pp. 128–136.

Lee, T. N., Yoder, J. and Atkinson, L. (1991). Gulf-stream frontal eddy influence on productivity of the southeast United-States continental-shelf, *Journal of Geophysical Research-Oceans* **96**, C12, pp. 22191–22205, doi:10.1029/91JC02450.

Li, J., Zhang, R. and Jin, B. (2011). Eddy characteristics in the northern South China Sea as inferred from Lagrangian drifter data, *Ocean Science* **7**, 5, pp. 661–669, doi:10.5194/os-7-661-2011.

Li, J., Jing, Z., Jiang, S., Wang, D. and Yan, T. (2015). An observed cyclonic eddy associated with boundary current in the northwestern South China Sea, *Aquatic Ecosystem Health & Management* **18**, 4, pp. 454–461, doi:10.1080/14634988.2015.1100959.

Lin, H., Hu, J., Liu, Z., Belkin, I. M., Sun, Z. and Zhu, J. (2017). A peculiar lens-shaped structure observed in the South China Sea, *Scientific Reports* **7**, p. 478, doi:10.1038/s41598-017-00593-y.

Mahadevan, A. (2014). Ocean science eddy effects on biogeochemistry, *Nature* **506**, 7487, pp. 168–169, doi:10.1038/nature13048.

Nan, F., Xue, H. and Yu, F. (2015). Kuroshio intrusion into the South China Sea: A review, *Progress in Oceanography* **137**, pp. 314–333, doi:10.1016/j.pocean.2014.05.012.

Nan, F., He, Z., Zhou, H. and Wang, D. (2011). Three long-lived anticyclonic eddies in the northern South China Sea, *Journal of Geophysical Research-Oceans* **116**, p. C05002, doi:10.1029/2010JC006790.

Nencioli, F., Dong, C., Dickey, T., Washburn, L. and McWilliams, J. C. (2010). A vector geometry-based eddy detection algorithm and its application to a high-resolution numerical model product and high-frequency radar surface velocities in the Southern California bight, *Journal of Atmospheric and Oceanic Technology* **27**, 3, pp. 564–579, doi:10.1175/2009JTECHO725.1.

Pohlmann, T. (1987). A three dimensional circulation model of the South China Sea, *Elsevier Oceanography Series*, **45**, pp. 245–268.

Storto, A., Masina, S. and Navarra, A. (2016). Evaluation of the CMCC eddy-permitting global ocean physical reanalysis system (C-GLORS, 1982–2012) and its assimilation components, *Quarterly Journal of the Royal Meteorological Society* **142**, 695, pp. 738–758, doi:10.1002/qj.2673.

Sun, Z., Zhang, Z., Zhao, W. and Tian, J. (2016). Interannual modulation of eddy kinetic energy in the northeastern South China Sea as revealed by an eddy-resolving OGCM, *Journal of Geophysical Research-Oceans* **121**, 5, pp. 3190–3201, doi:10.1002/2015JC011497.

Taburet, G. and Pujol, M.-I. (2018). Sea level tac-duacs products. Available online: https://catalogue.marine.copernicus.eu/documents/QUID/CMEMS-SL-QUID-008-032-051.pdf.

Tang, S., von Storch, H., Chen, X. and Meng, Z. (2019). "Noise" in climatologically driven ocean models with different grid resolution, *Oceanologia* **61**, 3, pp. 300–307, doi:10.1016/j.oceano.2019.01.001.

Tim, N., Zorita, E. and Hünicke, B. (2015). Decadal variability and trends of the Benguela upwelling system as simulated in a high-resolution ocean simulation, *Ocean Science* **11**, 3, pp. 483–502, doi:10.5194/os-11-483-2015.

Tuo, P., Yu, J.-Y. and Hu, J. (2018). The changing influences of ENSO and the Pacific meridional mode on mesoscale eddies in the South China Sea, *Journal of Climate* **32**, 3, pp. 685–700, doi:https://doi.org/10.1175/JCLI-D-18-0187.1.

von Storch, H. and Hannoschöck, G. (1984). Empirical orthogonal function-analysis of wind vectors over the tropical pacific region — comment, *Bulletin of the American Meteorological Society* **65**, 2, pp. 162–162.

von Storch, J.-S., Haak, H., Hertwig, E. and Fast, I. (2016). Vertical heat and salt fluxes due to resolved and parameterized meso-scale eddies, *Ocean Modelling* **108**, pp. 1–19, doi:10.1016/j.ocemod.2016.10.001.

von Storch, J.-S., Eden, C., Fast, I., Haak, H., Hernandez-Deckers, D., Maier-Reimer, E., Marotzke, J. and Stammer, D. (2012). An estimate of the Lorenz energy cycle for the world ocean based on the 1/10 degrees STORM/NCEP simulation, *Journal of Physical Oceanography* **42**, 12, pp. 2185–2205, doi:10.1175/JPO-D-12-079.1.

Wang, D., Xu, H., Lin, J. and Hu, J. (2008). Anticyclonic eddies in the Northeastern South China Sea during winter 2003/2004, *Journal of Oceanography* **64**, 6, pp. 925–935, doi:10.1007/s10872-008-0076-3.

Wang, D., Liu, Q., Huang, R. X., Du, Y. and Qu, T. (2006). Interannual variability of the South China Sea throughflow inferred from wind data and an ocean data assimilation product, *Geophysical Research Letters* **33**, 14, p. L14605, doi:10.1029/2006GL026316.

Wang, G. H., Su, J. L. and Chu, P. C. (2003). Mesoscale eddies in the South China Sea observed with altimeter data, *Geophysical Research Letters* **30**, 21, p. 2121, doi:10.1029/2003GL018532.

Wang, Q., Zeng, L., Zhou, W., Xie, Q., Cai, S., Yao, J. and Wang, D. (2015). Mesoscale eddies cases study at Xisha waters in the South China Sea in 2009/2010, *Journal of Geophysical Research-Oceans* **120**, 1, pp. 517–532, doi:10.1002/2014JC009814.

Weisse, R., Heyen, H. and von Storch, H. (2000). Sensitivity of a regional atmospheric model to a sea state dependent roughness and the need of ensemble calculations, *Monthly Weather Review* **128**, pp. 3631–3642.

Xiu, P., Chai, F., Shi, L., Xue, H. and Chao, Y. (2010). A census of eddy activities in the South China Sea during 1993–2007, *Journal of Geophysical Research-Oceans* **115**, p. C03012, doi:10.1029/2009JC005657.

Yi, X., Hünicke, B., Tim, N. and Zorita, E. (2017). The relationship between Arabian sea upwelling and Indian Monsoon Revisited in a high resolution ocean simulation, *Climate Dynamics* **50**, pp. 1–13.

Yuan, D., Han, W. and Hu, D. (2007). Anti-cyclonic eddies northwest of Luzon in summer-fall observed by satellite altimeters, *Geophysical Research Letters* **34**, 13, p. L13610, doi:10.1029/2007GL029401.

Zhan, P., Subramanian, A. C., Yao, F. and Hoteit, I. (2014). Eddies in the Red Sea: A statistical and dynamical study, *Journal of Geophysical Research-Oceans* **119**, 6, pp. 3909–3925, doi:10.1002/2013JC009563.

Zhang, M. and von Storch, H. (2017). Toward downscaling oceanic hydro-dynamics — suitability of a high-resolution OGCM for describing regional ocean variability in the South China Sea, *Oceanologia* **59**, 2, pp. 166–176, doi:10.1016/j.oceano.2017.01.001.

Zhang, M., von Storch, H. and Li, D. (2017). The effect of different criteria on tracking eddy in the South China Sea, *Research Activities in Atmospheric and Oceanic Modelling (WGNE Blue Book)* **2**.

Zhang, Z., Wang, W. and Qiu, B. (2014). Oceanic mass transport by mesoscale eddies, *Science* **345**, 6194, pp. 322–324, doi:10.1126/science.1252418.

Zhang, Z., Tian, J., Qiu, B., Zhao, W., Chang, P., Wu, D. and Wan, X. (2016). Observed 3D structure, generation, and dissipation of oceanic mesoscale eddies in the South China Sea, *Scientific Reports* **6**, p. 24349, doi:10.1038/srep24349.

Zu, T., Wang, D., Yan, C., Belkin, I., Zhuang, W. and Chen, J. (2013). Evolution of an anticyclonic eddy southwest of Taiwan, *Ocean Dynamics* **63**, 5, pp. 519–531, doi:10.1007/s10236-013-0612-6.

Zwiers, F. and von Storch, H. (1995). Taking serial-correlation into account in tests of the mean, *Journal of Climate* **8**, 2, pp. 336–351, doi:10.1175/1520-0442(1995)008<0336:TSCIAI>2.0.CO;2.

Chapter 4

"Noise" in Climatologically Driven Ocean Models with Different Grid Resolution[*]

S. Tang[†,‡], H. von Storch[†,‡], X. Chen[†], and M. Zhang[‡]

*†College of Oceanic and Atmospheric Sciences,
Ocean University of China, Qingdao, China*
*‡Institute of Coastal Research, Helmholtz-Zentrum Geesthacht,
Geesthacht, Germany*

The internally generated variability in the climate system, which is unrelated to any external factors, can be conceptualized as "noise". This noise is a constitutive element of high-dimensional nonlinear models of such systems. In a three-layer nested simulation, which is forced by climatological (periodic) atmospheric forcing and includes an (almost) global model, a West Pacific model, and South China Sea (SCS) model, we demonstrate that such "noise" builds also ocean models. They generate variability by themselves without an external forcing. The "noise" generation intensifies with higher resolution, which favors macroturbulence.

---

[*]This chapter was originally published in *Oceanologia*, 61, 300–307, doi: 10.1016/j.oceano.2019.01.001. Copyright (2019), with permission from Elsevier.

1. Introduction

The climate system is a high-dimensional macroturbulent system, which features many nonlinear processes. Based on the concept of the "stochastic climate model" (Hasselmann, 1976) the trajectory of the climate system can be described as that of an inert system subject to internally generated variations, which may be conceptualized as "noise". Here, we use the term "noise" to refer to variability which cannot be traced back to external "drivers". Instead, the variability is generated internally.

The net effect of very many interacting through many non-linearities is the creation of variability, which is well described by the of random processes. Whether this variability truly stems from stochastic processes, or, put differently, whether it's akin to God rolling a dice, is inconsequential. We're unable to unravel the complexities of the high-dimensional dynamics, but we observe that characterizing it as stochastic noise effectively serves its purpose.

Since macro-turbulence is an inherent part of the dynamics of the climate system then such "noise" should be present also in such models. Without such "noise", the climate system will be incompletely described and may lack significant features. Therefore, the recognition of the "noise" is helpful for scientists to explore the climate dynamics and modeling.

"Noise" represents for certain issues a nuisance (hiding real effects, as for instance when deriving eddy statistics from satellite data) but is also constitutive for the dynamical properties of the climate system (von Storch *et al.*, 2001). "Noise", i.e., unprovoked, internal variability, has significant implications for issues like "detection and attribution of climate change" (Hasselmann, 1993) and for numerical experimentation with climate models (Chervin and Schneider, 1976; Weisse *et al.*, 2000).

Since the nonlinear high-dimensional ocean system is part of the climate system, we suggest that the ocean system should also generate significant "noise", which is unrelated to any external factors (atmospheric forcing, lateral boundary conditions, and so on). Moreover, since "noise" takes place in climate models describing macro-turbulence, we suggest that the formation of "noise" in the ocean model may intensify with ocean model resolution increasing.

Some may find our suggestion about noise in ocean models almost trivial; indeed, in the framework of the stochastic climate model, it is mostly so. However, most climate modelers hardly know about the stochastic climate model, and it seems that many climate scientists are not aware of this unprovoked variability. There seem to be quarter in the climate science community, where efforts are made to find "explanations" for whatever what appears as not normal, but which may be simply the effect of this internal variability.

In the present study, we use a three-layer nested numerical simulation, which is subject to climatological atmospheric forcing, to test our hypothesis. The concerned region in this study is the South China Sea (SCS). The existing and intensity of "noise" in the SCS will be discussed in the framework of this three-layer nested simulation. The model resolutions change from coarse to fine, so the "noise" generation is conditioned by different model resolution.

The present chapter is organized as follows. A brief introduction the simulation setup is given in Section 2. The results of the simulation and "noise" in the simulation are presented in Section 3. Conclusions are summarized in Section 4.

The subject of this chapter is not finding out how well the simulations of the dynamics of the SCS are reproducing observed features. Such studies have been plentiful (e.g., Wang *et al.*, 2006; Zhang and von Storch, 2016), but the issue dealt with here is merely the conceptual issue of "noise" generation.

2. Simulation Setup

The ocean model used in this study is the Hybrid Coordinate Ocean Model (HYCOM). The HYCOM used in this study is a primitive equation ocean general circulation model. Its vertical coordinates are isopycnic in the open, stratified ocean, but smoothly change to z coordinates in the weakly stratified upper-ocean mixed layer, and change to terrain-following sigma coordinate in shallow water regions, and back to z-level coordinates in very shallow water (Bleck, 2002). The vertical mixing schemes chosen in this paper is the K-Profile Parameterization (KPP) scheme (Large *et al.*, 1994). The KPP scheme provides mixing throughout the water column with an abrupt but smooth transition between the vigorous mixing in the surface boundary and the relatively weak diapycnal mixing in the ocean interior.

A three-layer nested numerical simulation in HYCOM is performed, with an almost global model (60°S–54°N, 180°W–180°E) with 1° grid resolution, an embedded West-Pacific (WP) model (6°S–48°N, 95°E–146°E) with 0.2° grid resolution, an embedded South China Sea (SCS) model (4°N–24°N, 98.4°E–124.4°E) with 0.04° grid resolution. The different integration regions are shown in Figure 1.

The global model starts from the state of zero velocity and is run 50 model years. After 25 model years, the global model reaches a (cyclo) stationary state. The fields of last 25 model years in the global model are taken as the boundary forcing fields for the WP model. The fields in the 26th year of the global model are taken as the initial state of the WP model, which is run for 25 years. After 2 model years, the WP model trajectory becomes stationary. The fields of last 23 model years in the WP model are taken as the boundary forcing fields for the SCS model. The fields in the 3rd year of the WP model are taken as the initial state of the SCS model, and the SCS model is run 23 model years. After 2 model years, the SCS model reaches stationary.

The nested simulation is exposed to periodic climatological atmospheric forcing, with a fixed annual cycle, and without weather variability. The atmospheric forcing, including the net shortwave longwave radiation, precipitation, air relative humidity, air temperature, sea surface temperature, and wind speed, are

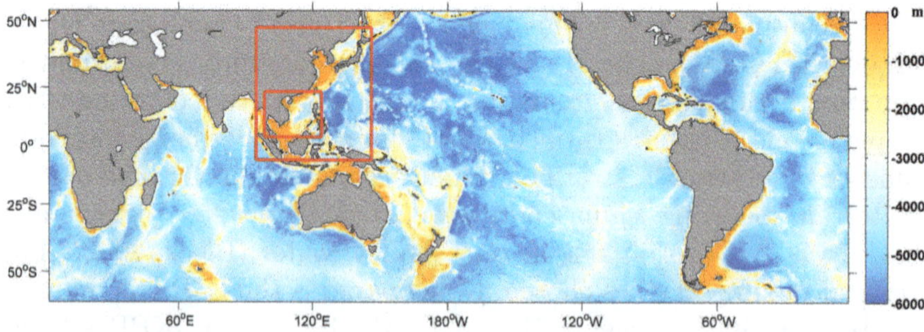

Fig. 1. The regions of the three-layer nested simulation, which includes an (almost) global model, a West-Pacific model, and a South China Sea (SCS) model.

all from the monthly Comprehensive Ocean-Atmosphere Data Set (COADS) climatology with 1 grid resolution.

The daily average data of the last 21 year in these three models are used to study the "noise". The barotropic streamfunction (BS) and sea surface height (SSH) simulated by these three models are discussed in this study.

3. Results

3.1. *The Amount of Variability of BS and SSH in These Three Models*

We measure the amount of variability by the variances of daily values at each grid point. The variance is calculated by subtracting the annual and semi-annual cycle. The annual and semi-annual cycle are fitted from the full-time series by harmonic analysis.

Table 1 lists variances of daily BS and SSH averaged across the SCS. The BS variances in the WP model are increased compared to the global model, and even larger variances can be found in the SCS model. The SSH variances in the WP model are approximately equal to that in the global model, and slightly larger variances are found in the SCS model.

The maps in Figure 2 show the spatial distributions of the logarithm of BS variances and SSH variances in the SCS in two seasons (summer and winter monsoon) simulated by these three models (global, WP, and SCS). From the global model to the SCS model, with the model resolution increasing, the BS variances in the whole

Table 1. Spatial averages of the daily BS variances and SSH variances in the SCS simulated by the three models.

Model	BS variance (Sv^2)			SSH variance (m^2)		
	Global	WP	SCS	Global	WP	SCS
Spring	0.5539	0.6141	1.1015	0.0004	0.0004	0.0005
Summer	1.0083	1.3374	2.2178	0.0010	0.0010	0.0012
Autumn	0.4382	0.6262	1.1739	0.0004	0.0004	0.0005
Winter	1.1281	1.3245	1.6055	0.0010	0.0009	0.0011

SCS strongly increase. In terms of SSH the changes are regionally different: While in the northern SCS, an increase from the global model to the SCS model is emerging, in the southern SCS, the differences are small. We suggest that the higher resolution model generates more or more intense eddies, which leads to intensified variability on all time scales (Hasselmann, 1976).

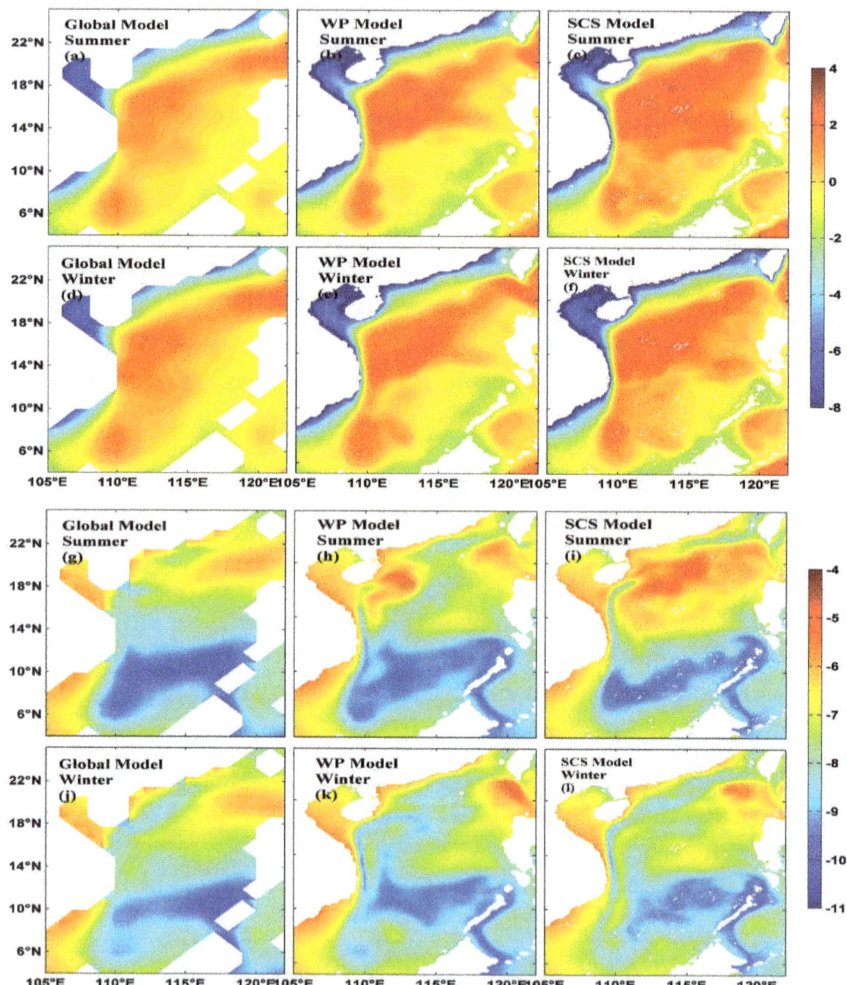

Fig. 2.   The spatial distributions of logarithm of daily variances of BS (top) and SSH (bottom) in the SCS simulated for summer and winter by the global model (a, d; g, j), the WP model (b, e; h, k) and the SCS model (c, f; i, l).

For comparing the eddies in the WP model and SCS model, we employed an eddy detection and tracking algorithm to search all eddies in the SCS. In order to compare, the SSH of the SCS model is interpolated from 0.04° resolution to 0.2° resolution. This algorithm only relies on the discrete SSHA (Zhang and von Storch, 2018). The potential eddy points are determined by the SSHA extrema in a moving 5×5 grid box according to the suggestion of Faghmous *et al.* (2015), with a relative intensity ≥ 5 mm. The relative intensity (RI) is defined by the absolute SSHA difference of the extrema and the mean SSHA of the other 24 neighbors in the box. The eddy centers at the consecutive time steps that are connected if their distance is ≤ 25 km (considering the eddy traveling speed ≤ 25 km per day) and the RI difference is ≤ 1.5 times of the RI in the previous time step. For the eddy tracks, the eddy should be tracked over at least 30 days.

Figure 3 shows that the annual eddy track numbers generated by the WP model are in all years smaller than those found in the SCS model. The eddy tracks numbers are comparable to those found by Chen *et al.* (2011) in satellite data.

Figure 4 shows the distribution of eddy occurrence for the 21 model years in the WP model and the SCS model, according to which the SCS model generates more eddies in the SCS, especially in the northern SCS.

We conclude that, since the atmosphere forcing of these three models is the same without any weather or interannual-variability,

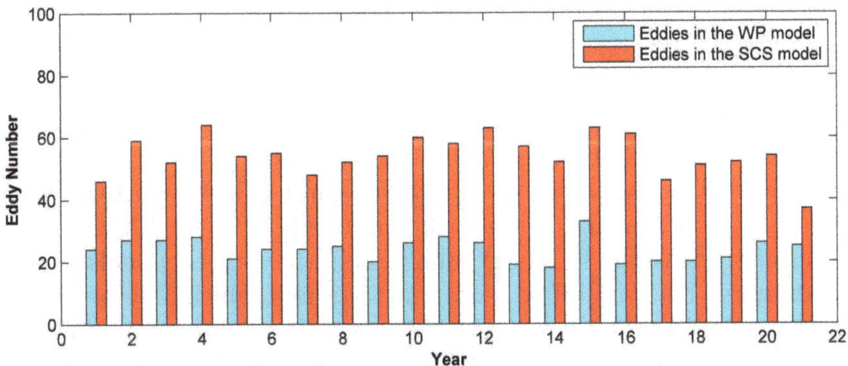

Fig. 3.   The annual number of eddy tracks in the WP simulation (blue bar) and in the SCS simulation (red bar).

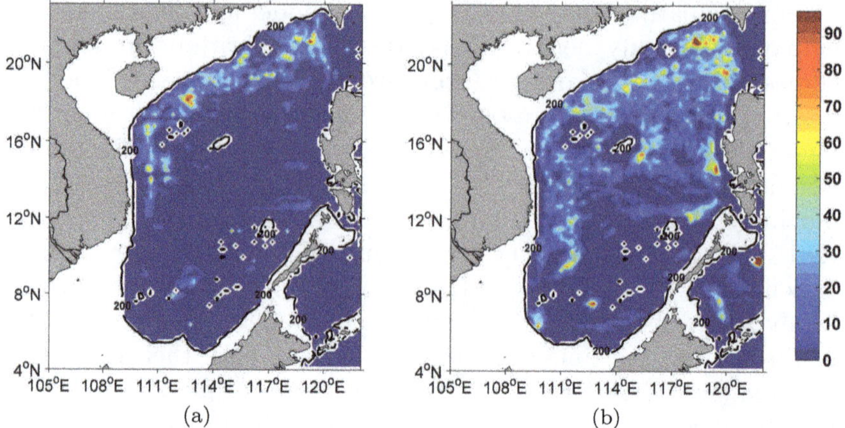

Fig. 4.   The total number of eddy occurrence for the 21 model years in the WP model (a) and SCS model (b). The units are the numbers of the eddy. The black lines indicate the 200 m isobaths (Tang *et al.*, 2019).

this increased variability in higher resolution models is internally generated by models.

3.2.   *Dominant Modes of BS and SSH in These Three Models*

Empirical Orthogonal Functions (EOF) decompose the time series of fields. A few orthogonal modes capture the main variability (Lorenz, 1956; von Storch and Zwiers, 1999). We apply the EOF decomposition to the BS and SSH fields in the South China Sea. The EOFs have been normalized so that the standard deviation of the time coefficients (principal component, PC) is 1 — so that the different intensity of the EOFs is given by the patterns.

Figures 5 and 6 show the first leading EOF patterns and associated standardized principal component (PC) for BS and SSH, respectively. All the PCs of BS and SSH in these three models appear stationary. The variability is not due to trends, which may be indicative for equilibrating from an initial state. Since the forcing in these three models is periodic and "without weather," these variations must be caused by internal dynamics, likely in the spirit of the "stochastic climate model."

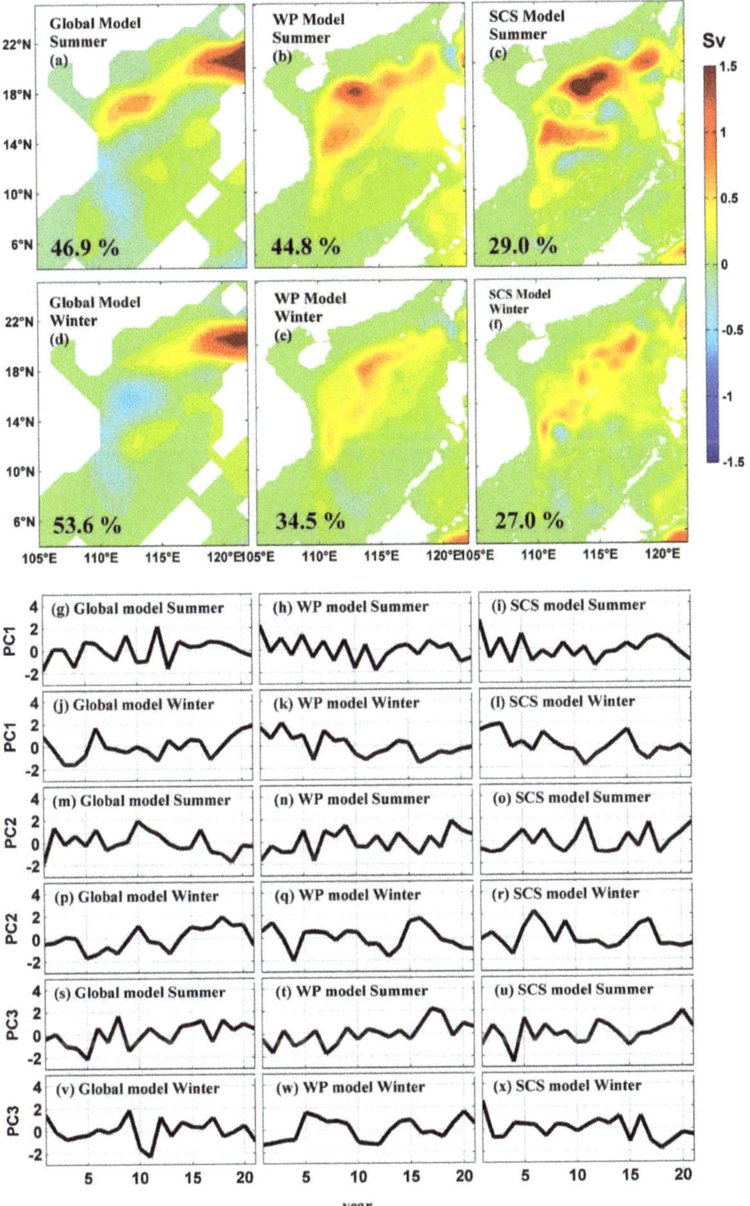

Fig. 5. The first leading EOF patterns (in Sv) (a–f) and associated standardized principal component (PC1) (g–l) of the summer and of winter mean BS in the SCS. The PC2 (m–r) and PC3 (s–x) of BS in the SCS in summer and winter are from the three-layer nested simulation. The numbers in the bottom left corner of (a–f) indicate the percentages of variance described by the first leading EOF.

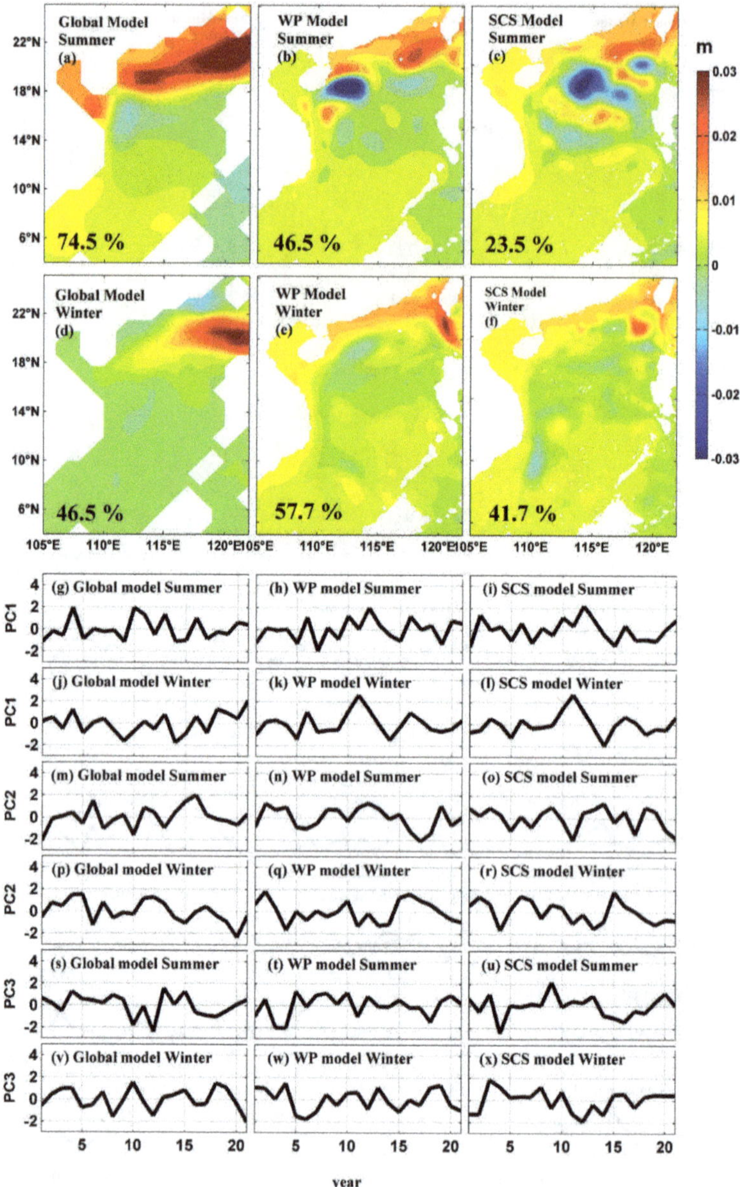

Fig. 6. The first leading EOF patterns of SSH (in m) (a–f) and associated standardized principal component (PC1) (g–l) of SSH in the SCS in summer and winter from the three-layer nested simulation. The PC2 (m–r) and PC3 (s–x) of SSH in the SCS in summer and winter are from the three-layer nested simulation. The numbers in the bottom left corner of (a–f) indicate the percentages of variance described by the first leading EOF.

Table 2. Cumulative percentages of variance described by the first three EOFs of BS (left) and SSH (right) in the SCS.

	BS			SSH		
	Global model	WP model	SCS model	Global model	WP model	SCS model
Spring	84.5	81.3	56.6	85.7	80.9	57.8
Summer	74.3	75.3	53.3	90.5	67.7	52.8
Autumn	82.2	64.7	54.1	90.8	68.2	67.0
Winter	76.5	72.1	48.9	80.8	77.8	63.2

Table 2 lists the sum of percentages of variance described by the first three EOFs for BS and SSH, respectively. We find that the leading EOFs of BS in the global model represent a higher percentage of cumulative variance than that in the WP model, and the leading EOFs of BS in the WP model explain the higher percentage of cumulative variance than that in the SCS model. From the global model to the SCS model, with model resolution increasing, the percentages of the sum of variances represented by the first three EOFs for BS decrease (except for summer).

The situations with of SSH are similar to that of BS, except for winter: The first leading EOF of SSH in the global model explains a lower percentage of variance than that in the WP model (Figure 6). We speculate that, because the resolution in the global model is too coarse, the Kuroshio invasion path in the global model is farther west and much broader than that in reality and in the WP and SCS simulations — which is reflected by a very strong first leading EOF (Figure 6). Therefore, the variability in the northern SCS in the global model is exaggerated compared to both the WP and SCS model.

We conclude that higher resolution models generate more "noise" so that a smaller percentage of the overall variability is represented by the dominant EOFs in the higher resolution models.

4. Conclusion

Basing on a three-layer nested simulation, which is forced by periodic climatological atmospheric forcing, featuring a global model, a West-Pacific model and South China Sea model, we find that high-dimensional nonlinear-systems like ocean dynamics generate variability by itself without an external forcing, and that "noise" generations are stronger in models with higher resolution, which favors the building of macroturbulence. Here "noise" is meant as variability which emerges in an "unprovoked" manner, i.e., which is unrelated to external drivers, and is not deterministically related to initial conditions.

It is important to note that we are not referring to low-dimensional non-linear systems which may generate beautiful attractors and other phenomena, but high-dimensional systems, whose variability maybe described by stochastic processes.

Ocean models can generate "noise" by internal nonlinear or stochastic dynamics, in the spirit of the "stochastic climate model." From the global model to the SCS model, with model resolution increasing, the "noise" generation increase. The higher resolution models can internally generate variability, which is absent in the lower resolution models. Because that higher resolution models generate more "noise," a smaller percentage of the overall variability is represented by dominant EOFs in the higher resolution models. The higher resolution models generate more "noise," which can motivate the generation of eddies in the ocean. This is important for scientists who study eddies in ocean models.

"Noise" may in some cases be a nuisance, but in dynamical simulations, it is a constitutive element of the dynamics of the system, which makes the dynamics richer, but creates the need of statistical efforts for determining if a change is beyond the range of internal variations. This becomes a significant issue when studying the effects of climate change or numerical experiments on the effect of formulating processes in models. So far, this practice is not widely recognized in ocean sciences (but, see for instance Leroux *et al.*, 2018), even if this mechanism is an almost trivial consequence of the concept of the stochastic climate model. In atmospheric sciences this is well-known, which may be related to the fact that atmospheric eddies have been part of the dynamics in most quasi-realistic atmospheric models, while global ocean models have for long operated with coarse resolution, without eddies but strong numerical viscosity and diffusion so that the late Ernst Maier-Reimer joked that older ocean models would be filled with mustard and not with water.

Another factor, which may limit the role of such internal variability in coastal seas, is the presence of tides, which acts as a kind of viscosity, erasing quickly and efficiently the emergence of non-forced long-living anomalies.

Funding

This work is supported by the Project "Oceanic Instruments Standardization Sea Trials (OISST)," (2016YFC1401300) of National Key Research and Development Plan, and Taishan scholars Program.

Acknowledgments

Thanks should be given to China National Super Computing Center in Jinan for providing the computing resources and Ms. Liu Xin for computing technology support.

References

Bleck, R. (2002). An oceanic general circulation model framed in hybrid isopycnic-cartesian coordinates, *Ocean Modelling* **4**, 1, pp. 55–88, doi:10.1016/S1463-5003(01)00012-9.

Chen, G., Hou, Y. and Chu, X. (2011). Mesoscale eddies in the South China Sea: Mean properties, spatiotemporal variability, and impact on thermohaline structure, *Journal of Geophysical Research-Oceans* **116**, p. C06018, doi:10.1029/2010JC006716.

Chervin, R. and Schneider, S. H. (1976). Determining statistical significance of climate experiments with general circulation models, *Journal of the Atmospheric Sciences* **33**, 3, pp. 405–412, doi:10.1175/1520-0469(1976)033<0405:ODTSSO>2.0.CO;2.

Faghmous, J. H., Frenger, I., Yao, Y., Warmka, R., Lindell, A. and Kumar, V. (2015). A daily global mesoscale ocean eddy dataset from satellite altimetry, *Scientific Data* **2**, p. 150028, doi:10.1038/sdata.2015.28.

Hasselmann, K. (1976). Stochastic climate models. Part I. Theory, *Tellus* **28**, pp. 473–485.

Hasselmann, K. (1993). Optimal fingerprints for the detection of time-dependent climate change, *Journal of Climate* **6**, 10, pp. 1957–1971, http://dx.doi:10.1175/1520-0442(1993)006<1957:OFFTDO>2.0.CO;2.

Large, W. G., Mcwilliams, J. C. and Doney, S. C. (1994). Oceanic vertical mixing — A review and a model with a nonlocal boundary-layer parameterization, *Reviews of Geophysics* **32**, 4, pp. 363–403, doi:10.1029/94RG01872.

Leroux, S., Penduff, T., Bessieres, L., Molines, J., Brankart, J., Serazin, G., Barnier, B. and Terray, L. (2018). Intrinsic and Atmospherically Forced Variability of the AMOC: Insights from a Large-Ensemble Ocean Hindcast, *Journal of Climate* **31**, 3, pp. 1183–1203, doi: 10.1175/JCLI-D-17-0168.1.

Lorenz, E. N. (1956). *Empirical Orthogonal Functions and Statistical Weather Prediction*, Vol. 1, Massachusetts Institute of Technology, Department of Meteorology Cambridge.

von Storch, H. and Zwiers, F. W. (1999). *Statistical Analysis in Climate Research*, Cambridge University Press.

von Storch, H., von Storch, J.-S. and Müller, P. (2001). Noise in the climate system — ubiquitous, constitutive and concealing, in B. Engquist and W. Schmid. eds., *Mathematics Unlimited — 2001 and Beyond. Part II*, pp. 1179–1194, Springer Verlag.

Wang, Y., Fang, G., Wei, Z., Qiao, F. and Chen, H. (2006). Interannual variation of the South China Sea circulation and its relation to El Niño, as seen from a variable grid global ocean model, *Journal of Geophysical Research* **111**, p. C11S14, doi:10.1029/2005JC003269.

Weisse, R., Heyen, H. and von Storch, H. (2000). Sensitivity of a regional atmospheric model to a sea state dependent roughness and the need of ensemble calculations, *Monthly Weather Review* **128**, pp. 3631–3642.

Zhang, M. and von Storch, H. (2016). Toward downscaling oceanic hydrodynamics — suitability of a high-resolution OGCM for describing regional ocean variability in the South China Sea, *Oceanologia* **59**, 2, pp. 166–176, doi:10.1016/j.oceano.2017.01.001.

Zhang, M. and von Storch, H. (2018). Distribution features of travelling eddies in the South China Sea, in *Research Activities in Atmospheric and Oceanic Modelling*, WGNE Blue Book, pp. 2–31.

Chapter 5

Atmospherically Forced Regional Ocean Simulations of the South China Sea: Scale Dependency of the Signal-to-Noise Ratio*

S. Tang[†,‡], H. von Storch[†,‡], and X. Chen[†]

[†]*Key Laboratory of Physical Oceanography,
Ocean University of China Qingdao, China*
[‡]*Institute of Coastal Research, Helmholtz Center Geesthacht, Germany*

When subjecting ocean models to atmospheric forcing, the models exhibits two types of variability — a response to the external forcing (hereafter referred to as signal) and inherently generated (internal, intrinsic, unprovoked, chaotic) variations (hereafter referred to as noise). Based on an ensemble of simulations with an identical atmospherically forced oceanic model that differ only in the initial conditions at different times, the signal-to-noise ratio of the atmospherically forced oceanic model is determined. In the large scales, the variability of the model output is mainly induced by the external forcing and the proportion

---

*This chapter was originally published in *Journal of Physical Oceanography*, 50(1), 133–144, doi:10.1175/JPO-D-19-0144.1. This chapter is licensed under the terms and conditions of the Creative Commons Attribution (CC BY) license (https://creativecommons.org/licenses/by/4.0/), which permits unrestricted use, distribution, and reproduction in any medium.

of the internal variability is small, so the signal-to-noise ratio is large. For smaller scales, the influence of the external forcing weakens and the influence of the internal variability strengthens, so the signal-to-noise ratio becomes less and less. Thus, the external forcing is dominant for large scales, while most of the variability is internally generated for small scales.

1.   Introduction

Based on the concept of the "stochastic climate model" (Hasselmann, 1976), the trajectory of the climate system can be described as that of an inert system subject to internally generated variations. That is, a relevant part of the variability of a system is not introduced by external factors, but internally generated. The internal variability (also called intrinsic variability, unprovoked variability, chaotic variability, or noise) is ubiquitous in the climate system and its oceanic and atmospheric components; it emerges at all locations and times, at all scales (von Storch *et al.*, 2001). The cause for the emergence of such noise is in part due to nonlinearities, but equally important is the presence of very many degrees of freedom, which transforms short term small-scale disturbances to large-scale variations — in the spectral domain, this morphs a white noise spectrum into a red noise spectrum (Hasselmann, 1976).

The presence of unprovoked variability was first recognized when doing numerical experiments with global atmospheric models, for instance, on the effect of anomalous sea surface temperature distributions (Chervin *et al.*, 1974; Laurmann and Gates, 1977). The noise is not just academically interesting but of great practical significance for the practice of numerical experimentation. For determining the effect of a modification, say of anomalous sea surface temperatures or changed parameterizations of physical processes, either very long simulations or ensembles of simulations are needed, and the null hypothesis "the modification has no effect" needs to be rejected with a statistical test. Thus, the discrimination between the signal, due to the modification, and the noise, due to internal processes, is needed.

The same problem comes to the forefront when analyzing the ongoing climate variability, if it would contain a component, which is forced by anthropogenic factors, in particular elevated greenhouse gas or aerosol presences in the atmosphere. Hasselmann coined the term "detection" for determining if a signal is present, and "attribution" for determining which external factor would plausibly be the source of the signal (Hasselmann, 1979, 1993). In the climatic system, the signal-to-noise (S/N) ratio problem has been studied since the 1980s (Alien *et al.*, 1994; Madden and Ramanathan, 1980; Santer *et al.*, 1995, 1994; Wigley and Jones, 1981; Wigley and Raper, 1990). Santer *et al.* (2011) estimated the timescale dependency of the S/N ratio in the climatic system.

The challenge of discriminating between the signal related to some forcing of interest, and the noise, which is generated internally, has been recognized in the global modeling communities for almost 50 years — with significant consequences for numerical experimentation and evaluation of observed variations. The regional atmospheric modeling communities, however, seem to have overseen the issue for a long time. Only in the late 1990s was the regional atmospheric modeling community confronted with the need of ensemble simulations (Ji and Vernekar, 1997; Rinke and Dethloff, 2000; Weisse *et al.*, 2000). In the global ocean community, the issue came up when eddy-resolving models came into use (e.g., Arbic *et al.*, 2014; Jochum and Murtugudde, 2004, 2005; Li and Han, 2015; Penduff *et al.*, 2018; Sérazin *et al.*, 2015).

In the regional ocean modeling community the issue went mostly unnoticed, possibly until 2016. In personal communication, we were confronted with the unsubstantiated claim that the presence of lateral boundary conditions would constrain the system, so that the emergence of noise would be suppressed. However, Büchmann and Söderkvist (2016) reported about the problem emerging in operational regional oceanography. Also, Waldman *et al.* (2017) addressed this issue for an analysis of Mediterranean Sea variability. Nevertheless, "noise" is rarely considered an issue for numerical experimentation, it seems. A simply designed numerical experiment demonstrated the problem for simulations with high grid resolution (Tang *et al.*, 2019).[1]

Indeed, ocean models with coarse resolution, as were common in early global climate models, behaved almost deterministically, but as soon as mesoscale variability, in particular eddies, enters the

---

[1]See Section 4.

dynamics, the situation changes, and the system becomes stochastic. The constraining by lateral boundary values does possibly reduce but not suppress the noise, if the considered region is not very small (cf. Schaaf *et al.*, 2017).

First systematic studies about the emergence of noise in global ocean models have been summarized by Penduff *et al.* (2018). They drew mostly on the Oceanic Chaos — Impacts, Structure, Predictability (OCCIPUT) project, within which an ensemble of 50 members of 56 years (1960–2015) "global ocean/sea ice simulations driven by the same atmospheric reanalysis, but with perturbed initial conditions" was built and examined. They showed that there is a massive presence of noise, mostly but not only on smaller spatial scales, and also at low frequencies.

The issue of noise in regional ocean simulations was addressed every now and then, such as the analysis of Jochum and Murtugudde (2004, 2005) who showed that internal variability can explain a significant part of the observed SST variability in the tropical Pacific Ocean and the western Indian Ocean. By studying the decadal sea level variations, Li and Han (2015) also indicated that ocean internal processes have considerable contribution along the Somali coast and in the western Bay of Bengal and the subtropical south Indian Ocean.

In this work, we study the presence of noise in simulations of the dynamics of the South China Sea (SCS). Plenty of previous works has shown that the main features of these dynamics of the SCS are influenced by the surface forcing and the water exchange at the boundary (e.g., Luzon Strait inflow). However, by using a three layer nested simulation with the climatological forcing without any short-term atmospheric variability, Tang *et al.* (2019) demonstrated that unprovoked variability, that is, noise, is formed, and that the noise generation becomes stronger in models with higher resolution. From this observation and the results reported by Penduff *et al.* (2018) we derive the hypothesis that the S/N ratio depends on the spatial scale. This paper is testing and quantifying this hypothesis and finds it valid.

Before we begin our analysis, a brief discussion of the term noise is needed. This term is used differently in different scientific and engineering quarters. Noise is unprovoked variability, but the term "provoked" may depend on which "signal" is studied. In climate change simulations, the signal is related to the anthropogenic drivers, and the weather variations and their oceanic responses are part of the noise. In our simulations, described below, the signal is the response to the atmospheric forcing, so that their oceanic response is not part of the noise.

An issue is, of course, where the unprovoked variations come from. There may be many sources; in case of the SCS, it is plausible that the mesoscale eddies and their dynamics contribute (cf. Zhang *et al.*, 2019). But various types of waves, variations in river discharge, spatially inhomogeneous radiative heating modulated by clouds may add to the small-scale and short-term cacophony of disturbances. Even unphysical interventions such as rounding errors, or miniscule differences in initial conditions, which no longer have any predictive significance, may become part of the source of noise, which then is transformed to larger and longer-living features via the mechanism suggested in the stochastic climate model (Hasselmann, 1976).

The significance of the presence of noise is that the effect of a modification in a numerical experiment may in some or maybe even most cases no longer be simply described as the plain difference between a modified simulation and a control simulation (e.g., Ådlandsvik, 2008; Luneva *et al.*, 2015; Verri *et al.*, 2018). Instead extended simulations of ensembles of simulations may be needed, and the experiments themselves must be considered a statistical challenge. If the expected response is mostly large scale, the effect may be minor, but when limited time windows, that is, events or episodes, are considered, the effect may intermittently be strong [cf. example by Weisse *et al.* (2000) and Weisse and Feser (2003)]. A second significant aspect is that the noise may have an effect of mixing and transport, and on marine ecosystems. To this end, knowledge about the scale dependency of the S/N ratio is helpful — large-scale features are likely robust and less affected by random variations, while at small scales all kind of variations unrelated to the forcing may take place [e.g., an example by Chen *et al.* (2019) on sediment dynamics on the South China Sea].

In present work, based on an ensemble of identical ocean model simulations which differ only in the initial conditions, the scale dependency of the S/N ratio of the atmospherically forced ocean model in the SCS is studied.

The novelty our study is, first a confirmation that noise forms in regional models in spite of the lateral boundary forcing; second, we embed the presence of noise on all spatial scales into the framework of the stochastic climate model, which implies that the noise is partly due to eddy dynamics, but also to other factor, such as wave dynamics. Indeed, for the stochastic climate model, neither nonlinearity is needed nor specific processes (such as eddies). Third, our approach of describing the complete spatial variability through empirical orthogonal functions (EOFs) circumvents problems related to complex geometries. A fourth aspect is that we emphasize the consequences for the practice of numerical experimentation, which will be significant also for ecological and morphodynamic studies on small scales.

The present chapter is organized as follows. A brief introduction to the model setup is given in Section 2. In Section 3, the methods used for this study are introduced. The results of the estimation of the scale length and the scale dependency of the S/N ratio are presented in Section 4. Conclusions are summarized in Section 5.

2.   Model Setup

The model used in this work is the Hybrid Coordinate Ocean Model (HYCOM), which is a primitive equation ocean general circulation model and has hybrid vertical coordinates (Bleck, 2002). It describes the dynamics of the SCS on a grid with 0.04° resolution Figure 1. The surface forcing used in this work comes from the Navy Operational Global Atmospheric Prediction System (NOGAPS) forcing products. The temporal resolution of the NOGAPS data is 3 hourly. The NOGAPS data are interpolated from a 0.5° resolution to the 0.04° SCS model resolution.

An ensemble of four simulations is generated with this model. The four simulations are identical, except for the initial condition and the integration time (see Table 1). They all cover the year 2008, and the evaluation of the simulations considers only this year.

In other studies, such as Penduff *et al.* (2018) and Waldman *et al.* (2017), larger ensembles of longer simulations are used. For our purpose, namely to demonstrate the emergence of noise in regional models, and the qualitative scale dependency of the S/N ratio, this is

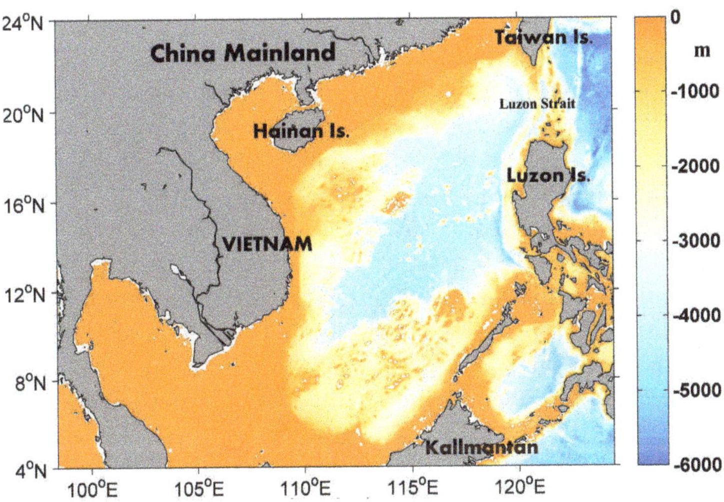

Fig. 1. The domain and bathymetry of the SCS model. There are 651 grid points in the zonal (east–west) direction and 541 in the meridional (north–south) direction. The grid includes 212 940 sea points.

Table 1. The setup of the four simulations.

	Start time	End time	Total run time	Initial condition
N1	1 Dec 2007	31 Dec 2008	13 months	1 Dec of 31st year
N2	1 Oct 2007	31 Dec 2008	15 months	1 Oct of 31st year
N3	1 Dec 2006	31 Dec 2008	25 months	1 Dec of 31st year
N4	1 Oct 2006	31 Dec 2008	27 months	1 Oct of 31st year

not needed. Indeed, the quantitative results will depend on the model used, on the region considered, on the time line (e.g., if an El Niño persists or not). These limitations do, however, not compromise our conclusions.

The simulations are initialized 1, 3, 13, and 15 months before the beginning of 2008 (Table 1) During the entire simulations the atmospheric forcing as given by NOGAPS is used. The initial conditions are taken from simulations done with the same model (HYCOM) in a nested setup, as described by Tang *et al.* (2019). The coarsest model (1°) is global, an intermediate model (0.2°) describes the west Pacific, and the finest the SCS model on the same 0.04° grid.

These simulations were done with climatological forcing, that is, a smooth annual cycle without weather variations.

The initial condition of the first model (N1) comes from the first day of December of the 31st year of the climatological SCS model and is run for 13 months (including December in 2007 and the whole year of 2008). The initial condition of the second model (N2) comes from the first day of October of the 31st year of the climatological SCS model and is run for 15 months (including from October to December in 2007 and the whole year of 2008). The initial condition of the third model (N3) comes from the first day of December of the 30th year of the climatological SCS model and is run for 25 months (including December in 2006, the whole year of 2007, and the whole year of 2008). The initial condition of the fourth model (N4) comes from the first day of October of the 30th year of the climatological SCS model and is run for 27 months (including from October to December in 2006, the whole year of 2007, and the whole year of 2008). The lateral boundary conditions of all these four simulations came from the intermediate model (or west Pacific model) in Tang *et al.* (2019), which is forced by climatological atmospheric forcing. Across the ensemble simulations, these boundary values were identical.

The four simulations differ only by the initial time and the initial conditions. However, the different initial conditions are fully consistent, taken from the timeline of the climatological HYCOM simulation. The time from initialization to analyzing the data is at least 1 month, a time which we consider sufficient to make sure that the details of the initial state do not matter (in the sense of predictability). But the small differences introduced by the different initial conditions will make sure that noise forms in the simulation.

We use barotropic velocity to make our case. Indeed, various variables of the ocean dynamics could be used to demonstrate our case, among them barotropic velocity or SSH. We decided on barotropic velocity because we did our demonstration of the presence of noise in the South China Sea, in Tang *et al.* (2019), with that variable. We found that the increase of daily variance of the barotropic streamfunction with finer model grid resolution is more obvious than when using SSH. We simply went on with the barotropic velocity because the noise in the barotropic velocity is more obvious than that in the SSH.

For demonstrating the presence of unprovoked variability, or of noise, in the set of simulations, we plot in Figures 2 and 3 the

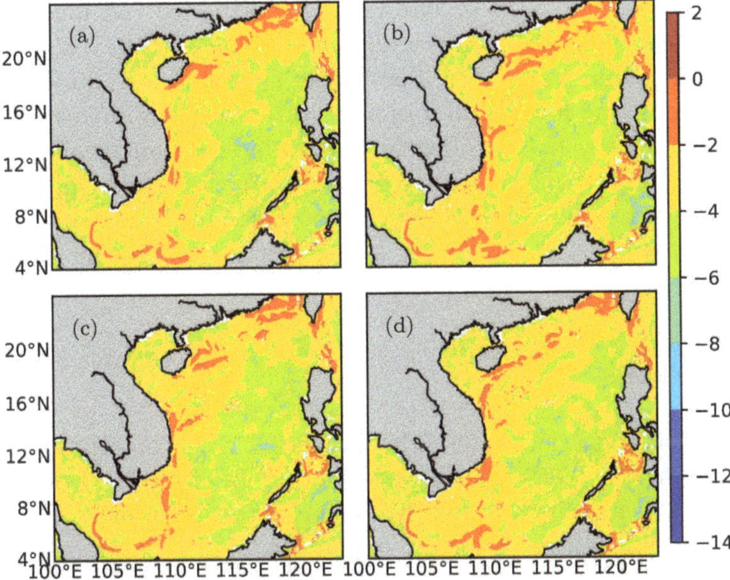

Fig. 2.   The logarithm of the speed of barotropic velocity for the four models (a) N1, (b) N2, (c) N3, and (d) N4 on the 87th day of the year 2008.

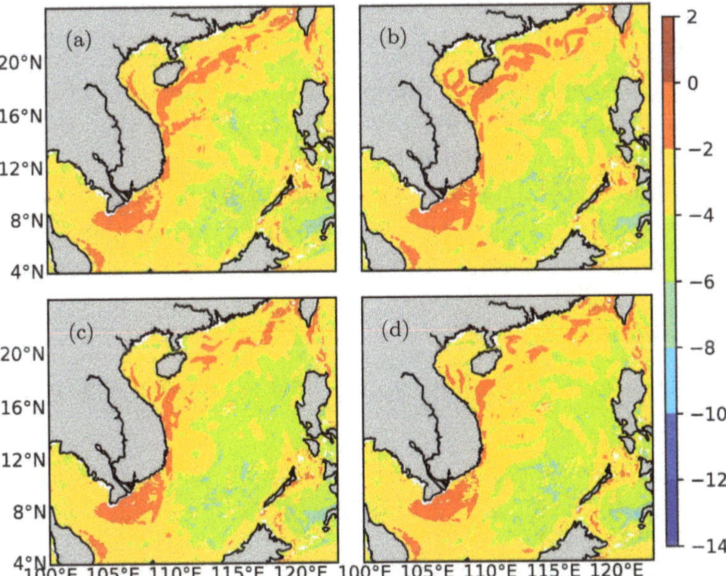

Fig. 3.   The logarithm of the speed of barotropic velocity for the four models (a) N1, (b) N2, (c) N3, and (d) N4 on the 255th day of the year 2008.

logarithm of the speed of barotropic velocity for the four simulations at two randomly chosen days — on the 87th day and on the 255th day of the year 2008. The figures show that the states on the same day in the four simulations are rather similar with respect to the large scales, but exhibit obvious differences on smaller scales. Thus, Figures 2 and 3 already demonstrate the validity of our base hypothesis, but we want to do the comparison not just in terms of two cases studies, but in a more systematic manner in the remainder of this paper.

3. Methods

We need first to decompose the simulated fields in components of different scales. We do so by considering EOFs in Section 3.1. In Section 3.2 we derive a measure of scale for the different EOFs, allowing for a quantitative assertion of the link of EOF rank and spatial scale, which is intuitively to be expected. In Section 3.3 we explain how to estimate S/N ratios from the four ensemble members.

3.1. *Decomposing According to Spatial Scales*

The EOF analysis is a multivariate analysis technique that can be applied for the separation of different dominant patterns of variability (von Storch and Zwiers, 1999). We will show below that the ranking of the EOFs is associated with a ranking of scales.

There are a total of four simulations in the year 2008 in this work. So, in order to get the scale involved for all the four model simulations, we determine the EOFs of the daily barotropic velocity across all four 2008 simulations, which is computed as

$$B(x,y,t) = \sum_{j=1}^{p} Q_j(t)e_j(x,y) \quad \text{for } t = 1\ldots 4 \times 366 \text{ days} \quad (1)$$

where B is the barotropic velocity across all four 2008 simulations, x and y denote the model's grid points in the zonal (east–west) direction and the meridional (north–south) direction, respectively, and t counts the days in the four simulations. The $e_j(x,y)$ is the jth EOF across all four 2008 simulations, the $Q_j(t)$ are the jth principal components (PCs), and j denotes the index of EOFs. Each simulation in 2008 has 366 days, so four simulations have $4 \times 366 = 1464$ days. Therefore, we can get $p = 1464$ nontrivial EOFs (i.e., with nonzero eigenvalues; see, e.g., von Storch and Zwiers, 1999).

Due to the problem related to the accuracy of computing and resulting inaccuracies of determining the last EOF, the first 1463 EOFs are used for the next analysis, and the last one is discarded. The EOFs are ranked as usual, with an index sorted according to the percentage of explained variance. A visual inspection of the patterns reveals that low-indexed EOFs go with large scales, medium-indexed EOFs with medium scales, and high-indexed ones with small scales. Thus, PCs of these EOFs are associated with scales that decrease with increasing indices of the EOFs, but we will quantify this link below.

3.2. *Estimating the Scale Lengths*

The spatial scale lengths of EOFs are estimated by the spatial auto-correlations (Wackernagel, 2013) of the EOFs derived earlier.

For explaining how this is done, we need to introduce some definitions. If M is a subset of gridpoints in the modeling domain, then $|M|$ is the number of elements of this set. Let G be the set of all sea points in the modeling domain. For calculating the spatial autocorrelation at lag k of a pattern e, we form all products of $e(x, y)$ and $e(x + k\Delta, y)$, with Δ being the distance between two points in the zonal or meridional direction, if both points (x, y) and $(x + k\Delta, y)$ are in G. Similarly, we form products of points with points neighboring in the meridional direction. Formally, when applied to the jth EOF e_j

$$
c_j(k) = \left[\sum_{(x,y)\in G^{kz}} [e_j(x + k\Delta, y) - \bar{e}_j][e_j(x, y) - \bar{e}_j] \right.
$$

$$
\left. + \sum_{(x,y)\in G^{km}} [e_j(x, y + k\Delta) - \bar{e}_j][e_j(x, y) - \bar{e}_j] \right] \tag{2}
$$

$$
\left/ \left[\left(|G^{kz}| + |G^{kz}|\right) \times \frac{\sum_{(x,y)\in G}[e_j(x, y) - \bar{e}_j]^2}{|G|} \right] \right.
$$

with the sets of points $G^{kz} = \{(x, y) \in G, (x + k\Delta, y) \in G\}$, containing all sea points with a sea point k grid points to the right (z like zonal displacement). The m in set G^{km} indicates that points displaced in the meridional direction are considered, and $\bar{e}_j$ is the spatial mean of the jth EOF e_j.

As the length scale of the EOF e_j we select $k^*\Delta$, with the largest k^* so that all $c_j(k)$ are larger than some threshold for all $k < k^*$. The gridmesh length is 4.0–4.4 km, and we set $\Delta = 4.2$ km.

3.3. *Computing the S/N Ratio*

Because the noise is represented by the differences between the simulations in this work, the signal and noise can be computed by the ensemble mean and ensemble standard deviation between the simulations, respectively. This is first done for each PCs of the 1463 EOFs at each day of the year 2008, of which the ensemble provides four samples. Obviously, a sample of four is not large, and noteworthy random variations prevail because of the limited sampling. However, as we will see, the results are plausible and qualitatively meaningful and consistent so that an expansion of the sample size, by adding more simulations of the year 2008, would lead to some quantitative changes but not to a need of questioning the conclusions of this paper.

If $Q^i(j, t)$ is the PC at the day t of the simulation N_i of jth EOF, then the signal at the day t is estimated to be the mean at that day across the four simulations

$$\mu(j, t) = \left[\sum_{i=1.4} Q^i(j, t) \right] \bigg/ 4 \tag{3}$$

and, consistently, the intensity of the noise at day t is measured by the standard deviation across the four simulations, that is, by

$$\sigma(j, t) = \sqrt{\frac{1}{4} \sum_{i=1,4} (Q^i(j, t) - \mu(j, t))^2} \tag{4}$$

Then, the S/N ratio is estimated by the ratio of the standard deviation of the signal across the year ($t = 1, \ldots, 366$) and of the standard deviation of the noise across the year for each of the PCs $j = 1, \ldots, 1463$.

All variations in the numerator, the temporal standard deviation of the ensemble mean of the PCs, are related to the common atmospheric variations acting upon the ocean in the four different samples of the year 2008. The variations in the denominator, the temporal standard deviation of the ensemble standard deviation of the PCs, on the other hand, reflect non-synchronous variations in the different realizations of 2008, and are thus not related to the common atmospheric forcing.

4. Results

4.1. *The Estimation of the Scale Lengths*

Figure 4 shows a few EOFs (viz., EOF1, EOF5, EOF50, and EOF 200) of the zonal u and meridional v components of the barotropic velocity, and the corresponding spatial autocorrelation functions. From the EOF patterns, we can see that with the increase of the EOF index the spatial scale becomes less and less. Consistently the spatial autocorrelations inform that a larger EOF index is associated with a steeper decline of the autocorrelation function. This confirms our hypothesis that the index of an EOF is indicative for their spatial scale.

As mentioned before, we may define a quantitative spatial scale by choosing a critical level for determining k^*. This choice is arbitrary but allows a consistent comparison across different patterns.

We choose 0.1 for this critical value. Figure 5 shows the resulting scale lengths for all EOFs. The scale lengths do not decline in a monotonic manner, which may be related to the fact that both, the EOFs themselves and the scales are estimated and subject to random variations. Indeed the sampling errors in EOF estimation are large for high index EOFs (North *et al.*, 1982). But, in spite of these limitations, the estimated scale lengths are a useful tool to characterize the patterns of the EOFs.

4.2. *The Scale-Conditioned S/N Ratios*

Using the formulas (3) and (4), we estimate the S/N ratios of the PCs of the EOFs. We consider the joint EOFs of the zonal and meridional components of the barotropic velocity. In one case we

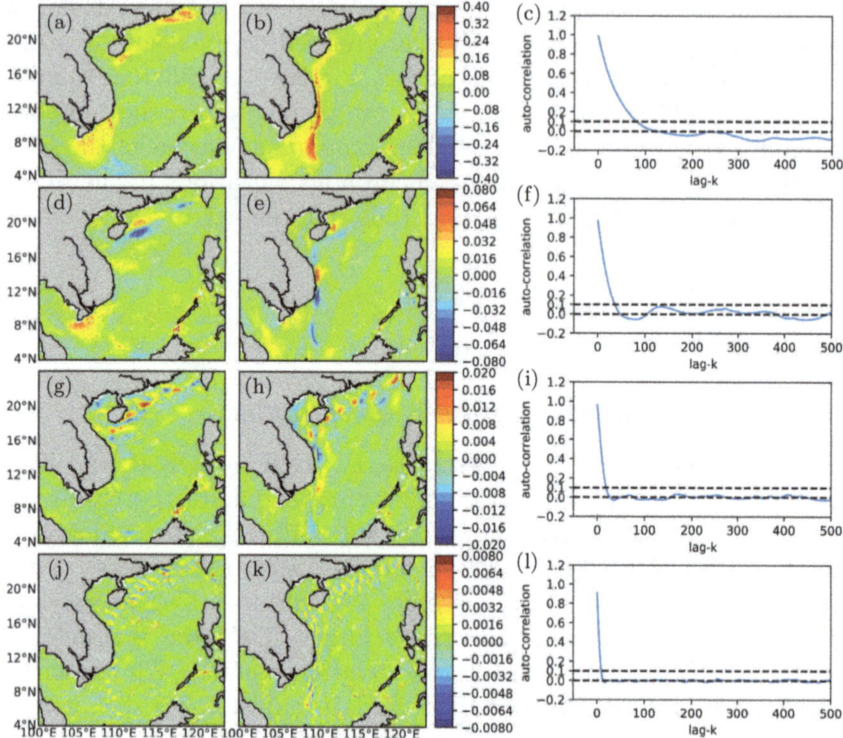

Fig. 4. The joint EOFs (a)–(c) 1, (d)–(f) 5, (g)–(i) 50, and (j)–(l) 200 of the zonal u and meridional (v components of the barotropic velocity. (left) The patterns of the zonal barotropic velocity u (ms^{-1}); (center) the patterns of the meridional barotropic velocity v (ms^{-1}); and (right) corresponding autocorrelation functions for EOF 1, EOF 5, EOF 50, and EOF 200.

consider the EOFs without prior subtraction of the annual cycle so that the signal is rooted in the atmospheric forcing and in the annual cycle. In the other case, we subtract the common annual cycle before doing the EOF analysis, and in this case the signal stems only from the atmospheric weather forcing. The annual cycle is fitted from the time series of the year 2008 by harmonic analysis.

Figures 6 and 7 show the scale distribution of the S/N ratio for the case with and without annual cycles. We find that with the increase of the EOF index the S/N ratio declines. Our base hypothesis is thus valid.

When looking more in detail (Figure 6), we find three ranges of EOF indices, namely, EOFs 1–10, 11–50, and 50–1463. Within the

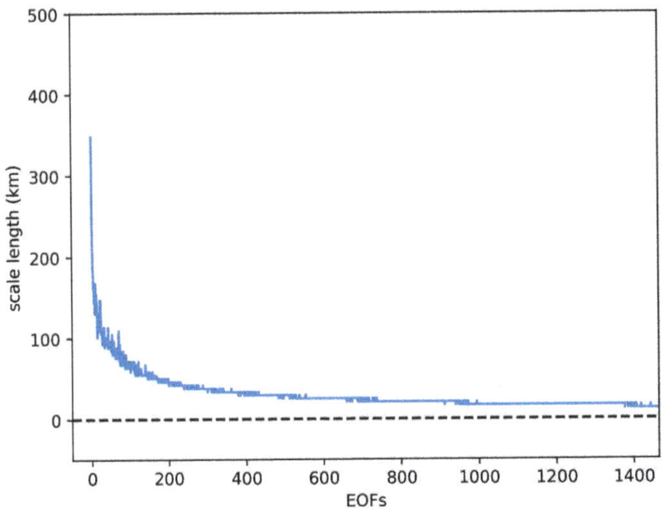

Fig. 5.   The estimated scale lengths (km) of the first 1463 EOFs.

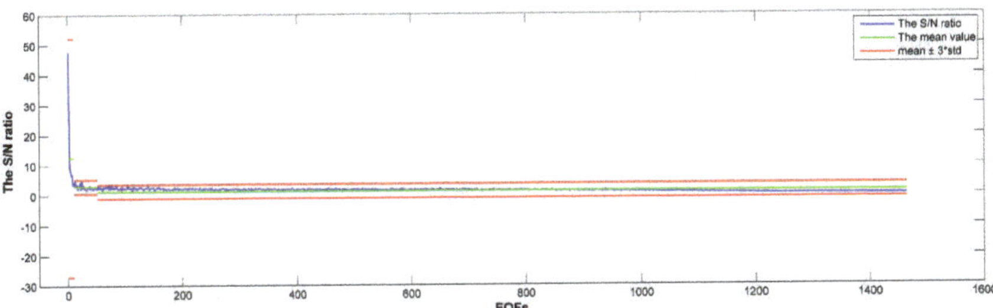

Fig. 6.   The S/N ratios as a function of the EOF index for the case when the annual cycle is included in the EOF analysis. Small scales are situated at the right end, and large scales are at the left end. Based on the distribution of the S/N ratio, the scales (EOFs) are divided into three blocks: EOFs 1–10, EOFs 11–50, and EOFs 51–1463. The green line in each block is the mean value of the ratio values, and the red lines above and below the green line in each box are the mean value ±3 standard deviations of the S/N ratio.

three ranges, the S/N ratios have characteristic distributions, as illustrated by 3 standard deviation blocks. The first block of EOFs 1–10 has large scales, namely on average about 220 km, the middle block of EOFs 11–50 only 110 km, and the third block of EOFs 51–1463 only 30 km. The S/N ratios in the first block decline quickly, in the

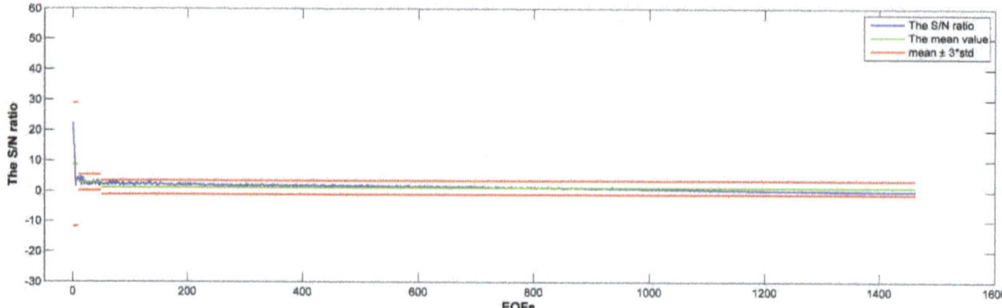

Fig. 7.   The S/N ratios as a function of the EOF index for the case when the annual cycle is excluded in the EOF analysis. Small scales are situated at the right end, and large scales are at the left end. Based on the distribution of the S/N ratio, the scales (EOFs) are divided into three blocks: EOFs 1–10, EOFs 11–50, and EOFs 51–1463. The green line in each block is the mean value of the ratio values, and the red lines above and below the green line in each box are the mean value ±3 standard deviations of the S/N ratio.

middle block the decrease is weaker, and in the third block corresponding to the small scale the S/N ratio declines more slowly, and the range of the high-indexed EOFs 51–1463 is not only very small but almost constant. When the annual cycle is taken out (Figure 7), the S/N ratios in the first block are smaller, but for the two other blocks almost unchanged.

In the same way, as for the time series of the PCs, formulas (3)–(4) we can calculate locally S/N ratios for any series of fields. We do so for the filtered fields of zonal and meridional barotropic velocity. The filtering is done by combining the EOF patterns and principal components of the three ranges mentioned above, that is, 1–10, 11–50, and 51–1463. Now, μ and σ are functions for each point (x, y) on the model grid.

Figure 8 shows the maps of the S/N ratios for the three spatial scale ranges, 1–10 (Figure 8(a)), 11–50 (Figure 8(b)), and 51–1463 (Figure 8(c)), for the case with annual cycle; the same without annual cycle is displayed in Figure 9.

For the large-scale component (Figure 8(a)) the largest S/N ratios (>16) mainly appear to the south of Vietnam and to the west of the island of Taiwan, where the ocean is shallow. Smaller S/N ratios (<4)

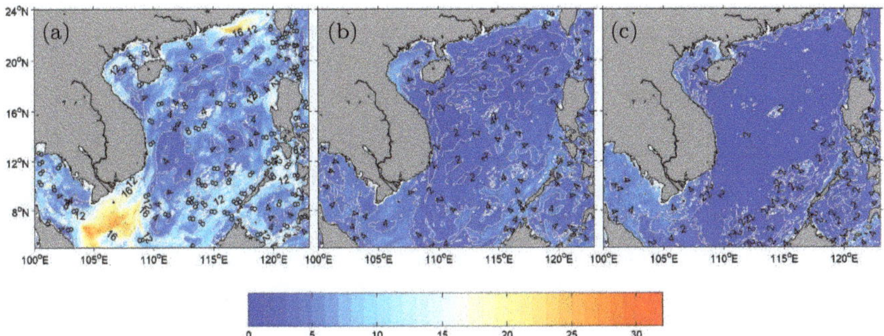

Fig. 8. Maps of the S/N ratio for the case that the annual cycle is included in the EOF analysis for (a) the first range of EOFs 1–10, (b) the second range of EOFs 11–50, and (c) the third range of EOFs 51–1463.

mainly appear in the deep ocean, north of 128°N. There are mainly three regions for S/N ratios in the large-scale component smaller than 4, which are to the east of Hainan Island, to the east of Vietnam, and to the west of Luzon Island. In Figure 8(b) describing medium scales, the spatial distribution of the S/N ratio varies irregularly with the value range from 2 to 4. In Figure 8(c) on small scales, the value of the S/N ratio in the shallow ocean (>2) is larger than that in the deep ocean (<2). Therefore, the results in Figure 8 demonstrate that with an EOF index increasing or the scale becoming smaller, the S/N ratio becomes less everywhere.

In general, the atmospherically triggered signals are best visible in the shallow parts of the SCS, whereas the internal dynamics dominate in the deep part of the SCS. This is not surprising as in the shallower parts of the ocean, the atmospheric forcing acts on a smaller water body, whereas eddies tend to form in deeper waters.

The results for the reconstructions without the annual cycle (Figure 9) differ only for the first, large scale range. In whole SCS the S/N ratios without the annual cycle in the large scale are smaller than that with the annual cycle. This is not surprising, since the annual cycle resides mostly in the low-indexed EOFs — and the patterns shows an obvious weakening of the maxima in Figure 9(a),

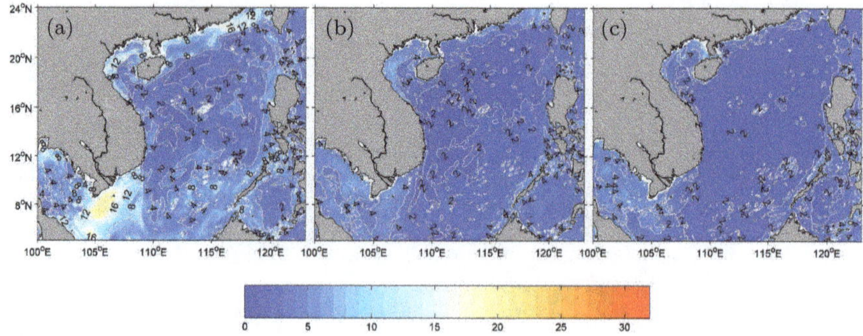

Fig. 9.   Maps of the S/N ratio for the case that the annual cycle is excluded in the EOF analysis for (a) the first range of EOFs 1–10, (b) the second range of EOFs 11–50, and (c) the third range of EOFs 51–1463.

south of Vietnam, where the effect of the seasonal monsoonal change is particularly strong.

5.   Conclusion and Comments

With an ensemble of identical ocean model simulations that differ only in the initial conditions at the different spin-up times, we test the hypothesis that the ratio of externally forced variability and of internally generated variability is strongly dependent on the spatial scale in a regional sea. This ratio is quantified with help of a signal-to-noise ratio. Here, the signal is the impact of atmospheric variability, and the noise is unprovoked variability.

By doing so, we demonstrate also that the presence of lateral boundary values is not eradicating the emergence of internal variability, which we call noise. A novel aspect of our analysis is that we define the spatial scales empirically with EOFs, so that we cover all spatial variability in a sea with irregular geometry. Two significant conceptual conclusions of our study refer to the applicability of the "stochastic climate model" for framing the issue of noise, and the implications for numerical experimentation and simulation of impacts with regional laterally constrained ocean models.

That intrinsic variability, that is, unprovoked variability or in our terminology noise, emerging in global eddy-resolving models has been known for about two decades. However, the regional oceanographic community mostly overlooked the problem until recently, possibly because of a tacit assumption that the regional constraints through lateral boundary conditions would transform the stochastic system into a deterministic one also for eddy-resolving dynamics. This perception is seemingly inaccurate.

Our study deals with this regional, laterally constrained system, specifically the dynamics of the South China Sea, described on a 0.04° grid in the year 2008. We do our analysis with the daily stored barotropic velocity. The choice of the year and the variable is ad hoc.

We find our hypothesis fully confirmed for the South China Sea. For spatial scales of about 220 km and more, the signal is stronger than the noise. This is particularly so in the shallower parts of the SCS. But there is noise at these large scales, as predicted by the stochastic climate model. The ratios become comparable for scales on the order of about 110 km, whereas for spatial scales of about 30 km and less, the noise dominates. This is particularly so in the deeper parts of the ocean.

Our results concerning the emergence and scale dependency of noise are consistent with what Penduff *et al.* (2018) found in their analysis on an ensemble of global ocean/sea ice models.

We know that the noisy mesoscale variability comes mostly from the accumulation of perturbations originating from instability mechanisms, in particular eddies but also waves. But what is driving the seeding of eddy formation and of wave formation? It is seemingly mostly not external factors (Zhang *et al.*, 2019), but a variety of random effects, related to nonlinearities or computational process details (e.g., Ludwig and Geyer, 2019), when the numerical viscosity is weak.

Of course, the S/N ratio depends on the forcing, which is the source of the signal. If this is weak, the S/N ratio will become small for all scales. If the forcing is of small scales and not dominated by large scales, the situation may change. But we expect that we will find the type of S/N spectra as in Figures 6 and 7, when the forcing is mostly large scale — which is the case in atmospherically driven regional ocean simulations or in climate change simulations. More cases, with different seas, need to be studied, but in our two cases, when the annual cycle is, or is not, considered part of the signal, we see the same type of S/N spectrum, albeit with smaller values for the leading EOFs.

The scale dependency of the S/N ratio is important to supplement for our recognition of oceanic hydrological dynamics, especially the small scales. For example, for the oceanic eddies, we see the hypothesis confirmed that the generation of some oceanic eddies is mostly caused by inherently internal variability and is hardly due to the external forcing (Zhang *et al.*, 2019). Consistently, Chen *et al.* (2019) rightly failed to connect small-scale deposition dynamics in the South China Sea to large-scale forcing.

The presence of noise is not only of such academic interest but has practical consequences. Namely, the effect of a modification in a model or in a simulation, be it a change in the forcing or in a parameterization, is not given by the mean difference between a control and an experimental simulation as the signal, but must be studied with an extended dataset, be it a very long simulation or an ensemble of simulations, for identifying a robust part of the signal. This is long known in atmospheric sciences, but seemingly hardly recognized in numerical experimentation with regional ocean models.

In certain cases, some argue, for instance when tides are very strong, the noise part may become less important if not outright negligible. Indeed, the tides may be efficient in limiting the predictive skill of the knowledge about an initial state but may not eradicate the formation of noisy components.[2] However, this needs to be tested in designed experiments. Additionally, the dynamical characters, as to which processes are mainly responsible for the formation of noise, need to be studied. Eddies are an obvious candidate, but internal wave activity or stochastic elements or stochastically varying driver discharge may have a share. Finally, one should keep in mind that the dynamics in the South China Sea may be substantially different from those in other regional seas, such as in the North Sea.

Given our results, one could assume that the stochastic character of the simulations is less important if variations that are large scale in space and slow in time are studied. But this may not always be so, as the examples of Penduff *et al.* (2018) demonstrate. Also, the effect of spurious signals, which are in reality an expression of the activity of noise, may become dominant if specific episodes of limited temporal extension are selected.

Our study is limited by the relatively short time window of 1 year for each member of our ensemble. Previous studies (Arbic *et al.*, 2014; Mikolajewicz and Maier-Reimer, 1990; Penduff *et al.*, 2018)

---

[2]This is done in Chapter 6.

have indicated that low frequency noisy large-scale variability unprovoked by specific forcing may emerge. Indeed, the theory behind Hasselmann's (1976) stochastic climate model does not imply that noteworthy unprovoked variability would prevail only at higher frequencies. The famous experiment by Mikolajewicz and Maier-Reimer (1990) operated with a coarse grid model without mesoscale eddies, but generated strong quasi-oscillatory variability after centuries as a response to space–time white noise freshwater fluxes. It certainly would be interesting to see if such large-scale variability, for instance, as in Mikolajewicz and Maier-Reimer (1990), would emerge in a millennial run with climatological (weather-free) forcing due to small-scale random disturbances in the ocean. However, this is beyond the scope of our present analysis.

Funding

This work is supported by the Project "Oceanic Instruments Standardization Sea Trials (OISST)", (2016YFC1401300) of National Key Research and Development Plan, and Taishan Scholars Program.

Acknowledgments

Thanks is given to German Climate Computing Center (DKRZ) and China National Super Computing Center in Jinan for providing the computing resources and Ms. Liu Xin for computing support. We thank Ralf Weisse and Zhang Wenyan from HZG and Frabk Janssen from BSH for valuable advice.

References

Ådlandsvik, B. (2008). Marine downscaling of a future climate scenario for the north sea, *Tellus A: Dynamic Meteorology and Oceanography* **60**, 3, pp. 451–458, doi:10.1111/j.1600-0870.2007.00311.x.

Alien, M. R., Mutlow, C. T., Blumberg, G. M. C., Christy, J. R., McNider, R. T. and Llewellyn-Jones, D. T. (1994). Global change detection, *Nature* **370**, 6484, pp. 24–24, doi:10.1038/370024b0.

Arbic, B. K., Müller, M., Richman, J. G., Shriver, J. R., Morten, A. J., Scott, R. B., Sérazin, G. and Penduff, T. (2014). Geostrophic turbulence in the frequency–wavenumber domain: Eddy-driven low-frequency variability, *Journal of Physical Oceanography* **44**, 8, pp. 2050–2069, doi:10.1175/JPO-D-13-054.1.

Bleck, R. (2002). An oceanic general circulation model framed in hybrid isopycnic-cartesian coordinates, *Ocean Modelling* **4**, 1, pp. 55–88, doi:10.1016/S1463-5003(01)00012-9.

Büchmann, B. and Söderkvist, J. (2016). Internal variability of a 3-D ocean model, *Tellus*, 68A, 30417, https://doi.org/10.3402/tellusa.v68.30417.

Chen, H., Zhang, W., Xie, X. and Ren, J. (2019). Sediment dynamics driven by contour currents and mesoscale eddies along continental slope: A case study of the Northern South China Sea, *Marine Geology* **409**, pp. 48–66, doi:10.1016/j.margeo.2018.12.012.

Chervin, R. M., Gates, W. L. and Schneider, S. H. (1974). The effect of the time averaging on the noise level of climatological statistics generated by atmospheric general circulation models, *Journal of the Atmospheric Sciences* **31**, pp. 2216–2219.

Hasselmann, K. (1976). Stochastic climate models. Part I. Theory, *Tellus* **28**, pp. 473–485.

Hasselmann, K. (1979). On the signal-to-noise problem in atmospheric response studies, in B. D. Shaw ed., *Meteorology Over the Tropical Oceans*, pp. 251–259, *Royal Meteorological Society*, Bracknell, Berkshire, England.

Hasselmann, K. (1993). Optimal fingerprints for the detection of time-dependent climate change, *Journal of Climate* **6**, 10, pp. 1957–1971, doi:10.1175/1520-0442(1993)006<1957:OFFTDO>2.0.CO;2.

Ji, Y. M. and Vernekar, A. D. (1997). Simulation of the Asian summer monsoons of 1987 and 1988 with a regional model nested in a global gcm, *Journal of Climate* **10**, p. 1965–1979.

Jochum, M. and Murtugudde, R. (2004). Internal variability of the tropical pacific ocean, *Geophysical Research Letters* **31**, 14, doi:10.1029/2004GL020488.

Jochum, M. and Murtugudde, R. (2005). Internal variability of Indian Ocean SST, *Journal of Climate* **18**, 18, pp. 3726–3738, doi:10.1175/JCLI3488.1.

Laurmann, J. A. and Gates, W. L. (1977). Statistical considerations in the evaluation of climatic experiments with atmospheric general circulation models, *Journal of Atmospheric Sciences* **34**, 8, pp. 1187–1199, doi:10.1175/1520-0469(1977)034<1187:SCITEO>2.0.CO;2.

Li, Y. and Han, W. (2015). Decadal sea level variations in the Indian Ocean Investigated with HYCOM: Roles of climate modes, ocean internal variability, and stochastic wind forcing, *Journal of Climate* **28**, 23, pp. 9143–9165, doi:10.1175/JCLI-D-15-0252.1.

Ludwig, T. and Geyer, B. (2019). Reproduzierbarkeit, *Informatik Spektrum* **42**, 1, pp. 48–52, doi:10.1007/s00287-019-01149-2.

Luneva, M. V., Aksenov, Y., Harle, J. D. and Holt, J. T. (2015). The effects of tides on the water mass mixing and sea ice in the Arctic Ocean, *Journal of Geophysical Research: Oceans* **120**, 10, pp. 6669–6699, doi: 10.1002/2014JC010310.

Madden, R. A. and Ramanathan, V. (1980). Detecting climate change due to increasing carbon dioxide, *Science* **209**, 4458, pp. 763–768, doi: 10.1126/science.209.4458.763.

Mikolajewicz, U. and Maier-Reimer, E. (1990). Internal secular variability in an ocean general circulation model, *Climate Dynamics* **4**, 3, pp. 145–156, doi:10.1007/BF00209518.

North, G. R., Bell, T. L., Cahalan, R. F. and Moeng, F. J. (1982). Sampling errors in the estimation of empirical orthogonal functions, *Monthly Weather Review* **110**, 7, pp. 699–706, doi:10.1175/1520-0493(1982)110<0699:SEITEO>2.0.CO;2.

Penduff, T. and Coauthors. (2018). Chaotic variability of ocean heat content: Climate-relevant features and observational implications, *Oceanography* https://doi.org/10.5670/oceanog.2018.210.

Rinke, A. and Dethloff, K. (2000). On the sensitivity of a regional Arctic climate model to initial and boundary conditions. *Climate Research* **14**, pp. 101–113.

Santer, B. D. and Coauthors. (1995). Ocean variability and its influence on the detectability of greenhouse warming signals, *Journal of Geophysical Research: Oceans* **100**, C6, pp. 10693–10725, doi: 10.1029/95JC00683.

Santer, B. D. and Coauthors. (2011). Separating signal and noise in atmospheric temperature changes: The importance of timescale, *Journal of Geophysical Research: Atmospheres* **116**, D22, doi:https://doi.org/10.1029/2011JD016263.

Santer, B. D., Brüggemann, W., Cubasch, U., Hasselmann, K., Höck, H., Maier-Reimer, E. and Mikolajewica, U. (1994). Signal-to-noise analysis of time-dependent greenhouse warming experiments, *Climate Dynamics* **9**, 6, pp. 267–285, doi:10.1007/BF00204743.

Schaaf, B., von Storch, H. and Feser, F. (2017). Does spectral nudging have an effect on dynamical downscaling applied in small regional model domains? *Monthly Weather Review* **145**, 10, pp. 4303–4311, doi:10.1175/MWR-D-17-0087.1.

Sérazin, G., Penduff, T., Grégorio, S., Barnier, B., Molines, J.-M. and Terray, L. (2015). Intrinsic variability of sea level from global ocean simulations: Spatiotemporal scales, *Journal of Climate* **28**, 10, pp. 4279–4292, doi:10.1175/JCLI-D-14-00554.1.

Tang, S., von Storch, H., Chen, X. and Zhang, M. (2019). "Noise" in climatologically driven ocean models with different grid resolution, *Oceanologia* **61**, 3, pp. 300–307, doi:10.1016/j.oceano.2019.01.001.

Verri, G., Pinardi, N., Oddo, P., Ciliberti, S. A. and Coppini, G. (2018). River runoff influences on the Central Mediterranean overturning circulation, *Climate Dynamics* **50**, 5, pp. 1675–1703, doi:10.1007/s00382-017-3715-9.

von Storch, H. and Zwiers, F. W. (1999). *Statistical Analysis in Climate Research*, Cambridge University Press.

von Storch, H., von Storch, J.-S. and Müller, P. (2001). Noise in the climate system — Ubiquitous, constitutive and concealing, in B. Engquist and W. Schmid. eds., *Mathematics Unlimited — 2001 and Beyond. Part II*, pp. 1179–1194, Springer Verlag.

Wackernagel, H. (2013). *Multivariate Geostatistics: An Introduction with Applications*, Springer, 388 pp., https://doi.org/10.1007/978-3-662-05294-5.

Waldman, R. and Coauthors. (2017). Modeling the intense 2012–2013 dense water formation event in the northwestern Mediterranean Sea: Evaluation with an ensemble simulation approach, *Journal of Geophysical Research: Oceans* **122**, 2, pp. 1297–1324, doi:10.1002/2016JC012437.

Weisse, R. and Feser, F. (2003). Evaluation of a method to reduce uncertainty in wind hindcasts performed with regional atmosphere models, *Coastal Engineering* **48**, 4, pp. 211–225, doi:10.1016/S0378-3839(03)00027-9.

Weisse, R., Heyen, H. and von Storch, H. (2000). Sensitivity of a regional atmospheric model to a sea state dependent roughness and the need of ensemble calculations, *Monthly Weather Review* **128**.

Wigley, T. M. L. and Jones, P. D. (1981). Detecting co2-induced climatic change, *Nature* **292**, 5820, pp. 205–208, doi:10.1038/292205a0.

Wigley, T. M. L. and Raper, S. C. B. (1990). Natural variability of the climate system and detection of the greenhouse effect, *Nature* **344**, 6264, pp. 324–327, doi:10.1038/344324a0.

Zhang, M., von Storch, H., Chen, X., Wang, D. and Li, D. (2019). Temporal and spatial statistics of travelling eddy variability in the South China Sea, *Ocean Dynamics* **69**, 8, pp. 879–898, doi:10.1007/s10236-019-01282-2.

Chapter 6

The Effect of Tides on Internal Variability in the Bohai and Yellow Sea*

L. Lin[†,‡], H. von Storch[‡], D. Guo[§], S. Tang[†], P. Zheng[†], and X. Chen[†]

[†] *College of Oceanic and Atmospheric Sciences,
Ocean University of China, Qingdao, China*
[‡] *Institute of Coastal Research, Helmholtz-Zentrum Hereon,
Geesthacht, Germany*
[§] *North China Sea Marine Forecasting Center of State Oceanic
Administration, Qingdao, China*

The hydrodynamics of marginal seas exhibit internal variability unprovoked by external forcing. To date, the role of tides in reducing this "noise" has not been evaluated. We investigated the effect of tides on internal variability in the Bohai and Yellow Sea. To do so, we conducted three ensembles of numerical experiments using the three-dimensional Finite-Volume Coastal Ocean Model (FVCOM) with tidal

---

*This chapter was originally published in *Dynamics of Atmospheres and Oceans*, 98, 101301, doi:10.1016/j.dynatmoce.2022.101301. This chapter is licensed under the terms and conditions of the Creative Commons Attribution (CC BY) license (https://creativecommons.org/licenses/by/4.0/), which permits unrestricted use, distribution, and reproduction in any medium.

forcing, with "half-tidal" forcing, and without tidal forcing, while everything else was unchanged, and determined the intensity of the signal-to-noise ratio (hereinafter referred to as the S/N ratio), with the "signal" represented by the variance of the coherent variations of the different simulations subject to the same atmospheric variability and the noise represented by the intra-ensemble variance. The S/N ratio was determined for depth-averaged velocities, surface temperature, and surface salinity. The first result was that in all three ensembles, noise emerged but with different intensities. In the ensemble with tidal forcing, unprovoked variability emerged primarily at smaller scales. When the tides were weakened or turned off, the S/N ratios were reduced, more so in the Yellow Sea than in the Bohai. The increase in the S/N ratio was largest for large scales and for depth-averaged velocity. The reduction in tidal forcing resulted in an approximately 30% increase in S/N ratios in the Bohai at large scales. Thus, the absence of tidal forcing favored the emergence of increased unprovoked variability at large and medium scales but not at small scales. A hypothesis for explaining this scale-selective effect of tides, based on the stochastic climate model, is suggested.

1. Introduction

Based on the concept of the "stochastic climate model" (Hasselmann, 1976), the trajectory of the climate system can be described as that of an inert system subject to internally generated variations. In other words, a relevant part of the variability of a system is not introduced by external factors but internally generated. This part of the internal variability is denoted intrinsic variability, unprovoked variability, chaotic variability, or simply noise. Climate variability, in particular the annual (Hirschi *et al.*, 2013) or diurnal cycles, is a mix of internally and externally generated variations (Hirschi *et al.*, 2013; Tang *et al.*, 2020). The external part (hereafter referred to as the signal) is a response to the external forcing, and the internal or noise part is mostly related to turbulence but likely also to other nonlinear processes. Noise is partly due to nonlinearities but is also due to the very many degrees of freedom in models, which transform short-term small-scale disturbances to low-frequency large-scale variations, as demonstrated in the Nobel Prize-recognized classical paper of Hasselmann (1976). The significance of the ubiquitous presence of noise is that only part of the slow variability may be traced back to external forcing, while some must be considered "unprovoked" and internally generated. Sensitivity studies must separate deterministic effects ("signal") from stochastic ones (noise), as demonstrated by Chervin *et al.* (1974) and the other Nobel Prize-recognized paper by Hasselmann (1979).

As often occurs in geosciences, the same term is used with different meanings both in the scientific community and in the public. A prominent example of the multiple uses of terms is the term "model", but "noise" is also used to refer to a variety of scientific features. It is important that the specific definition used in this article and in the line of research to which the present article contributes is recognized, namely, variability unrelated to specific external factors.

A number of model ocean dynamic studies have found that noise generates interesting dynamical features, for example, the noise-induced transitions of thermohaline circulation (Mikolajewicz and Maier-Reimer, 1990; Timmermann and Lohmann, 2000; Weisse *et al.*, 1994, 1999); thus, hydrodynamic noise is certainly not only a nuisance but interferes constructively with the dynamics.

Even if the external forcing is similar from year to year, noteworthy variability emerges when different years are compared (Hirschi *et al.*, 2013). This was also documented in simulations of global warming with ocean-atmosphere coupled models, which showed deterministic (externally forced) changes and stochastic (unprovoked) variability (Cai *et al.*, 2020).

The emergence of noise was first recognized in global atmosphere models (Chervin *et al.*, 1974). Approximately 30 years later, internal variability was recognized in ocean simulations with eddy-resolving models (Jochum and Murtugudde, 2004; Jochum *et al.*, 2004), showing that internal variability can explain a significant part of the observed sea surface temperature (SST) variability in the tropical Pacific Ocean and the western Indian Ocean. The Atlantic meridional overturning circulation, which modulates sea surface temperature and mediates the sea ice extent at high northern latitudes, is also masked by internal variability (Grégorio *et al.*, 2015; Kushnir, 1994; Mahajan *et al.*, 2011). Penduff *et al.* (2018) gave a holistic picture of oceanic noise in the Oceanic Chaos — Impacts, Structure, Predictability (OCCIPUT) project based on a large ensemble (50 members) of long-term (1960–2015) global ocean/sea ice 1/4° simulations with slightly different initial conditions. Their research showed that in the presence of mesoscale turbulence, oceanic signals cannot be entirely attributed to atmospheric variability because in several midlatitude areas, chaotic processes have more impact than atmospheric variability on both low-frequency variability and small-scale variability; substantial uncertainty related to chaotic internal variability should be kept in mind when interpreting modeled and observed time series.

Noise in regional models is a subject of ongoing research. Büchmann and Soderkvist (2016) and Tang *et al.* (2019) illustrated that noise occurs in regional ocean models. Büchmann and Söderkvist demonstrated that front and eddy activity increase the internal variability in salinity (2016).

Internal variability is related to turbulence (Callies, 2016) or other dynamics, such as Lagrangian coherent structures (Peacock and Haler, 2013) or internal waves and other nonlinear processes. Mesoscale and submesoscale eddies likely play an important role. The statistics of the number and intensities of mesoscale eddies in the ocean are not directly linked with external forcing, as shown by Zhang *et al.* (2019) in simulations of the deep South China Sea. Slightly different initialization times lead to different mesoscale eddy activities (Tang *et al.*, 2019; Zhang *et al.*, 2019). Slightly different initial conditions, such as model integration time, contribute to different mesoscale eddy activities (Tang *et al.*, 2019; Zhang *et al.*, 2019). On the other hand, noise at small scales may also be generated as the net effect of many nonlinear effects related to a variety of subgrid scale processes and then transformed to large scales through the memory of the system, as outlined in Hasselmann's stochastic climate model.

In shallow coastal seas such as the Bohai and Yellow Sea, the noise intensity and scales have not been analyzed systematically. Tang *et al.* (2020) analyzed the internal variability in the South China Sea with a regional model. However, the Bohai and Yellow Sea, with shallow water depths, seasonal stratification and strong tides, differ strongly from the South China Sea. The tidal mixing estimated in the Bohai is two orders of magnitude greater than the global background mixing on the order of $10^{-5}\mathrm{m^2s^{-1}}$, and the maximum reaches $5 \times 10^{-5}\mathrm{m^2s^{-1}}$ (Deng and Zhao, 2020).

A satellite image may serve as a demonstration of the ubiquity of small-scale features in the Bohai and the Yellow Sea with little cloud cover (Figure 1(b)). It shows the near-surface distribution of chlorophyll-a on a randomly chosen day, 28 April 2020.

In the following section, we first examine the characteristics of hydrodynamic noise in these marginal seas and then ask what role

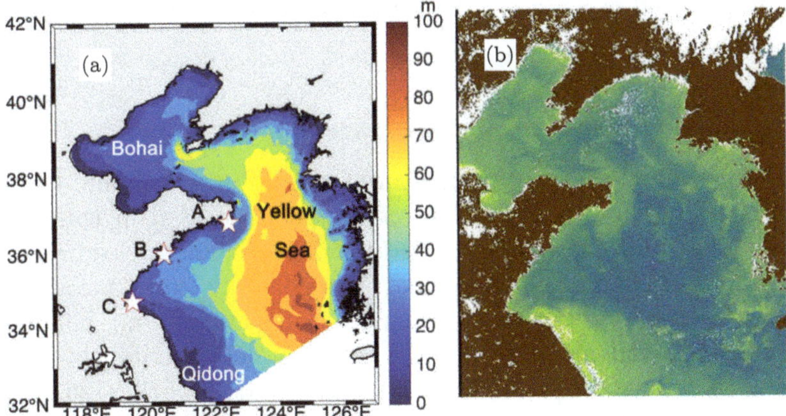

Fig. 1. The model topography and SENTINEL-3-derived sea surface chlorophyll-a concentration. (a) The model topography (the 3 white stars labeled A, B, and C represent the Shidao, Xiaomaidao, and Lianyungang hydrological stations, respectively). (b) SENTINEL-3-derived sea surface chlorophyll-a concentration on a randomly chosen day, 28 April 2020, from https://www.eumetsat. int/coda. For the white areas, data are missing because of cloudiness.

the tides are playing. In the conclusions, we speculate about a mechanism that links tides and noise formation. The chapter is organized as follows: Section 2 includes an introduction to the model, its experimental configuration, and methods used to calculate internal variability and "signal-to-noise" ratios (hereinafter referred to as S/N ratios). In Section 3, the model validation, scales, and scale dependency of the S/N ratios in the Bohai and Yellow Sea are presented. In Section 4, the effects of the intensity of tides (full, half, and no tides) on internal variability are analyzed separately for the Bohai and Yellow Sea. Discussion, some speculations, and summaries are presented in Section 5.

2. Model, Data, and Methods

Section 2 provides the model configuration (Section 2.1) and the setup for three ensembles of numerical simulations, with tidal forcing, with "half-tidal" forcing, and without tidal forcing (Section 2.2). Observational datasets are introduced in Section 2.3. Our definitions of "signal", "noise", and "signal-to-noise ratio" used for this study are introduced in Section 2.4, and the scale separation is introduced in Section 2.5.

2.1.   *Model Simulation Configuration*

The three-dimensional unstructured grid Finite-Volume Coastal Ocean Model (FVCOM) (Chen *et al.*, 2006) is used in this study. The Mellor and Yamada level 2.5 (MY-2.5) turbulent closure model (Mellor and Yamada, 1982) and Smagorinsky eddy parameterization method are used to calculate the vertical and horizontal diffusion coefficients, respectively. The model domain covers the Bohai and Yellow Sea (31.885°–40.942°N and 117.572°–126.915°E) (Figure 1(a)). The Bohai and Yellow Sea are shallow, with average depths of 18 m and 44 m, respectively. The open boundary across the Yellow Sea extends from Qidong in China eastward to the southern tip of the Korean Peninsula, and radiation boundary conditions are used. The grid resolution is approximately 4 km in the Bohai and 8 km in the Yellow Sea horizontally and 30 layers vertically.

Six-hourly surface forcing data are from the National Centers for Environmental Prediction (NCEP) Climate Forecast System Version 2 (CFSv2) data, with a global resolution of $0.20° \times 0.20°$, including the sea surface temperature, cloud cover, air pressure, wind, specific humidity, evaporation, precipitation, and heat flux. The model data output interval is three hours, and daily average data are used for calculations. The first 60 days of the 2019 hourly output elevation data are also saved for harmonic analysis and comparison.

2.2.   *The Ensembles*

We generate three ensembles of four numerical simulations each, one with tidal forcing, another with "half-tidal" forcing, and a third without tidal forcing. Apart from the different treatments of the tides and different dates of the initializations, all simulations are identical. The surface fluxes in all twelve simulations are identical, derived from NCEP (CFSv2).

The ensemble with tidal forcing is regarded as a control/standard run (hereafter referred to as "with tides" (wT)). The ensembles with half-tidal forcing and without tidal forcing are hereinafter referred to as hT and nT, respectively. Each of the simulations covers 2019, but the integration time is longer (see Table 1), as each simulation is

Table 1.   The model setup for the twelve simulations.

	NCEP forcing				Tidal forcing		
	Start time	End time	Run time	Initial condition	Full	Half	None
1	01/11/2017	31/12/2019	26 months	1 Sep year 7	1/wT	1/hT	1/nT
2	01/01/2018	31/12/2019	24 months	1 Jan year 8	2/wT	2/hT	2/nT
3	01/03/2018	31/12/2019	21 months	1 Mar year 8	3/wT	3/hT	3/nT
4	01/11/2018	31/12/2019	14 months	1 Sep year 8	4/wT	4/hT	4/nT

allowed to equilibrate for several months. Each ensemble consists of four simulations forced by the same surface fluxes but with different initialization times. In the case of the control ensemble, the four simulations are labeled 1/wT, 2/wT, 3/wT, and 4/wT (see Table 1). All twelve simulations cover 2019, and we choose only this year's model output data for evaluation.

The effect of gravitational tidal potential is usually ignored for the region, as gravitational tides account for only ~3% in the Yellow Sea (An, 1977). Thus, most models use global/basin-scale tidal models to provide open boundary conditions. The tidal forcing is introduced by adding a time-dependent elevation at the open boundary across the Yellow Sea. The tidal elevation forcing is composed of 8 major tidal components ($M_2, S_2, N_2, K_2, K_1, O_1, P1,$ and Q_1) derived from the TPXO8 database (Egbert and Erofeeva, 2002). In the half-tidal ensemble, the tidal elevation forcing at the open boundary is halved compared with that in the control run. In the "no tide" simulations, the tidal elevation is turned off.

Of the simulations, three are initialized for fourteen months (1/wT, 1/hT, and 1/nT), and another three are initialized for twelve months (2/wT, 2/hT, and 2/nT), nine months (3/wT, 3/hT, and 3/nT), and two months (4/wT, 4/hT, and 4/nT) prior to 1 January 2019 (Table 1). Thus, when analyzing the variations in 2019, the model has had at least 2 months to adjust. The initial conditions of the ocean state are taken from a separate 9-year climatological FVCOM simulation: 1/wT, 1/hT, and 1/nT are run for twenty-six months (consisting of fourteen months before 1 January 2019 and twelve months in 2019), with initial conditions from 1 September of the 7th year of the climatological 9-year simulation. 2/wT, 2/hT, and 2/nT are run for twenty-four months (consisting of twelve months before 1 January 2019 and twelve months in 2019), with initial conditions from 1 January of the 8th year of the climatological 9-year simulation. Consistently, 3/wT, 3/hT, and 3/nT are run for twenty-one months (consisting of nine months before 1 January 2019 and twelve months in 2019), with initial conditions from 1 March of the 8th year of the climatological 9-year simulation. The last group, 4/wT, 4/hT, and 4/nT, are run for fourteen months (consisting of two months before 1 January 2019 and twelve months in 2019), with initial conditions from 1 September of the 8th year of the climatological 9-year simulation.

Thus, within each ensemble, the four simulations only differ in the times of the initial conditions and integration lengths. This setup of shifting the initial conditions, which are fully consistent, along the timeline of the nine-year climatological simulation is sufficient to "seed" noise generation (Geyer *et al.*, 2021).

The deterministic variability caused by different initial conditions will be more obvious a few days after the model initialization time but will be weak after a period of time (Büchmann and Soderkvist, 2016). Since the time from initialization to analyzing the data are at least two months, we think the details of the initial conditions do not matter; also, the time delay since initialization is insignificant for the formation of noise. To demonstrate this, we calculate the variance $\delta_{ij} = \frac{1}{T}\Sigma_{t=1}^{T}(f_i(t) - f_j(t))^2$ with the ith and jth simulations with tidal forcing, namely, the four simulations: 1/wT, 2/wT, 3/wT, and 4/wT (see Table 1). $f(t)$ is the time series of the spatial mean of the depth-averaged velocities; T represents the number of times. We will show that δ_{ij} does not show an increase with increasing differences in the initialization times.

We calculate δ_{ij}^{Jan} for the first month (thus T = 1), i.e., January 2019, and δ_{ij}^{whole} for all twelve months of 2019 (T = 12). As there are four simulations in the control run, δ_{ij} is a 4×4 matrix, which is shown in Equation (1a); the results of δ_{ij} are shown in Equation (1b). The upper and lower triangular matrices show δ_{ij}^{Jan} and δ_{ij}^{whole}:

$$
\begin{bmatrix}
\delta_{11}^{Jan} & \delta_{12}^{Jan} & \delta_{13}^{Jan} & \delta_{14}^{Jan} \\
\delta_{21}^{whole} & \delta_{22}^{Jan} & \delta_{23}^{Jan} & \delta_{24}^{Jan} \\
\delta_{31}^{whole} & \delta_{32}^{whole} & \delta_{33}^{Jan} & \delta_{34}^{Jan} \\
\delta_{41}^{whole} & \delta_{42}^{whole} & \delta_{43}^{whole} & \delta_{44}^{whole}
\end{bmatrix}
=
\begin{bmatrix}
0 & 0.1073 & 0.1081 & 0.0997 \\
0.1002 & 0 & 0.1342 & 0.1376 \\
0.0975 & 0.0664 & 0 & 0.0970 \\
0.0591 & 0.1252 & 0.0705 & 0
\end{bmatrix}
$$

$$\times 10^{-5} \tag{1}$$

There is no obvious trend showing that the variance decreases when the initialization times of the simulation decrease. The variances based on only January are similar to those derived from all twelve months, which indicates that the model is stable after two months.

2.3. *Observational Datasets*

The simulated water level and temperature are compared with hydrological station data derived from the China Science and Technology Resources Sharing service platform — China Ocean Science Data Center. To determine the model performance in simulating tides, the harmonic constants of the M_2 and K_1 model results are compared with harmonic constants calculated from tide gauge data (Hu *et al.* 2012). To do so, root-mean-square errors (RMSEs) and the correlation skill between the model results and observations are determined. This skill evaluates the correlation between the model results and observations, with a value of 1 indicating perfect agreement and a value of 0 indicating complete disagreement. Formulations and detailed descriptions of RMSEs and skill are given by von Storch and Zwiers (1999).

The ensemble averages of the control run of temperature and astronomical tide level are used for further comparison.

2.4. *Method of Scale Separation*

To separate fields into components associated with different scales, we apply the method introduced by Tang *et al.* (2020). To obtain a uniform distribution of grid points, the FVCOM 65969 unstructured grid points are interpolated to 800 × 800 regular grid points with a grid mesh length of 1 km. All daily fields of the four simulations in the 2019 control ensembles (i.e., 4 × 365 grids) are expanded into a series of empirical orthogonal functions (EOFs) $e_j(x, y)$ and principal components (PC) $Q_j(t)$. This is not done with the intention of identifying "modes" with certain statistical or dynamical properties in the considered field (which is often a questionable effort to begin with, cf. von Storch and Zwiers, 1999). Instead, we want to transform the fields from representation in the point space to another algebraically equivalent pattern representation. All daily fields of the four simulations in the 2019 control ensembles (i.e., 4 × 365 grids) are expanded into a series of empirical orthogonal function representations in a pattern space, which allows sorting according to scale. To have no loss of information, all possible EOFs must be considered (see below). Additionally, since EOFs optimize the distribution of variance in the considered dataset, we do not calculate the EOFs from only one simulation but from all simulations.

The presentation in the EOF space is

$$B(x, y, t) = \Sigma_{j=1}^{p} Q_j(t) e_j(x, y)$$

where B is the depth-averaged velocity across all four 2019 simulations, x,y representing the model grid points in the zonal direction and meridional direction, respectively, and t counts the days in four simulations. $p = \min(n,T)$ is the number of maximum number of possible EOFs, with the number of regular grid points $n = 800 \times 800$ and T the number of samples, which are all 365 daily fields from the four simulations. Thus, $p = 1460$, i.e., either the number of daily samples or the number of grid points. There are a total of n EOFs, but p-n of them have zero eigenvalues and do not contribute to the presentation of B (for details, refer to von Storch and Zwiers, 1999).

For each of the EOFs, we determine a spatial scale using the spatial autocorrelation function (cf. Wackernagel, 2003): the spatial scale of the jth EOF is set to be that length when this spatial autocorrelation function passes the 0.14 level. It turns out, as in the analysis of Tang *et al.* (2020),[1] that the scale decreases with increasing j.

The spatial scales are large in low-indexed EOFs and small in high-indexed EOFs. Figure 2 shows four selected EOFs (namely, EOF 1, EOF 10, EOF 300, and EOF 1300) of the depth-averaged velocities for the control run and their corresponding spatial autocorrelation functions. Corresponding spatial autocorrelation functions indicate the spatial scales quantitatively: Figure 2(e–h) reflects the correlations between two grid points at different distances. The horizontal coordinate represents the distance in the zonal or meridional direction. The autocorrelation functions decline smoothly in Figure 2(e) and (f), which represent a small EOF index, and the functions in Figure 2(g) and (h) decline steeply, especially in Figure 2(h). Thus, the correlations between two grid points at long distances are

---

[1]Reprinted in Chapter 5.

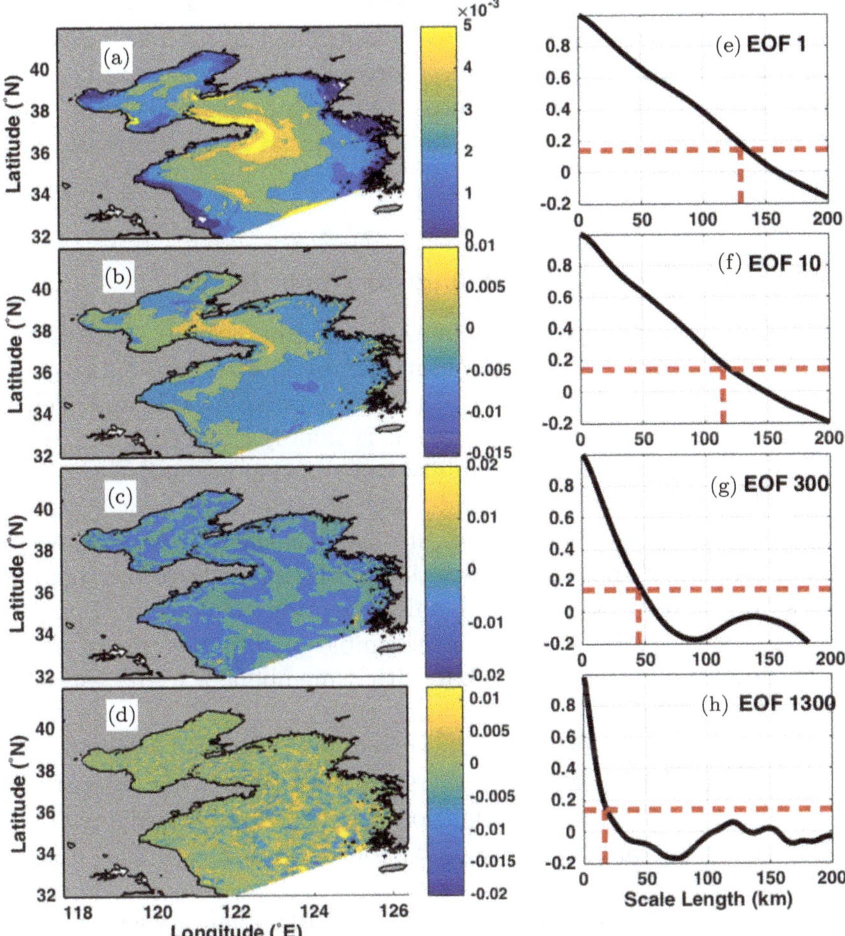

Fig. 2.　The four selected EOF spatial patterns and their corresponding auto-correlation functions of depth-averaged velocities for the control ensemble. The spatial patterns for EOF 1 (a), EOF 10 (b), EOF 300 (c), and EOF 1300 (d) of the depth-averaged velocities (m/s) (left). The scales are selected to be the lengths when the autocorrelation functions (right) pass the 0.14 level (red dashed line).

higher in large (low-indexed) EOFs than in small (high-indexed) EOFs. We define a quantitative spatial scale by selecting a critical level for determining the scale, which is an arbitrary but consistent value for different patterns. In the Bohai and Yellow Sea, we choose 0.14 as the critical value (Tang *et al.*, 2020). In Figure 2(e–h), the red horizontal dashed lines represent the critical value of 0.14, and the abscissa value of the intersection point of the red horizontal dashed line and black solid line represents the corresponding EOF scale. Both EOF spatial patterns and corresponding spatial auto-correlation functions indicate that the spatial scales decrease with increasing EOF indices.

2.5. *Determination of Signal-to-Noise Ratio*

The S/N ratio compares the intensity of the "signal", i.e., of the coherent variations across the four members of the ensemble, with the intensity of the "noise", i.e., the intra-ensemble variations. When the S/N ratio is large, the variation is mainly induced by external forcing, and the internal variability contribution is small.

To formally define signal and noise, consider a variable $x(j, t)$ in the jth ensemble at time t. Then, the signal is the time series consisting of the ensemble mean at each day t. The intensity of the signal is the standard deviation of the signal across time. Obviously, a perfect determination would require an ensemble of infinitely many members; however, it is assumed that with 4 members, a useful estimate is derived.

The noise, on the other hand, is given by the standard deviation of $x(j, t)$ across the ensemble members j at each time t. The intensity of the noise is the time mean of this noise, i.e., of the standard deviation. Both the noise and the signal have the same units as the original variable x.

The signal-to-noise ratio is the ratio of the signal intensity to that of the noise. It can be determined at each grid point but also as the average across the integration region.

We choose PCs of depth-averaged velocities $Q^i(j, t)$ as x. Additionally, for sea surface temperature and surface salinity, the intensity of the noise is determined, as well as the S/N ratio.

3.   Results

In this section, we consider the ensemble of control simulations, i.e., all runs with full tidal forcing.

3.1.   *Validations of Control Simulations*

We use the ensemble mean of the four simulations of the control run for validation. We define the ensemble mean at each time t as the mean of the four simulations at the same time t; the dispersion of the data is defined as the anomaly of each simulation to the ensemble mean.

The time series of the ensemble mean water level is compared with the astronomical tide level observed at the Shidao hydrological station (Figure 3) during January 2019. The simulation results (red) follow the measurements (black) very well in phase; however, the magnitudes are slightly underestimated. The RMSE is 0.24 m, and the correlation skill is 0.91. The magnitude underestimation is partly because of water depth deviation: the water depth in the model is not from the high-resolution measured data, which is not easily available, but from historical nautical charts. If the model water depth at Shidao is shallower than the real water depth, then the water level will be underestimated in the model. Compared with the tide gauge observation data and model results, the RMSEs of the amplitudes of M_2 and K_1 are 0.09 m and 0.04 m and the RMSEs of the phase lags are 6.8° and 6.10°, respectively; we conclude that the model results agree well with the observations.

The time series of the ensemble mean temperature and of the observed temperature at three stations are compared for the whole year of 2019. The RMSEs for Xiaomaidao, Shidao, and Lianyungang are 0.84°C, 0.98°C, and 0.94°C, respectively; the correlations are

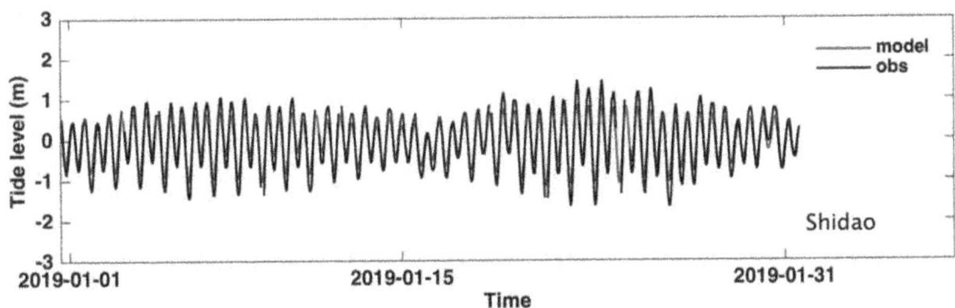

Fig. 3.   Comparison of model-predicted (red line) and observed (black line) astronomical tide levels at Shidao (A in Figure 1) for January 2019.

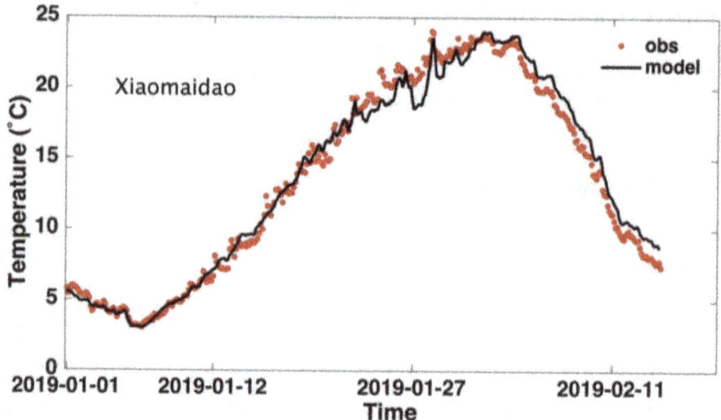

Fig. 4. Comparison of model-predicted (black line) and observed (red dots) temperatures at Xiaomaidao (C in Figure 1) in 2019.

0.929, 0.998, and 0.999, respectively. Figure 4 shows the temperature recorded at the Xiaomaidao hydrological station as an example: the seasonal temperature signal is in good agreement with the actual measurements; the comparisons for the other two stations show similar results (not shown).

3.2. *Variability at Large, Medium, and Small Scales in the Control Ensemble*

To illustrate the internal variability in the Bohai and Yellow Sea, we show barotropic flow maps on a randomly chosen day, namely, the 100th day of 2019, for the four members of the control ensemble (Figure 5). To see the variability more clearly, we use the logarithm of the speed of depth-averaged velocities because the magnitude of the depth-averaged velocities is small, approximately 0.1 m/s. Among the four ensemble members, the depth-averaged velocity patterns are similar at large scales, but obvious differences emerge at small scales in the centers of the northern and southern Yellow Sea.

To explore the internal variability more thoroughly and systematically, we calculate scales and S/N ratios with respect to depth-averaged velocities after separating the flow into large, medium and small scales. Using the method described in Section 2, we first derive the scale lengths for each EOF (Figure 6). The scale lengths decrease

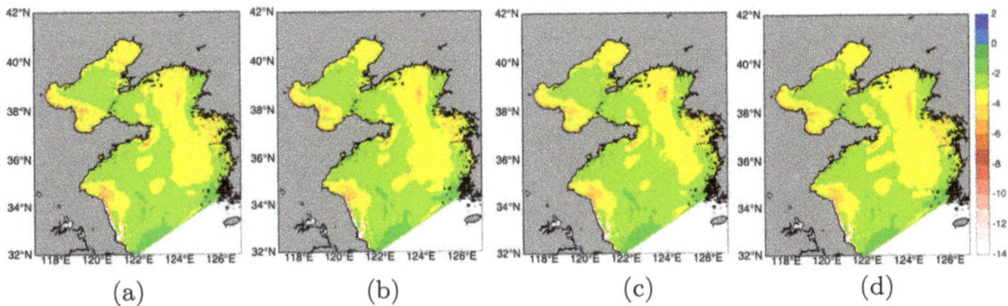

(a)   (b)   (c)   (d)

Fig. 5. Snapshots of the logarithm of the speed of depth-averaged velocity on the 100th day of 2019. (a–d) refer to the four members of the control ensemble (1/wT, 2/wT, 3/wT, and 4/wT). (a–d) are similar at large scales, but at small scales, they show some differences, e.g., in the middle of the South Yellow Sea. These results are consistent with our conclusion that the signal is dominant at large scales, while the relative importance of noise increases with decreasing scale.

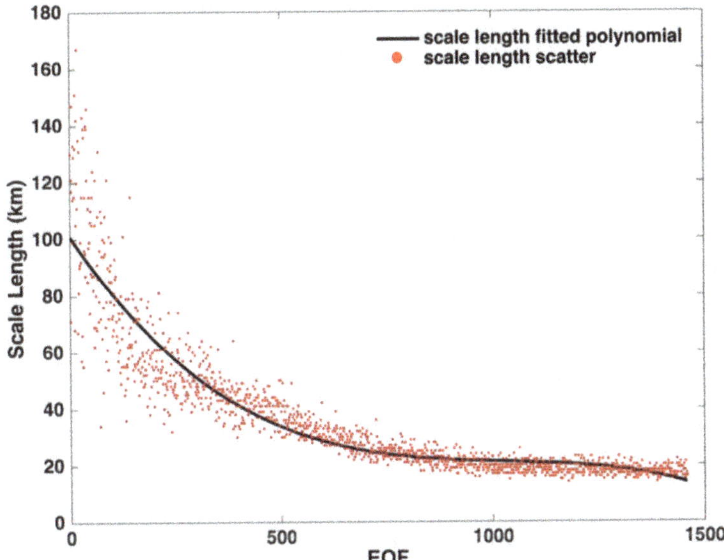

Fig. 6. The estimated scale lengths (km) of all 1459 EOFs. The red dots are the scales of each mode. The black line is a smoothed interpolation.

from approximately 140 km to 20 km. The scale lengths do not decrease in a monotonic way, but the trend is clear.

There is a clear decreasing trend in the S/N ratio as the EOF index increases, which can be seen from the S/N ratios for all EOFs

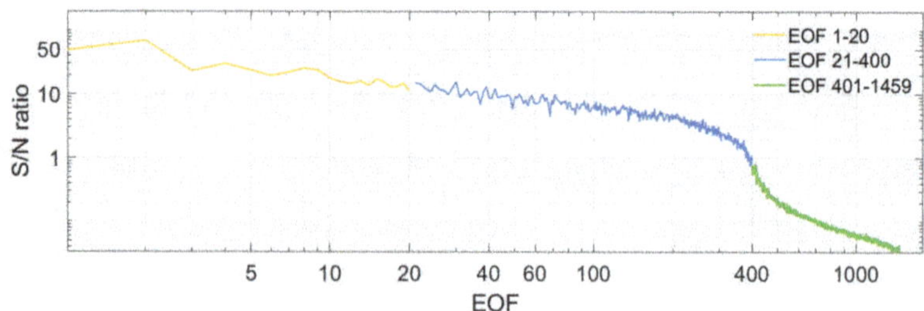

Fig. 7.   The depth-averaged velocities S/N ratios as a function of the EOF index. Based on the distribution of the S/N ratios, the scales (EOFs) are divided into three blocks: EOFs 1–20 (yellow line), EOFs 21–400 (blue line), and EOFs 401–1459 (green line).

of depth-averaged velocities (Figure 7). In particular, there is a dramatic decrease in the first few EOFs. We split the full range of EOFs into three subranges, namely, EOFs 1–20, 21–400, and 401–1459. The S/N ratios in the first segment (yellow) decline sharply; in the middle segment (blue), they decline slowly; and in the third segment (green), the S/N ratios are almost zero. The yellow line in Figure 7 represents the "large-scale" range, which has an average length of 100 km, and the S/N ratio declines sharply from a very high value of 68 to approximately 18. The blue line represents "medium scales" with an average length of approximately 60 km and reduced but nonzero S/N ratios. The green line in Figure 7 shows the close-to-zero S/N ratios of the last "small scale" (23 km) segment.

To illustrate the S/N ratio spatial distribution of different scales, we show maps of the S/N ratios for depth-averaged velocities in the control run for three different scale ranges, EOFs 1–20, 21–400, and 401–1459 (Figure 8). At large scales, the largest S/N ratios (>20) appear on the coast of the Bohai, where the depth is shallow. At medium scales, the S/N ratios are smaller everywhere than those at large scales, but the spatial distribution characteristics of larger S/N ratios along the coasts of the Bohai and Yellow Sea are visible. At small scales, the S/N ratios are small everywhere.

Internal variability is also found in surface temperature and surface salinity, as shown by the S/N ratios for temperature, depth-averaged velocities and salinity in Figure 9. At large scales, the temperature S/N ratios are larger than those for depth-averaged velocities and salinity, which indicates that externally forced signals

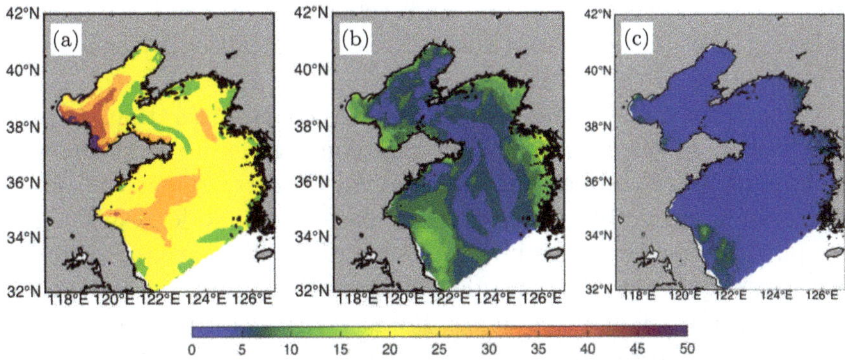

Fig. 8.   The S/N ratio spatial distributions of the control run at large (EOFs 1–20; (a)), medium (EOFs 21–400; (b)), and small scales (EOFs 401–1459; (c)).

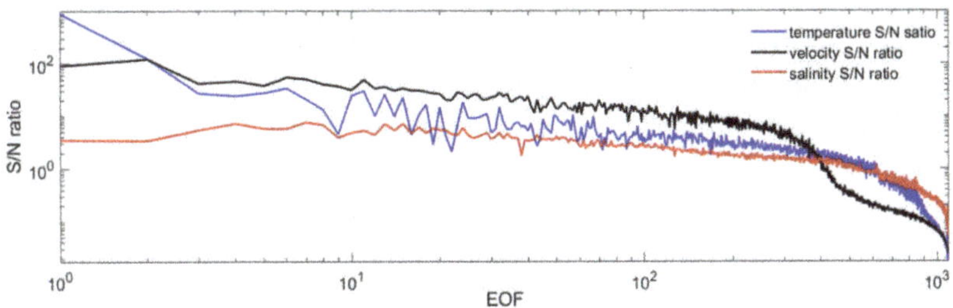

Fig. 9.   The S/N ratios of temperature, salinity, and depth-averaged velocities as a function of the EOF index.

are relatively important. We suggest that the S/N ratios of temperature at large scales are related to the marked seasonal signal. The S/N ratios are lower for depth-averaged velocities at large scales; the S/N ratios for salinity are the lowest.

4.   Change in Internal Variability in the Ensembles with Reduced Tidal Forcing

In this section, the changes in internal variability in the ensembles with reduced tidal forcing are analyzed separately in the Bohai and Yellow Sea. This is motivated by the difference in the intensity of tides in the Bohai and Yellow Sea. First, we analyze the S/N ratios in the simulations with reduced (half, no) tidal forcing, particularly the differences between the Bohai and Yellow Sea. Second, we analyze whether the tide effects on the S/N ratio are caused by changes in the signal or in the noise.

The intensity of tides in the Yellow Sea is larger than that in the Bohai, as one can conclude from comparing the amplitudes of dominant tidal components M_2 and S_2 (Figure S1[2] in the online supplementary material, which is calculated from model results). The amplitude represents the potential energy of tides. From Figure S1, we can see that the amplitudes in most of the area in the Yellow Sea are larger than those in the Bohai. With regard to tidal range, another indicator of tidal strength, Sun (2006) found that the Yellow Sea has the largest tidal range among the coastal seas of China, whereas Bohai's range is smaller. Yao *et al.* (2012) also found that based on simulation results and observations, the largest M_2 tidal range is seen along the west coast of the Korean Peninsula; inside the Bohai, the M_2 tide is largely damped by the enhanced bottom friction associated with its shallow bathymetry.

When repeating the analysis of Section 2, we calculate new EOFs for the two seas separately after combining all members of the 1–4/wT, 1–4/hT, and 1–4/nT ensembles for EOF analyses, resulting in 4380 (12 × 365) modes for both the Bohai and Yellow Sea.

With no tidal forcing, the S/N ratio is notably reduced at large scales, slightly reduced at medium scales, and not reduced at small scales. In other words, the relative importance of internal variability is increased at large and medium scales without tidal forcing. Figures 10 and 11 show the spatial distributions of the S/N ratios for the depth-averaged velocities in the Bohai and Yellow Sea, respectively. The rows in Figures 10 and 11 are the S/N ratios of the wT, hT, and nT ensembles; the columns in Figures 10 and 11 are the S/N ratios at large scales, medium scales, and small scales. Spatially averaged estimates are given in Table 2.

There is a clear decrease in S/N ratios at large and medium scales in both the Bohai and Yellow Sea when tidal forcing decreases. In the Bohai, the spatially averaged S/N ratios decrease from 32.63, 8.44, and 0.20–23.40, 7.11, and 0.20 at large, medium, and small

---

[2]Refer to https://ars.els-cdn.com/content/image/1-s2.0-S0377026522000203-mmc1_lrg.jpg.

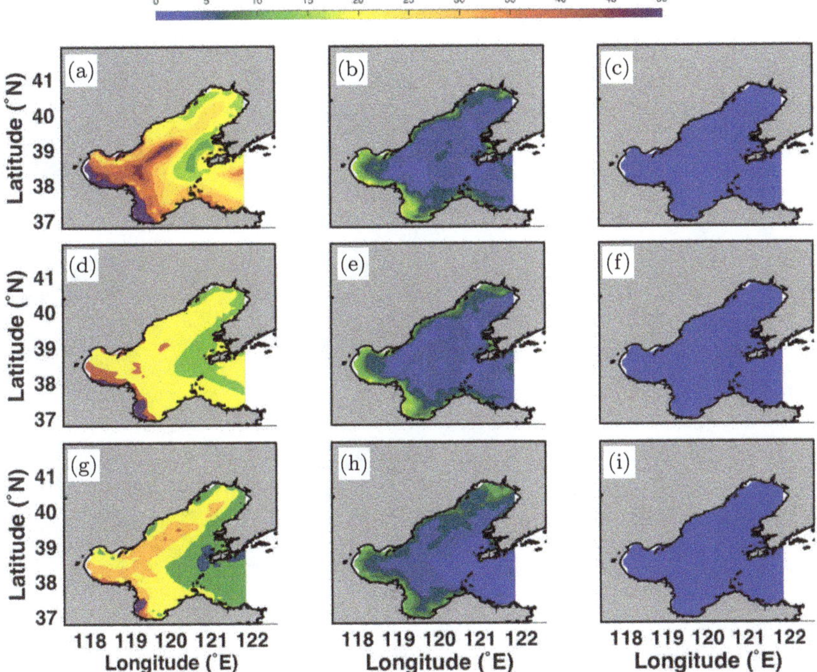

Fig. 10.   The S/N ratio spatial distributions of wT, hT, and nT ensembles at large (EOFs 1–20), medium (EOFs 21–2000), and small scales (EOFs 2001–4379) in the Bohai. The columns are S/N ratio spatial distributions at large scales (a, d, g), medium scales (b, e, h), and small scales (c, f, i). The rows are S/N ratio spatial distributions of wT (a, b, c), hT (d, e, f), and nT ensembles (g, h, i). For example, (a) shows the S/N ratio spatial distribution of the wT ensemble at large scales in the Bohai; (e) shows the S/N ratio spatial distribution of the hT ensemble at medium scales.

scales, respectively, when tidal forcing is turned off (Table 2). Percentagewise, this amounts to losses of 28.3%, 15.8%, and 0 in terms of large, medium, and small scales, respectively. In the case of the Yellow Sea, this decline is much larger, namely, 83.6%, 68.2%, and 0 at large, medium, and small scales, respectively (Table 2). When comparing Figures 10 and 11, the decrease in S/N ratios in response to reduced tidal forcing is much smaller in the Bohai than in the Yellow Sea.

We suggest that this difference is related to the different intensities of the tides in the two marginal seas. The tide intensity is stronger in the Yellow Sea than in the Bohai; therefore, the tidal effect on the

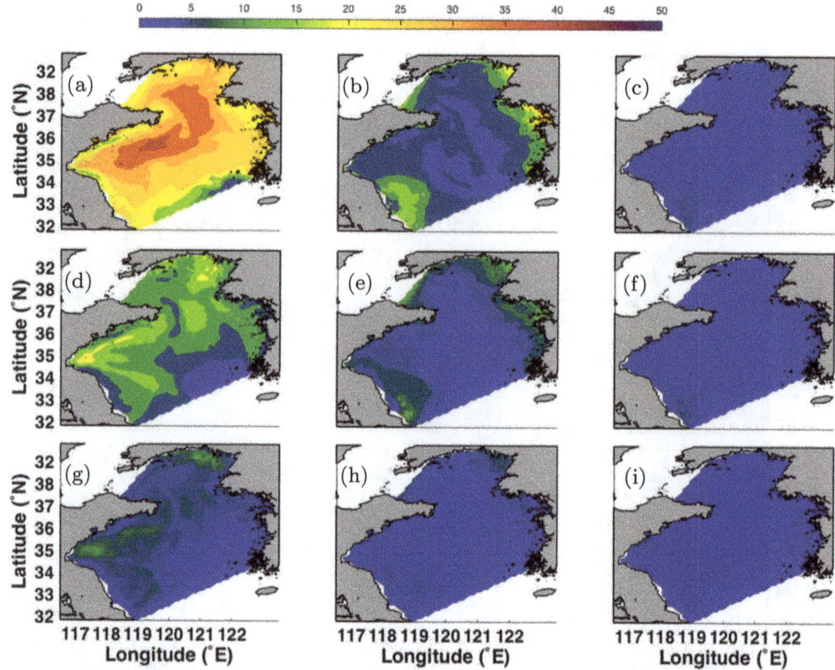

Fig. 11.   The S/N ratio spatial distributions of wT, hT, and nT ensembles at large (EOFs 1–20), medium (EOFs 21–2000), and small scales (EOFs 2001–4379) in the Yellow Sea. The columns are S/N ratio spatial distributions at large scales (a, d, g), medium scales (b, e, h), and small scales (c, f, i). The rows are S/N ratio spatial distributions of wT (a, b, c), hT (d, e, f), and nT ensembles (g, h, i). For example, (a) shows the S/N ratio spatial distribution of the wT ensemble at large scales; (e) shows the S/N ratio spatial distribution of the hT ensemble at medium scales.

internal variability in depth-averaged velocities in the Yellow Sea is stronger than that in the Bohai.

When examining the EOF-spectra of the S/N ratios of depth-averaged velocities for the wT, hT, and nT ensembles as a function of the EOF index in the Bohai and Yellow Sea (Figure 12), consistent differences between the Bohai and Yellow Sea and the different scale ranges emerge, which are in agreement with the S/N ratio spatial distribution results in which S/N ratios decrease with reduced tidal forcing, especially at large and medium scales in the Yellow Sea. In the Bohai, the S/N ratios of the first EOFs of the wT ensemble are larger than those of the hT ensemble, and those of the hT ensemble

Table 2.   Spatially averaged S/N ratios of depth-averaged velocities in the wT, hT, and nT ensembles.

	Large scale	Medium scale	Small scale
Bohai			
S/N ratio wT	32.63	8.44	0.20
S/N ratio hT	30.11	8.07	0.20
S/N ratio nT	23.40	7.11	0.20
Yellow sea			
S/N ratio wT	27.92	8.93	0.23
S/N ratio hT	11.37	5.31	0.23
S/N ratio nT	4.57	2.84	0.23

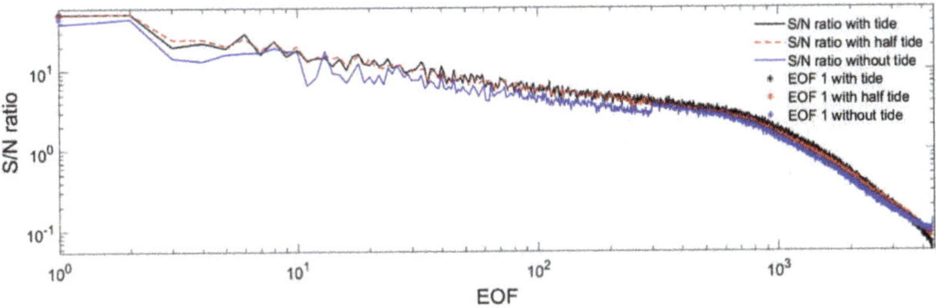

Fig. 12.   The depth-averaged velocity S/N ratios of the wT, hT, and nT ensembles as a function of the EOF index in the Bohai. The black solid line, red dashed line, and blue solid line represent the S/N ratios of the wT, hT, and nT ensembles, respectively.

are larger than those of the nT ensemble, but these differences among the three ensembles are relatively small (Figure 12). However, in the Yellow Sea, while the same pattern prevails, the differences are considerably larger (Figure 13).

Is the change mostly due to an increase of the signal, or a reduction of the noise? Figure 14 shows the intensities sorted after the spatial scale of the EOFs, which span from approximately 20 km to 200 km. The black and yellow dots display the signals, and both clouds are rather similar. The green and pink clouds represent noise, and obvious differences emerge. The general shapes are similar, with very high values at the shortest scales and a gradual weakening for large scales. However, for most scales, except for the smallest ones,

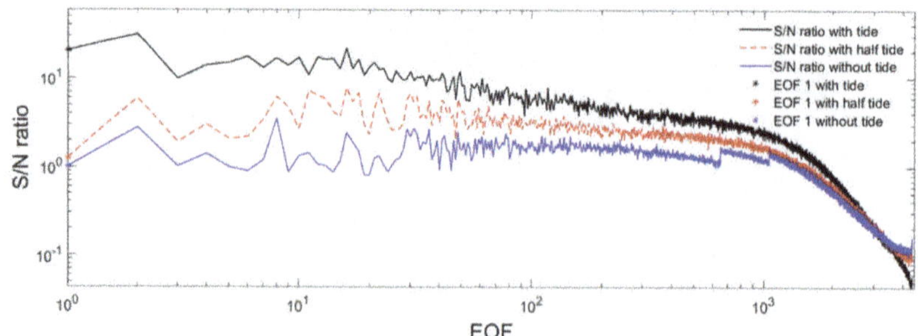

Fig. 13.   The depth-averaged velocity S/N ratios of wT, hT, and nT ensembles as a function of the EOF index in the Yellow Sea. The black solid line, red dashed line, and blue solid line represent the S/N ratios of the wT, hT, and nT ensembles, respectively.

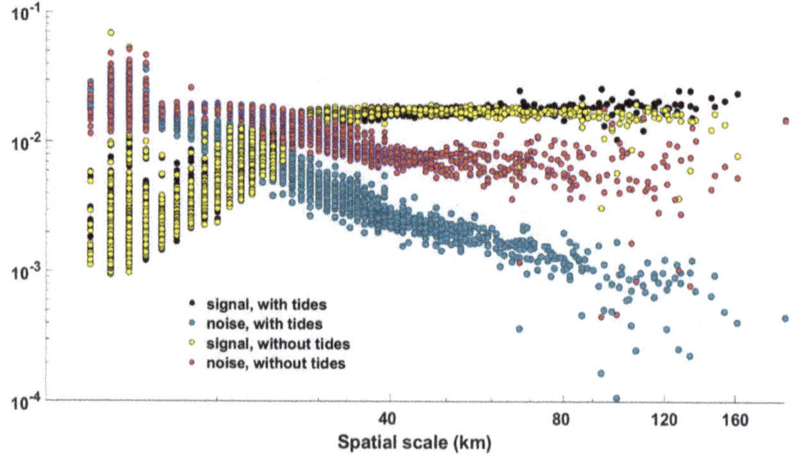

Fig. 14.   Noise and signal levels of depth-averaged velocities in the nT and wT ensembles sorted after the spatial scale of the EOFs.

the pink dots are located above the green dots — the noise level is reduced at all other scales.

To quantify the tidal forcing effects on the Bohai and Yellow Sea, we list the levels of noise for the different spatial scale ranges separately for the Bohai and Yellow Sea in Table 3. In the Bohai, tidal forcing results in 47% and 39% decreases in spatially averaged noise at large and medium scales, respectively; in contrast, tidal forcing results in an increase of 88% and a decrease of 75% at large and medium scales in the Yellow Sea, respectively. At small scales, the

Table 3. Spatially averaged noise of depth-averaged velocity in the wT, hT, and nT ensembles.

	Large scale	Medium scale	Small scale
Bohai			
noise wT	0.0001	0.0023	6.60e-04
noise hT	0.0012	0.0024	6.90e-04
noise nT	0.0019	0.0038	6.70e-04
Yellow Sea			
noise wT	0.0011	0.0026	8.09e-04
noise hT	0.0033	0.0043	8.63e-04
noise nT	0.0098	0.0103	8.20e-04

noise levels are approximately the same in the wT, hT, and nT ensembles.

We draw two conclusions concerning the effect of tides: first, tidal forcing reduces the emergence of noise at large and medium scales but not at small scales; second, the tidal forcing effect on noise is more obvious in the Yellow Sea than in the Bohai, likely because tides are weaker in the Bohai than in the Yellow Sea.

5.   Discussion and Conclusion

The temporal and spatial scales in the abiotic climate system are linked — small scales are associated with quickly changing variables, and large scales are associated with slowly changing variables (e.g., von Storch and Zwiers, 1999). The variations are related to external forcing from tidal potentials, the annual cycle, and changes in the atmospheric composition. Some variations are internally generated through myriads of nonlinear components and processes, among them fronts and eddies. These variations may be conceptualized as stochastic. Even though the external signals act mostly on larger scales, their effects cascade down to smaller scales ("downscaling"). However, noisy components are integrated, leaving their traces at larger scales (Hasselmann, 1976; a good example is given by Mikolajewicz and Maier-Reimer (1990)).

At large scales, external factors are commonly dominant — this is why scientists can construct scenarios of possible future climate change as responses to modifications of the atmospheric composition; from this follows the very practical understanding that in mid-latitudes, the annual cycle is considered trivial. However, a minor part of the large-scale variability also commonly appears to be stochastic. At small scales, the contributions are different — most variability is noisy and thus unprovoked by external factors, and a smaller part of the variability is externally forced.

All this has long been known in the paradigm of climate dynamics; it may be the key innovation in the field when physics took over the field from geography (Hasselmann, 1976) — and was recognized in 2021 by the Nobel Prize in Physics. This view was illustrated and demonstrated to be valid first in global models of the atmosphere, later in global oceanic models, and recently in regional atmospheric and oceanic models (see references in the Introduction). This is the context for the present study.

The results of Tang *et al.* (2020) for the deep South China Sea have shown that the S/N ratio of external signals and internal variations gradually changes from large at large scales to almost nil at small scales, also applying to shallower marginal seas. Penduff *et al.* (2018) also found such a scale dependency.

An obvious question that we can only answer speculatively at this time is as follows: What generates internal variability in shallow marginal seas? Stratification may modulate the tendency to form internal variability. The development of the standard deviation of the annual cycle runs parallel to the build-up of the seasonal stratification (not shown). Additionally, the usage of surface fluxes generated in reanalysis may allow for more internal variability compared with a setup when the fluxes are derived from simulated SST and air temperature via bulk formulae. In hydrodynamic models, small-scale disturbances generated at the smallest scale grow over time and accumulate noise on larger time and spatial scales (e.g., Roeckner and von Storch, 1980, and references therein), conditional upon the degree of diffusion applied. This accumulation reminds us of the accumulation of noise on larger scales in Hasselmann's (1976) stochastic climate model.

In our study, we asked two specific questions. First, what are the characteristics of internal variability in the Yellow Sea and in the Bohai? Second — what is the effect of tides on the intensity of the formation of internal variability?

Internal noise is a significant issue in the Bohai and Yellow Sea. We demonstrated the scale dependency and sensitivity of internal variations to the presence of external tides, similar to the South China Sea. Most of the analysis was performed with depth-averaged velocities.

Comparing the S/N ratios of depth-averaged velocities in the Bohai and Yellow Sea with those in the South China Sea, the internal variability in the Bohai and Yellow Sea is smaller. For the first EOF in the Bohai and Yellow Sea, the S/N ratio is approximately 70, whereas in the South China Sea, the S/N ratio of the first EOF is approximately 40. The reason for this may be that the hydrodynamics of the South China Sea and the Bohai and Yellow Sea are different. Substantial mesoscale eddies exist in the South China Sea (Zhang *et al.*, 2019)[3] and are a significant source of noise in the South China Sea. In the Bohai and Yellow Sea, eddies decrease and tend to be of the submesoscale type.

The study of internal variability is important for model results to be more realistic and to exclude accidental factors. In different regions, the level and scale of internal variability are various, and internal variability is reflected in different variables because dominant dynamic processes, topography, latitude, and open boundaries are various in different seas. When studying or simulating a physical process, it is necessary to clarify whether internal variability masks its effect, as originally worked out by Chervin *et al.* (1974); see also von Storch and Zwiers (1999). If internal variability is important in an area, for instance, mesoscale eddies in the South China Sea, an ensemble setup will be necessary (Zhang *et al.*, 2019). If statistics of variations on large spatial scales are of interest, the internal variability may often be ignored. Another avenue is opened by the ubiquity of small-scale noise: how stochastic does this make ecological and morphodynamical developments?

The second question, on the role of tides, was suggested by Schrum (2019, pers. communication). The analysis of three ensembles of long model integrations with different intensities of enforced tides verifies the Schrum hypothesis: the relative importance of internal variability increases when tidal forcing decreases; tidal forcing suppresses

---

[3]Reprinted as Chapter 3.

the internal variability at large and medium scales but not at small scales. Consistent with this finding is that the noise-reducing effect is smaller in the weaker-tidal Bohai and larger in the stronger-tidal Yellow Sea.

We cannot identify a process that leads to a reduction in the noise when the tides occur. However, we hypothesize which mechanism may play a role.

It is possible that the theory of the stochastic climate model points to a solution — this model transforms white noise, short-scale forcing, into a red spectrum, with enhanced variance at longer scales. The strength of the transformation depends on the memory of the system. A longer memory, i.e., a system within which once formed anomalies persist for a longer time, has a redder spectrum compared with a short-memory system, when such anomalies decay faster. Our hypothesis is that tides tend to shorten the lifetime of anomalies, resulting in less internal variance at long scales, even though the driving short-scale forcing is hardly changed.[4]

Funding

The modeling research of this work is supported by the Oceanic Instruments Standardization Sea Trials (OISST) project (2016YFC1401300) of the National Key Research and Development Plan and Taishan Scholars Program.

Acknowledgments

We thank the China Scholarship Council for enabling Xueen Chen to build constructive cooperation between the Ocean University of China and the Helmholtz-Zentrum Hereon in Germany. A major motivation for our work was provided by Corinna Schrum of Helmholtz-Zentrum Hereon, who suggested that the intensity of the noise in the system would strongly depend on the intensity of the tides. Thanks should be given to Deutsches Klimarechenzentrum GmbH (DKRZ) for high-performance computing system support. We thank

---

[4]At a later time, this hypothesis was examined in more detail, and found consistent with the data, see Lin *et al.* (2023).

the China National Science and Technology Resources Sharing service platform at the National Marine Science Data Center (http://mds. nmdis.org.cn/) for providing data support. Roland Doerffer provided us with Figure 1(b).

Constructive exchanges with Emil V. Stanev of the Helmholtz-Zentrum Hereon, Uwe Mikolajewicz from MPI of Meteorology in Hamburg, and Hans Burchard of Institute of Baltic Research in Warnemünde are acknowledged.

References

An, H. S. (1977). A numerical experiment of the M2 tide in the Yellow Sea, *Journal of Oceanography* **33**, 2, pp. 103–110.

Büchmann, B. and Soderkvist, J. (2016). Internal variability of a 3-D ocean model, *Tellus Series a-Dynamic Meteorology and Oceanography* **68**, p. 30417, doi:10.3402/tellusa.v68.30417.

Cai, W., Ng, B., Geng, T., Wu, L., Santoso, A. and McPhaden, M. J. (2020). Butterfly effect and a self-modulating El Niño response to global warming, *Nature* **585**, 7823, pp. 68–73, doi:10.1038/s41586-020-2641-x.

Callies, J. (2016). *Submesoscale Turbulence in the Upper Ocean*, Phd thesis, Massachusetts Institute of Technology and the Woods Hole Oceanographic Institution.

Chen, C., Beardsley, R. C., Cowles, G., Qi, J., Lai, Z., Gao, G., Stuebe, D. *et al.* (2006). An unstructured grid, finite-volume coastal ocean model: Fvcom user manual, *SMAST/UMASSD* **6**, p. 78.

Chervin, R., Gates, W. and Schneider, S. (1974). The effect of the time averaging on the noise level of climatological statistics generated by atmospheric general circulation models, *Journal of the Atmospheric Sciences* **31**, pp. 2216–2219.

Deng, Z. and Zhao, Y. (2020). Impact of tidal mixing on water mass properties and circulation in the Bohai Sea: A typhoon case, *Journal of Marine Systems* **206**, p. 103338, doi:10.1016/j.jmarsys.2020.103338.

Egbert, G. D. and Erofeeva, S. Y. (2002). Efficient inverse modeling of barotropic ocean tides, *Journal of Atmospheric and Oceanic Technology* **19**, 2, pp. 183–204, doi:10.1175/1520-0426(2002)019 <0183:EIMOBO>2.0.CO;2.

Geyer, B., Ludwig, T. and von Storch, H. (2021). Reproducibility and regional climate models — seeding noise by changing computers and initial conditions, *Communications Earth & Environment* **2**, p. 17, doi:10.1038/s43247-020-00085-4.

Grégorio, S., Penduff, T., Serazin, G., Molines, J.-M., Barnier, B. and Hirschi, J. (2015). Intrinsic variability of the Atlantic Meridional overturning circulation at interannual-to-multidecadal time scales, *Journal of Physical Oceanography* **45**, 7, pp. 1929–1946, doi:10.1175/JPO-D-14-0163.1.

Hasselmann, K. (1976). Stochastic climate models part I. Theory, *Tellus* **28**, pp. 473–485, https://doi:10.1111/j.2153-3490.1976.tb00696.x|.

Hasselmann, K. (1979). On the signal-to-noise problem in atmospheric response studies. In: Shaw, B.D. (Ed.), Meteorology over the tropical oceans. *Royal Meteorological Society*, Bracknell, Berkshire, England, pp. 251–259.

Hirschi, J. J.-M., Blaker, A. T., Sinha, B., Coward, A., de Cuevas, B., Alderson, S. and Madec, G. (2013). Chaotic variability of the meridional overturning circulation on subannual to interannual timescales, *Ocean Science* **9**, 5, pp. 805–823, doi:10.5194/os-9-805-2013.

Hu, S., Chen, C., Gao, G., Lai, Z., Ge, J., Lin, H. and Qi, J. (2012). Preliminary analysis of tide simulation of the East China Sea of the Global-FVCOM model, *Journal of Shanghai Ocean University* **21**, pp. 621–629.

Jochum, M. and Murtugudde, R. (2004). Internal variability of the tropical Pacific Ocean, *Geophysical Research Letters* **31**, 14, doi:10.1029/2004GL020488.

Jochum, M., Malanotte-Rizzoli, P. and Busalacchi, A. (2004). Tropical instability waves in the Atlantic Ocean, *Ocean Modelling* **7**, 1–2, pp. 145–163, doi:10.1016/S1463-5003(03)00042-8.

Kushnir, Y. (1994). Interdecadal variations in North-Atlantic sea-surface temperature and associated atmospheric conditions, *Journal of Climate* **7**, 1, pp. 141–157, doi:10.1175/1520-0442(1994)007 <0141:IVINAS>2.0.CO;2.

Mahajan, S., Zhang, R. and Delworth, T. L. (2011). Impact of the Atlantic meridional overturning Circulation (AMOC) on Arctic surface air temperature and sea ice variability, *Journal of Climate* **24**, 24, pp. 6573–6581, doi:10.1175/2011JCLI4002.1.

Mellor, G. L. and Yamada, T. (1982). Development of a turbulence closure-model for geophysical fluid problems, *Reviews of Geophysics* **20**, 4, pp. 851–875, doi:10.1029/RG020i004p00851.

Mikolajewicz, U. and Maier-Reimer, E. (1990). Internal secular variability in an ocean general circulation model, *Climate Dynamics* **4**, 3, pp. 145–156, doi:10.1007/BF00209518.

Peacock, T. and Haler, G. (2013). Lagrangian coherent structures the hidden skeleton of fluid flows, *Physics Today* **66**, 2, pp. 41–47, doi:10.1063/PT.3.1886.

Penduff, T., Sérazin, G., Leroux, S., Close, S., Molines, J.-M., Barnier, B., Bessières, L., Terray, L. and Maze, G. (2018). Chaotic variability of ocean heat content: Climate-relevant features and observational implications, *Oceanography* https://doi.org/10.5670/oceanog.2018.210.

Roeckner, E. and von Storch, H. (1980). On the efficiency of horizontal diffusion and numerical filtering in an arakawa-type model, *Atmosphere-Ocean* **18**, 3, pp. 239–253.

Sun, X. China, (2006). Tides and Storm. In: Yang, C.X. (Ed.), China offshore regional oceanography. *Ocean Press*, pp. 130–140.

Tang, S., von Storch, H. and Chen, X. (2020). Atmospherically forced regional ocean simulations of the South China Sea: Scale dependency of the signal-to-noise ratio, *Journal of Physical Oceanography* **50**, 1, pp. 133–144, doi:10.1175/JPO-D-19-0144.1.

Tang, S., von Storch, H., Chen, X. and Zhang, M. (2019). "Noise" in climatologically driven ocean models with different grid resolution, *Oceanologia* **61**, 3, pp. 300–307, doi:10.1016/j.oceano.2019.01.001.

Timmermann, A. and Lohmann, G. (2000). Noise-induced transitions in a simplified model of the thermohaline circulation, *Journal of Physical Oceanography* **30**, 8, pp. 1891–1900, doi:10.1175/1520-0485(2000)030<1891:NITIAS>2.0.CO;2.

von Storch, H. and Zwiers, F. W. (1999). *Statistical Analysis in Climate Research*, Cambridge University Press.

Wackernagel, H. (2003). *Multivariate Geostatistics: An Introduction with Applications*, Springer Science & Business Media.

Weisse, R., Mikolajewicz, U. and Maier-Reimer, E. (1994). Decadal variability of the North-Atlantic in an Ocean General-Circulation Model, *Journal of Geophysical Research-Oceans* **99**, C6, pp. 12411–12421, doi:10.1029/94JC00524.

Weisse, R., Mikolajewicz, U., Sterl, A. and Drijfhout, S. (1999). Stochastically forced variability in the Antarctic Circumpolar Current, *Journal of Geophysical Research-Oceans* **104**, C5, pp. 11049–11064, doi:10.1029/1999JC900040.

Yao, Z., He, R., Bao, X., Wu, D. and Song, J. (2012). M-2 tidal dynamics in Bohai and Yellow Seas: A hybrid data assimilative modeling study, *Ocean Dynamics* **62**, 5, pp. 753–769, doi:10.1007/s10236-011-0517-1.

Zhang, M., von Storch, H., Chen, X., Wang, D. and Li, D. (2019). Temporal and spatial statistics of travelling eddy variability in the South China Sea, *Ocean Dynamics* **69**, 8, pp. 879–898, doi:10.1007/s10236-019-01282-2.

https://doi.org/10.1142/9781800615816_0010

Part III
Regional Climatologies of Cyclones

In this part of the anthology, we deal with studies on North Pacific Polar Lows and on statistics of millennial changes of regional mid-latitude storminess.

Our study on North Pacific Polar Lows was inspired by Matthias Zahn's research on similar cyclones in the sub-Arctic North Atlantic. Zahn, a scientist at HZG, reconstructed the formation and statistics of Polar Lows using the dynamical downscaling methodology. In a subsequent step, scenarios were developed to project the possible future frequency of these relatively small but severe marine storms in a changed climate at the end of the 21st century. The reprinted articles entitled "Trends and variability of North Pacific Polar Lows" and "Polar Low genesis over North Pacific under different global warming scenarios" addressed similar challenges but focused on the less frequently studied Polar Lows in the North Pacific.

There are four more papers that would also fit this chapter, but due to the limited space, they are not reprinted. The first article "Towards a multidecadal climatology of North Pacific Polar Lows employing dynamical downscaling" was on testing the methodology and found the following: "The tracks of the simulated Polar Lows closely follow the tracks derived from satellite imagery. We conclude that the suggested method is suitable for constructing multi-decade climatologies, including trends and variability, of Polar Lows in the North Pacific by dynamically downscaling NCEP re-analyses." (Chen *et al.*, 2012).

In order to obtain statistics on polar lows, such as their tracks, intensities, and lifetimes, it is necessary to be able to identify and track these disturbances. This was the focus of the study (Xia *et al.*, 2012) "A comparison of two identification and tracking methods for polar lows". In this study, two methods make use of digital band-pass filtering and of discrete cosine transforms. The latter is found to be "more precise in scale separation than the digital filter. The detection and tracking parts also influence the numbers of tracks although less critically. After a selection process that applies criteria to identify tracks of potential polar lows, differences between both methods are still visible though the major systems are identified in both."

Two other articles (Xia *et al.*, 2013, 2016) pursued a different line of analysis, namely the centennial variability of baroclinic storms on both the northern and southern midlatitudes. The first article "Quasi-stationarity of centennial Northern Hemisphere midlatitude winter storm tracks" focuses on Northern Hemisphere midlatitude winter storm tracks as simulated in a millennial simulation with varying external forcing. The study tracked cyclones, binned them into regional clusters, and examined their temporal variability. The companion paper "A study of quasi-millennial extratropical winter cyclone activity over the Southern Hemisphere" explores the same analysis for Southern Hemisphere baroclinic storms. A "very strong year-to-year variability for Southern Hemispheric winter extratropical cyclone numbers and larger variations on centennial time scale, more so than for its Northern Hemispherical counterparts" was found. "However, no obvious trend can be found. The mean tracks of clusters over the Southern Indian Ocean and near New Zealand shift poleward from the eleventh to the twentieth century while the clusters in the central Southern Pacific shift equatorward."

Chapter 7

Trends and Variability of North Pacific Polar Lows*

F. Chen and H. von Storch

*Institute of Coastal Research, Helmholtz-Zentrum Geesthacht,
Germany*

The 6-hourly 1948–2010 NCEP1 reanalyses have been dynamically downscaled for the region of the North Pacific. With a detecting-and-tracking algorithm, the climatology of North Pacific Polar Lows has been constructed. This derived climatology is consistent with the limited observational evidence in terms of frequency and spatial distribution. The climatology exhibits strong year-to-year variability but weak decadal variability and a small positive trend. A canonical correlation analysis describes the conditioning of the formation of Polar Lows by characteristic seasonal mean flow regimes, which favor, or limit, cold air outbreaks and upper air troughs.

---

*This chapter was originally published in *Advances in Meteorology*, 170387, doi:10.1155/2013/170387. This chapter is licensed under the terms and conditions of the Creative Commons Attribution (CC BY) license (https://creativecommons. org/licenses/by/4.0/), which permits unrestricted use, distribution, and reproduction in any medium.

1. Introduction

"Polar Low" has been defined as the generic term for all mesoscale cyclonic vortices poleward of main polar front. It should be used for intense maritime mesocyclones with scales less than 1000 km with strong wind speeds (Heinemann and Claud, 1997).

Since the availability of comprehensive observations was supported by satellite imagery, a couple of authors have dealt with space-time statistics of Polar Lows in the North Pacific (Businger, 1987; Ninomiya, 1989; Yarnal and Henderson, 1989). However, these studies using satellite data cover only a few years of Polar Low occurrences. Also, a combination of subjective detections methods and inhomogeneities in data leads to the possibility that derived trends and variability may not be robust.

A number of authors have applied global reanalysis data for investigating conditions, which are favorable or unfavorable for the formation of Polar Lows and mesocyclones statistics in both Atlantic and Pacific (Condron *et al.*, 2006; Kolstad, 2006). Kolstad identified low static stability and reverse-wind shear conditions in the 40-year period of ERA-40 reanalyses as favorable conditions. By computing the space-time statistics (i.e., climatology) of favorable conditions for Polar Lows over the North Atlantic, the North-West Pacific and over Southern Ocean, statistics of Polar Low occurrences could be estimated in this indirect manner.

A suitable method for constructing climatology is using long-term dynamical downscaling with regional climate models (Zahn and von Storch, 2008a). Various authors have demonstrated that Polar Lows and other mesoscale windstorms can be described by high-resolution numerical models, such as Fu *et al.* and Yanase *et al.* in Japan Sea (Fu *et al.*, 2004; Yanase *et al.*, 2004; Yanase and Niino, 2007); Bresch *et al.* in Bering Sea (Bresch *et al.*, 1997); Businger and Blier in Gulf of Alaska (Blier, 1996; Businger, 1987); Chen *et al.* in North Pacific (Chen *et al.*, 2012); Cavicchia and von Storch for "medicanes" in the Mediterranean Sea (Cavicchia and von Storch, 2012); and Zahn and von Storch in North Atlantic (Zahn and von Storch, 2008a). Dynamical mechanisms and synoptic conditions of Polar Low formation as well as life cycles have been discussed in various case studies. Reed supported the baroclinic theory in 1979, but he did not reject the possibility of other instabilities for the Polar Low mechanism

(Reed, 1979). The barotropic shear instability has also been put forward to explain the genesis of meso-scale Polar Lows over the polar-air mass convergence zone of Japan Sea (Nagata, 1993). High-resolution simulations have also been applied for some idealized experiments to explain mechanisms of Polar Low development such as baroclinicity (Yanase and Niino, 2007). A theoretical balanced axisymmetric model has been applied to investigate the wind-induced surface heat exchange intensification mechanism in Polar Low development by Gray and Craig (Gray and Craig, 1998). Various case studies indicate that usually several mechanisms together lead to the formation of Polar Lows (Businger, 1985; Mailhot *et al.*, 1996; Nordeng, 1990). A recent study also pointed out that Polar Lows influence the large-scale ocean circulation and deep water transport over the Nordic sea (Condron and Renfrew, 2013).

In this paper, we employ the method developed by Zahn and von Storch (2008a, 2008b) to dynamically downscale the gridded large-scale synoptic fields, as provided by large-scale component of reanalyses — here: NCEP 1 (Kalnay *et al.*, 1996). In a first step, we have shown in a series of cases that downscaling generates Polar Lows with sufficient accuracy (Chen *et al.*, 2012). Now, in this study, a regional climate model (RCM) has been run continuously for 63 years during which the NCEP 1 reanalyses are available. During the integration, the model is constrained in its large-scale components to be similar to the driving NCEP reanalysis, using the method of spectral nudging (von Storch *et al.*, 2000). The output of this multidecadal simulation is used for investigating trends and variability of Polar Low occurrences in the North Pacific and their linkage to the seasonal mean large-scale circulation situations over the last 63 years.

We are sometimes confronted with the request that such an analysis should result in new insights into the dynamics of Polar Low formation and life cycles. However, this is not the intention of the present study. Such case studies have been done in many cases (as stated previously), and we do not intend to extend this large number of dynamical analyses. Instead, we are interested in space-time statistics, including the conditioning by large-scale dynamical configurations, of Polar Lows, in particular, on the differences between regions and years and decades, and in systematic changes. Thus, the present analysis does not contribute to dynamical meteorology but to climatology.

In Section 2, we describe briefly the model and the detection and tracking methodology used in this study. This methodology differs a bit from the previous study of Chen *et al.* (2012) as well as the North Atlantic study by Zahn and von Storch (2008b). The derived statistics of the formation of Polar Lows in space and time is the subject of Section 3. In Section 4, we determine linkages of seasonal mean large-scale flow in the North Pacific and the number of storms formed in different subregions of the North Pacific. The chapter is concluded with a discussion and the recapitulation of major results finally in Section 5.

2.   Data and Methodology

The RCM we applied in this study is the COSMO-CLM 4.8 (COSMO model in CLimate Mode) (Rockel *et al.*, 2008; Steppeler *et al.*, 2003). This model is the climate version of the operational weather prediction model of the Deutscher Wetterdienst (German National Weather Service) and the COnsortium for Small-scale MOdeling (COSMO), adapted to climate simulation purposes by the CLM-Community (http://www.clm-community.eu/). The model domain covers the whole North Pacific (Figure 3(b)). We used NCEP 1 reanalysis data as initial and lateral boundary conditions. In particular, the sea surface temperature (SST) and sea ice extent were prescribed as lower boundary according to the NCEP 1 reanalyses. Spectral nudging (von Storch *et al.*, 2000) of tropospheric wind components was applied for enforcing the NCEP large-scale situation in the model region in order to prevent COSMO-CLM from significantly deviating from the analyzed large-scale state. The nudging is applied at 850 hPa and above, with the nudging becoming stronger with height. Only spatial scales larger than 800 km are constrained; smaller features are unconstrained. A rotated grid with 0.4° grid resolution and 220 and 80 points is employed for the longitudinal and latitudinal grid map. The 8-grid sponge zone is introduced to avoid reflection of waves at the boundaries. Boundary data is prescribed by this zone with decreasing influence for the inner grid points. The simulation results from the sponge zone are not usable for further analysis. The number of vertical levels is 40. The simulation period began on 1 January 1948 and ended on 31 December 2010. For further details about the model setup, refer to Chen *et al.* (2012).

An automatic detection-and-tracking procedure was applied to determine the presence and tracks of Polar Lows in the North Pacific. The detection procedure searches for the minima in the band-pass filtered mean sea level pressure (MSLP) fields and concatenates the minima in consecutive time steps to tracks. Along these tracks, the fulfillment of further criteria is requested for categorizing an event as Polar Low:

(1) strength of the minimum band-pass filtered MSLP: ≤-2 hPa once along the track;
(2) wind speed: $\geq$13.9 m/s once along the track;
(3) sea surface versus mid troposphere temperature difference: SST-T500 hPa $\geq$ 39 K;
(4) average direction of the track: a north-to-south component;
(5) limits to allowable adjacent grid boxes. No land: those tracks are excluded when over 50% of the positions near the coastal grid boxes along the track.

When the minimum of the band-pass filtered MSLP along these tracks decreases below -6 hPa once and there are no coastal grid boxes close to that location, a Polar Low is recorded irrespective of the other criteria.

This is mostly identical to the North Atlantic setting of but some parameters (like the air–sea temperature difference) have been modified in order to meet the conditions in the North Pacific better (Chen *et al.*, 2012). However, we used now a cosine (discrete cosine transforms: DCT) band-pass filter of Denis *et al.* (2002) instead of the digital filter used by after Xia *et al.* (2012) pointed out the superiority of the first one. DCT is more precise in scale separation than the original digital filter.

Some Polar Lows in the North Pacific move on average in zonal direction, like the case of 22 March 1975 described by Chen *et al.* (2012). Therefore, the request for a north-to-south movement may be too strict. However, without this criterion, also smaller baroclinic storms may be categorized as Polar Lows. In order to examine the significance of this zonal movement criterion for variability and long-term trend, we did the analysis with both criteria, with and without N-S direction criterion (see Figure 1).

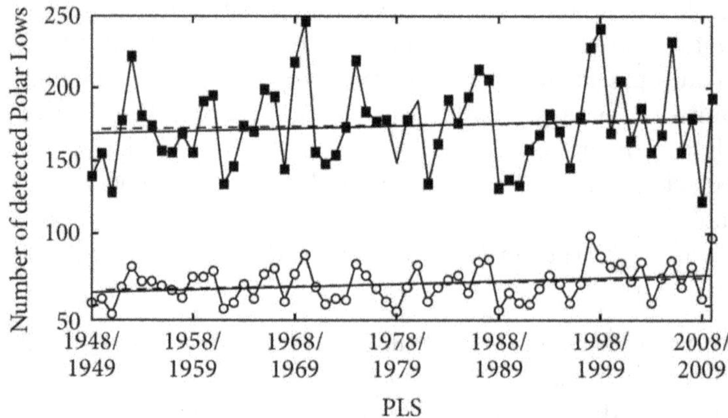

Fig. 1.   Number of detected Polar Lows in the North Pacific per Polar Low season (October to April). Top curves: the detected number of Polar Lows without N-S criterion (marked with ■). The solid line represents the trend from 62 PLSs, from 1948/1949 to 2009/2010, which is 0.17 cases/year; the dashed line represents the trend from only 60 years, from 1949/1950 to 2008/2009, which is 0.09 cases/year. Bottom curves: the detected number of Polar Lows with N-S criterion (marked with ○). The solid line represents the trend from 1948/1949 to 2009/2010, which is 0.2 cases/year; the dashed line represents the trend from 1949/1950 to 2008/2009, which is 0.14 cases/year.

3.   Results: Space Time Statistics of Tracks

Polar Lows are a phenomenon forming in the "cold season;" therefore, the "Polar Low season" (PLS) in the North Pacific is defined as the time from October through April the following year. This is a bit different from Zahn and von Storch (2008a) in the case of the North Atlantic: they count the PLS from July to next June. As we know, there are few Polar Lows in summer (see Figure 1); the Polar Low season is addressed by the first and second year; for example, the PLS 1950/1951 begins in October 1950 and ends in April 1951. The statistical analysis excluded the first half year of 1948 and last half year of 2010, as they represent only part of a Polar Low season. So we have 62 PLSs, with the first from October 1948 to April 1949 and the last from October 2009 to April 2010.

Figure 1 shows the time series of the number of detected Polar Lows per PLS both without and with N-S criterion. In all the following analyses, the criterion of an average movement with a north-to-south component is applied.

When looking only for cases without a directional constraint of the movement, a total of 10812 Polar Lows were detected by the tracking algorithm during the 62 Polar Low seasons. On average, 174 Polar Lows were found per PLS, with a strong year-to-year variability indicated by a standard deviation of 29 ($\pm17\%$ of the long-term mean). The decadal variability is weak. The overall trend, from the first PLS in 1948/1949 to the last PLS in 2009/2010, in the frequency of Polar Low is positive with 0.17 cases/PLS, which yields about 11 Polar Lows more in the end than in the beginning of the series (11 cases correspond to 6% of the long-term mean). We have to point out that the slope of the trend depends on the number of cases of the first and last PLSs. When disregarding the last and the first PLSs, the trend of Polar Lows from PLS 1949/1950 to 2008/2009 is smaller, with only 0.09 cases/PLS.

For the result with the N-S criterion, fewer Polar Lows are detected, namely, only 4052 Polar Lows during the 62 Polar Low seasons. On average, 65 Polar Lows were found per PLS, with a strong year-to-year variability indicated by a standard deviation of 12 ($\pm18\%$ of the long-term mean). Maximum number of detected cases is found in PLS 1997/1998 with 98 cases; the minimum number is in PLS 1950/1951 with 44 cases. The overall trend, from the first PLS in 1948/1949 to the last PLS in 2009/2010, in the frequency of Polar Low is positive with, on average, additional 0.2 cases per PLS, which yields about 12 Polar Lows more in the end than in the beginning of the series (corresponding to 21% of the mean total). The trend of Polar Lows from PLS 1949/1950 to 2008/2009 is positive with 0.18 cases/PLS. When calculating 10 trends from 1948/1949–2009/2010, 1949/1950–2008/2009 to 1957/1958–2000/2001, the mean trend is 0.16 cases/PLS; the standard deviation of these 10 trends is 0.03/PLS, so that the estimate of the trend appears relatively insensitive to the early and late values.

There is no acceleration of a trend towards the end of the time series. Furthermore, the trend seems mostly uniform and rather small. According to the prewhitened Mann-Kendall trend test (Kulkarni and von Storch, 1995), the 62 PLSs trend of the number of detected Polar Low is significant (5% risk; result not shown) for the configurations with directional constraining, but insignificant when examining the curve derived without directional constraint.

The annual cycle of monthly numbers of detected Polar Lows with N-S criterion exhibits the highest frequency in winter with maxima in December and January and almost no Polar Low activity in summer (Figure 2), consistent with the observation that Polar Lows form in the cold season. Furthermore, we determined the annual cycle of the days with Polar Lows during the same time period as Businger (Businger, 1987), namely, 1975–1983, and found our results consistent with Businger's results. The results differ with respect to a secondary peak in January in our 1975–1983 climatology but in February in Businger's climatology (figure not shown). In view of the very different methods applied by us and by Businger, namely, satellite observations versus downscaled reanalyses, the differences may be considered acceptable. When examining the set of Polar Lows derived without the directional constraint, a very similar annual cycle, apart of a uniform bias in the magnitude, is found (not shown).

There is no suggestion in the literature about the general number and trend of Polar Lows in the North Pacific that we should expect. Therefore, we find it difficult to decide if the directional constraint is really helpful or not. Clearly, the total number depends strongly on this criterion, but the overall trends are similar in both cases. This indicates that the N-S criterion does not much influence the general characteristic of long-term trend. The correlation coefficient between

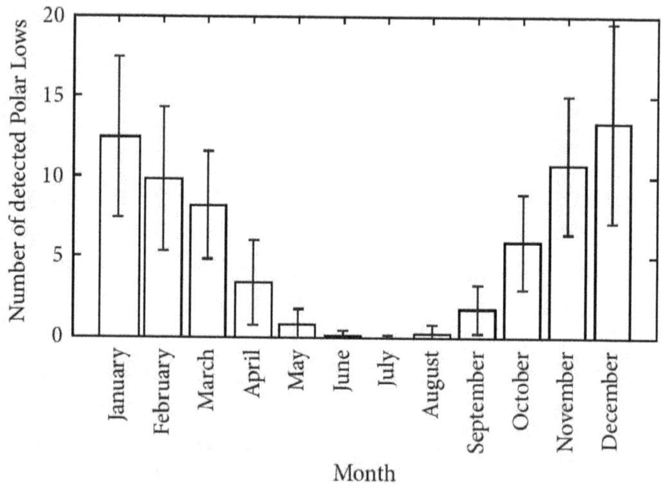

Fig. 2.   Number of detected Polar Lows in an average calendar month for the whole 62 PLSs. Tracking with N-S criterion. The error bar is the standard deviation of Polar Low variance in each month during the 62 PLSs.

the two curves is 0.82, in both cases, with and without N-S criterion. Under the consideration that the N-S criterion is more reasonable from the dynamical view of Polar Low generation, we have chosen the result with N-S criterion for further discussions.

The spatial distribution of Polar Low density is shown in Figure 3(a). The Polar Low density counts the frequency per grid

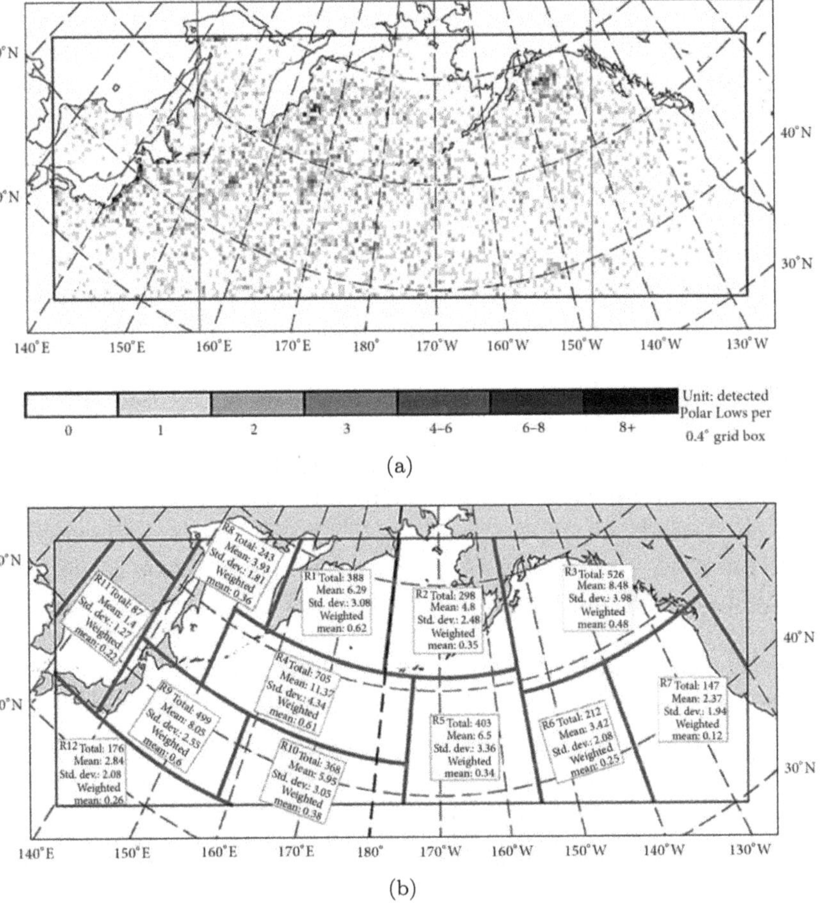

Fig. 3. (a) Density distribution of first appearance of Polar Lows. Unit: detected Polar Lows per 0.4° grid box. (b) Subregions R1–12, for which statistics of Polar Lows were aggregated: the total number, the mean number per Polar Low season (mean), year-to-year standard deviation (std dev), and number of Polar Lows per number of no land grid points in the subregion (weighted mean).

box of a first detection of a Polar Low. Highest densities are found in the region east of Japan in accordance with the results of Yarnal and Henderson (1989). By analyzing defense meteorological satellite program (DMSP) infrared imagery from 7 winter seasons, Yarnal and Henderson concluded that the most active Polar Low cyclogenesis takes place in the western extratropical North Pacific. Our peak area of Polar Low density is just off the east coast of Japan Island, while Yarnal and Henderson's result has its peak a little north, near the island of Hokkaido. In both analyses, there is much less Polar Low activity in the eastern North Pacific than in the western part.

Yarnal and Henderson (1989) pointed out that there are two bands that extend from northern Japan through the Kamchatka Peninsula into the western Bering Sea; another one extends eastward into the open waters of the North Pacific to just east of the international date line. Such two bands are also present in our climatology of the first detected position of the Polar Lows. Additionally, we find also high values for the Gulf of Alaska, where Yarnal and Henderson have detected only "a couple of small, weak pockets of formation."

In order to compare the information of Polar Lows over different regions of North Pacific, we divided the domain into 12 subregions (Figure 3(b)). Some of the subregions are based on the oceanographic feature, such as the Bering Sea (R1 and R2; they are divided by the international date line), the Gulf of Alaska (R3), the Okhotsk Sea (R8), and the Japan Sea (R11). The others are dependent on the climatology cyclogenesis by Yarnal and Henderson (1989). The two bands with the highest frequency of Polar Low occurrence are referred to as R4 and R9. The eastern North Pacific with much less activity is represented by R5, R6, and R7. R10 and R12 are the subregions in the south with seldom distribution. For the different regions in Figure 3(b), small trends and high year-to-year variability for Polar Low frequency were found. Some regions (especially for R1, R3, and R4 with high density) have a higher year-to-year standard deviation than others (R6, R7, R8, R11, and R12) (not shown).

4. Linkage to Large-Scale Pressure Patterns

We begin with investigating how mean sea level pressure (MSLP) fields, averaged across a Polar Low season, are related to the distribution of the numbers of Polar Low occurrences in that season. Here, we emphasize that the analysis is not about short-term synoptic situation or the instantaneous air pressure field directly related with the probability of a Polar Low to form. Instead, we compare two statistics during the same Polar Low season, namely, the geographical distribution of Polar Lows, aggregated to the twelve subregions shown in (Figure 3(b)) and the gridded time-mean MSLP field.

The link between MSLP and the number of Polar Lows is done through a canonical correlation analysis (CCA, von Storch and Zwiers (1999)). CCA is a method for calculating correlation structures between two fields of variables. To reduce noise in each set, we projected the full fields on the first 5 empirical orthogonal functions (EOFs) of the Polar Low time series (representing 77% of the variance) and of the PLS-mean MSLP (87% of the variance). Prior to the CCA, the multiyear mean field is subtracted; that is, the analysis is done with anomalies.

Figure 4 shows the resulting time series and spatial patterns of the two most important linkages between the regional distribution of Polar Lows and the time-mean MSLP field.

The first canonical pattern (CCA1, Figure 4(c)) describes a unipolar pressure distribution. When the CCA coefficient is positive, then, on average, there will be a west-eastward cold air flow across the Bering Sea and south-eastward across the Gulf of Alaska and, consistently, more Polar Lows in that region. These time-mean flows are characteristic for more or less, short-term marine cold air outbreaks in these regions. It is indicated that there is a close relationship between Polar Low formation and the presence of a trough in winter over East Asia and the nearby North Pacific (Businger, 1987). The strong land-sea thermal contrast along the marginal ice zone pulls cold continental polar or Arctic air over the Bering Sea and the Japan Sea. The relatively warm waters in the open ocean lead to the formation of Polar Lows through convective instability and baroclinicity.

The time series of canonical correlation coefficient (Figure 4(a) and (b)) explains the significance of the two fields of variables over the 62 PLSs. Dashed lines represent the pattern of Polar Low occurrence over the 12 subregions; meanwhile, the solid lines represent the MSLP pattern. When examining the time series in Figure 4(a), it is

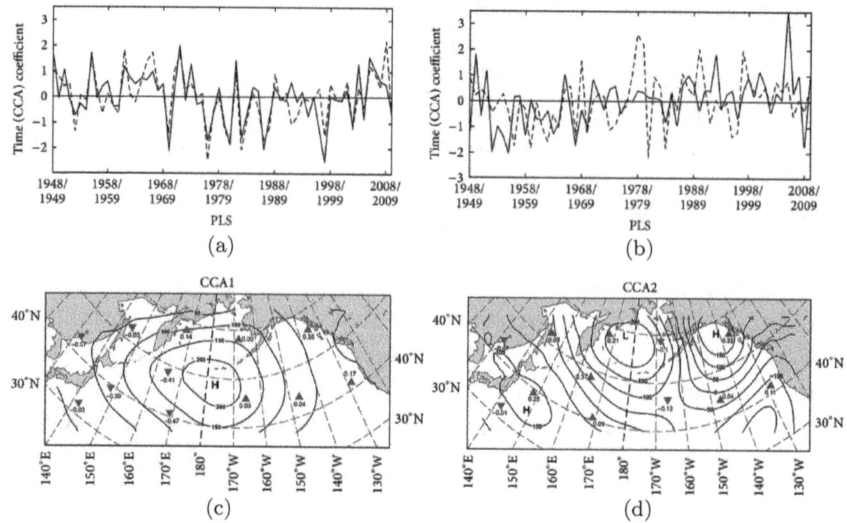

Fig. 4. ((a) and (b)) First two canonical correlation coefficient time series (dashed lines represent the Polar Low occurrence and solid lines represent the MSLP pattern) and ((c) and (d)) corresponding canonical correlation patterns between regional time series of Polar Low occurrences per PLS in subregions R1–R12 ($\triangle$ for positive values, ∇ for negative values) and mean sea level pressure fields in Pa. The first CCA pair ((a) and (c)) shares a correlation coefficient of 0.89. The second CCA pair ((b) and (d)) shares a correlation of 0.72.

evident that CCA1 dominates the Polar Low seasons in 1954/1955, 1971/1972, and 1981/1982.

The second canonical pattern (CCA2, Figure 4(d)) shows a bipolar pressure distribution — there is a negative anomaly (below -2 hPa) on the Bering Sea at the same time as two positive anomalies on the Gulf of Alaska and Japan Island (over 1 hPa). Consequently, there will be a south-eastward time-mean flow starting from Siberia, across the Sea of Okhotsk, and then the southwest of Bering Sea where, in subregion R4, consistently more Polar Lows are detected on average. A time-mean pressure contrast of about 3 hPa between the Bering Sea and the west and east part of North Pacific (the region around Japan Island and the west coast of North American continent) is associated with 0.2 to 0.3 more Polar Lows per PLS in the corresponding regions (R1, R3, R4, and R9).

By examining the time series in Figure 4(b), we found that CCA2 dominates the Polar Low seasons of 1951/1952, 1975/1976,

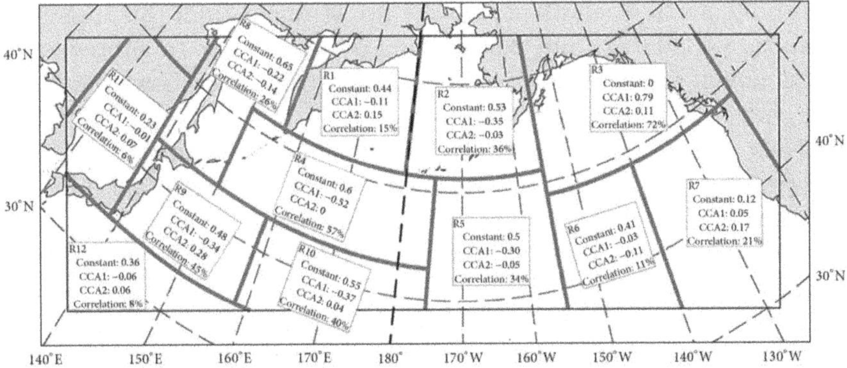

Fig. 5. Result of multiple regression analyses of the first two CCA coefficients on the number of Polar Lows in subregions R1–12: coefficient α_0 (constant), α_1 (connected to CCA1), α_2 (connected to CCA2), and the correlation of number of Polar Lows and trough regressions estimated number of Polar Lows.

and 1998/1999, which means there were more Polar Lows in regions R1, R3, R4, and R9; the negative signs in 1949/1950 and 1978/1979 point to remarkably less Polar Lows in these regions.

To determine the relative importance of the two CCA patterns, we have built for each subregion R1–R12 a regression model, of the form $\mathrm{PL}_i(t) = a_{0,i} + a_{1,i} \times \mathrm{CCA}_1(t) + a_{2,i} \times \mathrm{CCA}_2(t) + \varepsilon_i$. Here, $\mathrm{PL}_i(t)$ is the number of Polar Lows in season t and in subregion i, $a_{0,i}$, $a_{1,i}$, and $a_{2,i}$ are the regression coefficients, which are determined by a least square fit, $CCA_1(t)$ and $CCA_2(t)$ are the CCA coefficients of patterns 1 and 2 in season i, and ε_i is the residual, the unexplained part. In Figure 5, we list the regression coefficients and the correlation between the number of Polar Lows and through these regressions estimated number of Polar Lows for every subregion.

The two series, $CCA_1(t)$ and $CCA_2(t)$, are about equally important, when counting how often $|a_{1,i}| > |a_{2,i}|$. They are particularly successful to describe the variability in all but one of the highest frequency occurrence area which we showed in Figure 3 — for subregions 3, 4, 9, and 10, the correlation is over 40%, while in the far eastern (subregion 6 and 7) and western (11 and 12), the correlations are less than 25%. When comparing with Figure 3(b), we see that the frequency of occurrence of Polar Lows in the subregions 6, 7, 11, and 12 is relatively low: the mean number of Polar Lows in 7, 11, and 12 together is 6.61 Polar Lows per PLS, while the occurrences in each of subregions 3, 4, and 9 separately are stronger than this intensity, namely, 8.48, 11.37, and 8.05 Polar Lows per PLS. In subregion 1, frequency of occurrence is relatively high (6.29 Polar Lows per PLS) but the correlation is low (15%).

Next, we examine the link to time-mean distribution throughout the troposphere. Therefore, we derive associated correlation patterns (von Storch and Zwiers, 1999) to describe the linkage between Polar Low occurrence and geopotential height in different pressure levels (Figure 6).

Associated correlation patterns are designed to describe the relationship between time series of an "index" variable and a physical field. By calculating the correlation coefficient between the time series of CCA coefficient for Polar Low occurrence pattern (dashed lines in Figure 4(a) and (b)) and the time series of geopotential height anomalies for each grid point on each level, we derive typical configurations on every pressure level associated with the two Polar Low patterns shown in Figure 4(c) and (d) (the triangles). The anomalies were formed by subtracting the time-mean fields for the 62 considered Polar Low seasons.

The maps of correlation coefficients present similar patterns to the corresponding MSLP field of the CCA results (Figure 4(c) and (d); isolines). It indicates that the relationship between the atmosphere circulation and Polar Low occurrence is mostly barotropic, even if in the lower stratosphere of 100 hPa the pattern is somewhat shifted. The cold flow which is inducing the Polar Low occurrence is uniform throughout the troposphere, from sea level to the upper troposphere and the lower stratosphere.

On the upper levels at 100, 300, 500, and 700 hPa, the isolines are smoother. On the lower level at 850 and 1000 hPa, the solid lines are more wiggly: the orography and sea surface temperature attain a stronger influence on the generation of more or less Polar Lows.

In order to investigate the link with sea surface temperature (SST) and respective sea ice temperature in case of ice, a pair of associated correlation patterns is presented (Figure 7) to describe the linkage between the CCA-time series of Polar Low occurrence (dashed lines in Figure 4(a) and (b)) and the surface temperature. These temperatures were taken from the same NCEP1 reanalysis, which was used to force the regional model simulation.

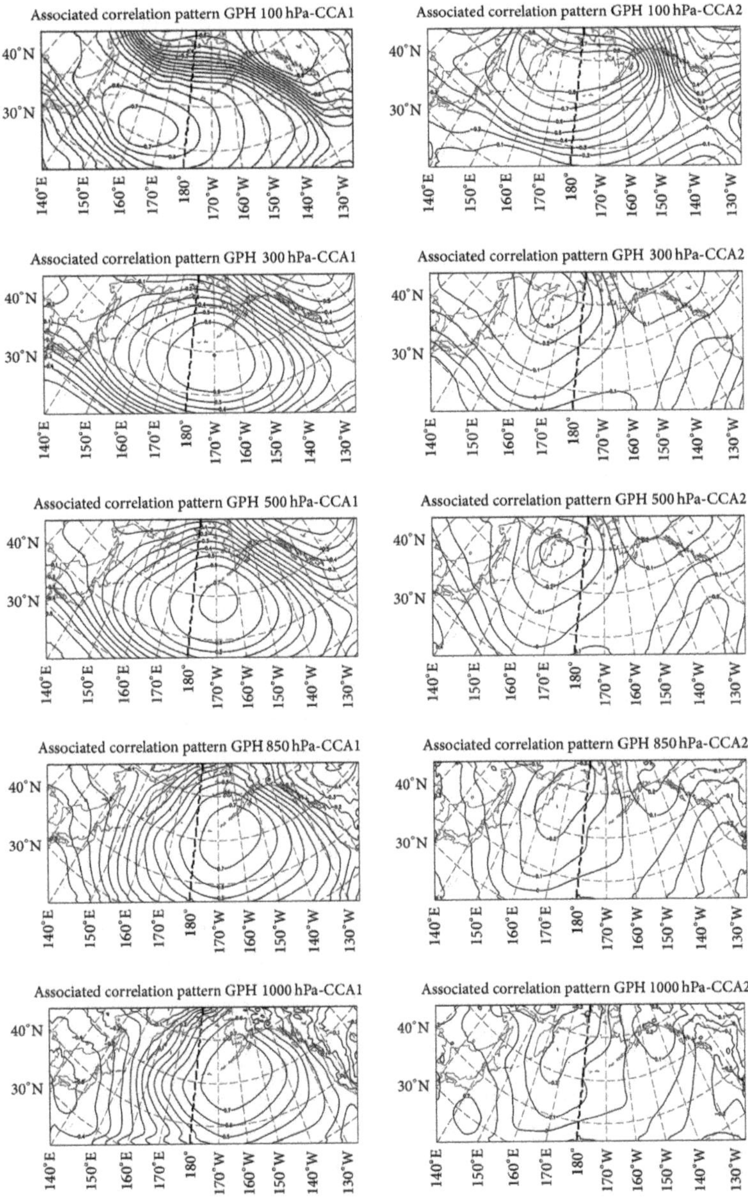

Fig. 6. Associated correlation patterns between time series of geopotential height over 62 PLSs and time coefficient of the Polar Low occurrence from the first two CCA patterns (1st CCA pair, left column, 2nd right column). From top to bottom, 100, 300, 500, 700, 850, and 1000 hPa. All variables are averaged across a Polar Low season, that is, from October to April.

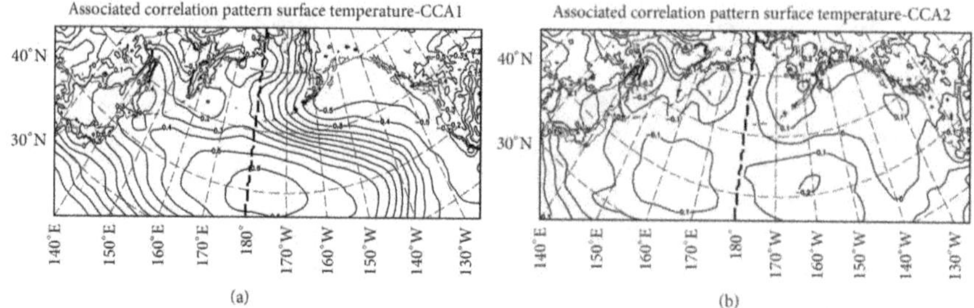

Fig. 7.　Associated correlation patterns between time series of ocean surface temperature anomalies over 62 PLSs and time coefficient of the Polar Low occurrence from the first two CCA patterns (1st CCA pair, (a), 2nd (b)). All variables are averaged across a Polar Low season, that is, from October to April. The surface temperature was taken from the NCEP1 reanalysis data, over the ocean. In case of no ice, it indicates the sea surface temperature (SST); else, it indicates the ice surface temperature.

Both associated correlation patterns of SST are consistent with the flow patterns of the CCA result. Temperatures tend to be lower, where more cold air is advected, and higher than on average, when the flow advects less cold or more warm air. Previous modeling studies have shown that the mechanism behind these patterns is that of an oceanic response to anomalous atmospheric flow (Luksch and von Storch, 1992; Luksch *et al.*, 1990), first suggested by Bjerknes (Bjerknes, 1964) for the Atlantic. We conclude that anomalous mean flow is responsible for both, the formation of anomalous SST as well as the formation of more, or less Polar Lows, in the North Pacific.

In order to investigate the large-scale dynamical environment of changing Polar Low occurrences, we analyze the correlation between the time series of detected Polar Low number in the different subregions and several climate variability indexes, namely, the Pacific decadal oscillation (PDO, Hare and Francis, 1995; Mantua, 2000; Zhang *et al.*, 1997), Pacific-North American tele-connection pattern (PNA, Wallace and Gutzler, 1981), and El Niño/La Niña-Southern oscillation (ENSO multivariate ENSO index; see http://www.esrl.noaa.gov/psd//people/klaus.wolter/MEI/table.html). The time series for the normalized Polar Low numbers PDO, PNA, and ENSO index are shown in Figure 8.

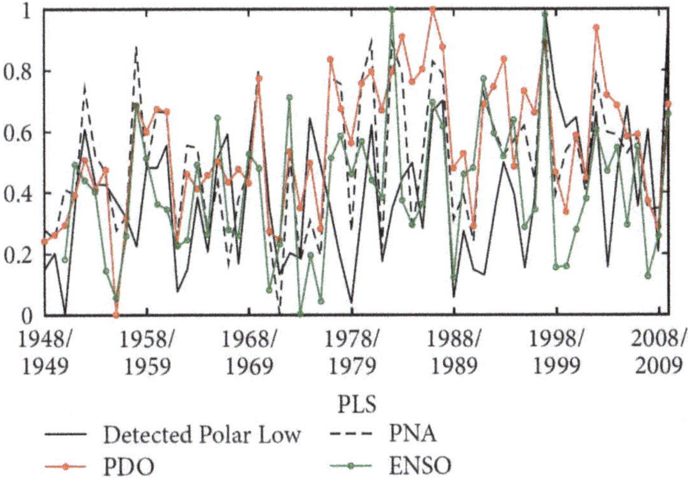

Fig. 8.   Time series of normalized detected Polar Low number (black line), PDO index (red, marked with ○), PNA index (black, dashed), and ENSO index (green, marked with ○). The indices are determined as average across the Polar Low seasons.

The correlation coefficient of detected Polar Low number in simulation area with PDO index is 0.39, with PNA index is 0.45, and with ENSO index is 0.22. It indicates that the Polar Low formation over the North Pacific has the highest relationship with PNA and PDO; there is limited influence by ENSO.

Multiple regression analyses of the number of Polar Lows in the 12 subregions (Figure 3(b)) with the time-mean circulation indices for PDO, PNA, and ENSO reveal which regions are mostly affected by the state of the three circulation systems. The regressions results are listed in Figure 9 for every subregion as well as the correlations between the number of Polar Lows and the through regression estimated number of Polar Lows. Obviously, the indices are not independent, so that there is no strict separation of the effect of the circulation systems. As an example, the result of the regression is shown for two subregions in Figure 10 — one is R1 in the North close to the date line, with a very low correlation of only 15%, and the other is R4 south of R1 in the central part of the North Pacific, with the largest correlation, namely, 55%. Obviously, the circulation indices

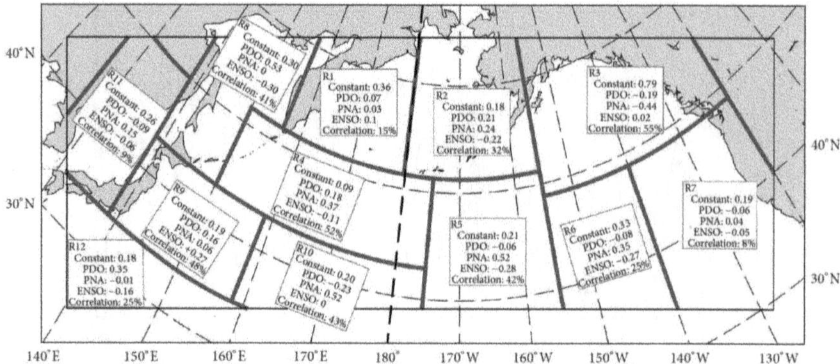

Fig. 9.   Results of multiple regression analyses of number of Polar Low subregions R1–12: The constant parts and the coefficients associated with the PLS time-mean circulation indices PDO, PNA, and ENSO.

cannot be associated with year-to-year variability in R1, while the skill of the indices in describing this variability is remarkable in R4.

We find only two regions, R3 and R4, where the total influence of the indices amounts to a correlation of more than 50%. R3 is in the Northeast, south of Alaska, while R4 is at middle latitudes west of the date line; in both cases, PNA is associated with the largest coefficient, which is consistent with the pattern of PNA (not shown). A weak link, as expressed by a correlation of 30% or less, is found for R1, R6, R7, R11, and R12, all subregions at the boundaries of the model domain.

To characterize the variability of the CCA patterns and the large-scale state indices of PDO, PNA, and ENSO, we have also established regression models. They relate the time series of the MSLP pattern with the three indices. The success and relative importance of the three indices are summarized in Table 1. Additionally, the correlations between the time series of MSLP pattern and the one estimated through the regression are listed.

The first CCA pattern is strongly linked to, primarily, PNA and also PDO (which are of course not independent) but hardly to ENSO; this is different for the second pattern, which is described as being negatively linked to ENSO, equally strongly linked to PNA but

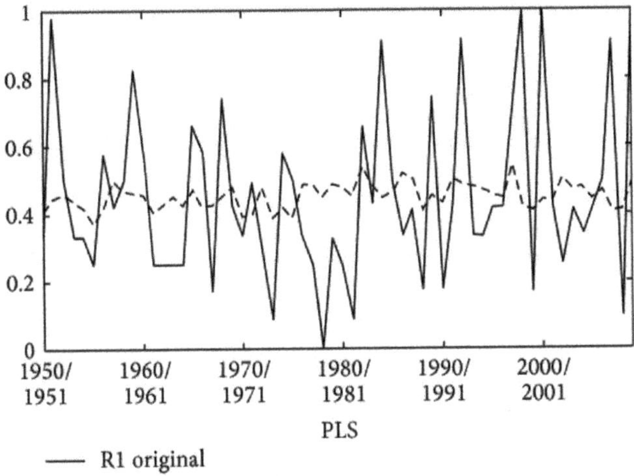

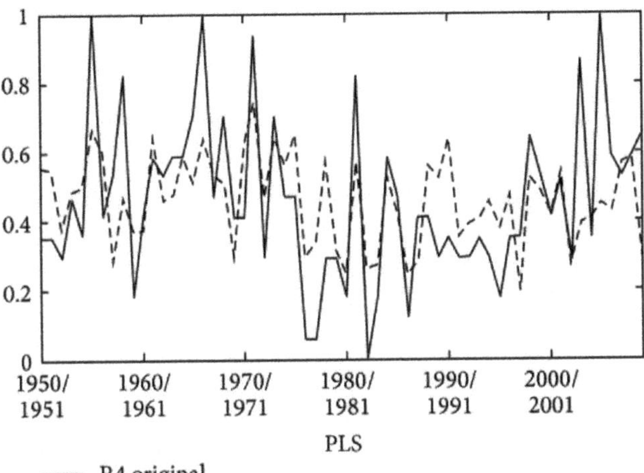

Fig. 10. Results of multiple regression analyses of number of Polar Low sub-regions R1–12: the constant parts and the coefficients associated with the PLS time-mean circulation indices PDO, PNA, and ENSO.

hardly to PDO. The overall link is much stronger for the first pattern (as indicated by a correlation of 71%), while the link for the 2nd pattern is weak (correlation of 26%).

Table 1.   Result of multiple regression analysis of the first
two time series of MSLP patterns of CCA between MSLP
and Polar Low occurrence and of PLS time mean circulation
indices PDO, PNA, and ENSO.

	Constant	PDO	PNA	ENSO	Correlation
CCA1	0.07	0.19	0.54	−0.01	71%
CCA2	0.48	−0.03	0.36	−0.37	26%

5.   Summary and Outlook

For the first time, a multidecadal climatology, including trends, of
Polar Low formation in the North Pacific has been constructed.
Because of an insufficient database of local observations and homogeneous high-resolution analyses, a dynamical downscaling strategy
has been employed, following the concept developed by Zahn and
von Storch (2008).

The main result is the presence of large interannual variability
in 1948–2010, but little decadal variability of Polar Low occurrences
and positive long-term changes in the North Pacific region. No obvious change was detected in recent decades, which indicates that the
large natural variability in the region is still dominating over possible
effects of global warming.

The tendency of forming more, or less, Polar Lows has been
related to the time-mean circulation in the North Pacific, in terms
of mean sea level pressure and geopotential height throughout the
troposphere. Anomalous flows from cold surfaces were found to support the large-scale synoptic environment of Polar Low formation.
Two major patterns are detected, which are correlated with the variations described by the indices of PDO, PNA, and, to a lesser extent,
ENSO.

Funding

The work was done with the support of the Chinese Scholarship
Council CSC, and it is a contribution to the Helmholtz Climate Initiative REKLIM (Regional Climate Change), a joint research project
of the Helmholtz Association of German Research Centers.

Acknowledgments

The authors thank Beate Gardeike for preparing most of the diagrams.The technical and scientific support, the various comments and suggestions by Dr. Beate Geyer and Dr. Matthias Zahn have greatly improved this chapter.

References

Bjerknes, J. (1964). Atlantic air-sea interaction, in *Advances in Geophysics*, Vol. 10, pp. 1–82, Elsevier, ISBN 0065-2687.

Blier, W. (1996). A numerical modeling investigation of a case of polar airstream cyclogenesis over the gulf of Alaska, *Monthly Weather Review* **124**, 12, pp. 2703–2725, doi:10.1175/1520-0493(1996)124 <2703:Anmioa>2.0.Co;2.

Bresch, J. F., Reed, R. J. and Albright, M. D. (1997). A polar-low development over the Bering Sea: Analysis, numerical simulation, and sensitivity experiments, *Monthly Weather Review* **125**, 12, pp. 3109–3130, doi:10.1175/1520-0493(1997)125<3109:Apldot>2.0.Co;2.

Businger, S. (1985). The synoptic climatology of polar low outbreaks, *Tellus Series — Dynamic Meteorology and Oceanography* **37**, 5, pp. 419–432.

Businger, S. (1987). The synoptic climatology of polar-low outbreaks over the gulf of Alaska and the Bering Sea, *Tellus A: Dynamic Meteorology Oceanography* **39**, 4, pp. 307–325.

Cavicchia, L. and von Storch, H. (2012). The simulation of medicanes in a high-resolution regional climate model, *Climate Dynamics* **39**, 9, pp. 2273–2290, doi:10.1007/s00382-011-1220-0.

Chen, F., Geyer, B., Zahn, M. and von Storch, H. (2012). Toward a multi-decadal climatology of North Pacific Polar Lows employing dynamical downscaling, *Terrestrial Atmospheric and Oceanic Sciences* **23**, 3, pp. 291–301, doi:10.3319/tao.2011.11.02.01(a).

Condron, A. and Renfrew, I. A. (2013). The impact of polar mesoscale storms on Northeast Atlantic Ocean circulation, *Nature Geoscience* **6**, 1, pp. 34–37, doi:10.1038/ngeo1661.

Condron, A., Bigg, G. R. and Renfrew, I. A. (2006). Polar mesoscale cyclones in the Northeast Atlantic: Comparing climatologies from ERA-40 and satellite imagery, *Monthly Weather Review* **134**, 5, pp. 1518–1533, doi:10.1175/mwr3136.1.

Denis, B., Côté, J. and Laprise, R. (2002). Spectral decomposition of two-dimensional atmospheric fields on limited-area domains using the discrete cosine transform (dct), *Monthly Weather Review* **130**, 7, pp. 1812–1829, doi:10.1175/1520-0493(2002)130<1812:SDOTDA>2.0.CO;2.

Fu, G., Niino, H., Kimura, R. and Kato, T. (2004). A polar low over the Japan sea on 21 January 1997. Part i: Observational analysis, *Monthly Weather Review* **132**, 7, pp. 1537–1551, doi:10.1175/1520-0493(2004)132<1537:Aplotj>2.0.Co;2.

Gray, S. L. and Craig, G. C. (1998). A simple theoretical model for the intensification of tropical cyclones and polar lows, *Quarterly Journal of the Royal Meteorological Society* **124**, 547, pp. 919–947, doi:10.1002/qj.49712454713.

Hare, S. R. and Francis, R. C. (1995). Climate change and salmon production in the northeast pacific ocean, *Climate Change Northern Fish Populations* **121**, p. 357.

Heinemann, G. and Claud, C. (1997). Report of a workshop on "theoretical and observational studies of polar lows" of the European Geophysical Society polar lows working group, *Bulletin of the American Meteorological Society* **78**, 11, pp. 2643–2658, doi:10.1175/1520-0477-78.11.2643.

Kalnay, E., Kanamitsu, M., Kistler, R., *et al.* (1996). The NCEP/NCAR 40-year reanalysis project, *Bulletin of the American Meteorological Society* **77**, 3, pp. 437–472.

Kolstad, E. W. (2006). A new climatology of favourable conditions for reverse-shear polar lows, *Tellus Series a-Dynamic Meteorology and Oceanography* **58**, 3, pp. 344–354, doi:10.1111/j.1600-0870.2006.00171.x.

Kulkarni, A. and von Storch, H. (1995). Monte Carlo experiments on the effect of serial correlation on the Mann-Kendall test of trend, *Meteorologische Zeitschrift* **4**, 2, pp. 82–85.

Luksch, U. and von Storch, H. (1992). Modeling the low-frequency sea surface temperature variability in the North Pacific, *Journal of Climate* **5**, 9, pp. 893–906, doi:10.1175/1520-0442(1992)005<0893:Mtlfss>2.0.Co;2.

Luksch, U., von Storch, H. and Maier-Reimer, E. (1990). Modeling North Pacific SST anomalies as a response to anomalous atmospheric forcing, *Journal of Marine Systems* **1**, 1, pp. 155–168, doi:10.1016/0924-7963(90)90204-n.

Mailhot, J., Hanley, D., Bilodeau, B. and Hertzman, O. (1996). A numerical case study of a polar low in the Labrador Sea, *Tellus Series a-Dynamic Meteorology and Oceanography* **48**, 3, pp. 383–402, doi:10.1034/j.1600-0870.1996.t01-2-00003.x.

Mantua, N. (2000). Comparison of typical warm PDO/El Niño SST, SLP, and wind anomalies. Available online: http://research.jisao. washington.edu/pdo/graphics.html.

Nagata, M. (1993). Meso-β-scale Vortices Developing along the Japan-Sea Polar-Airmass Convergence Zone (JPCZ)Cloud Band : Numerical Simulation, *Journal of the Meteorological Society of Japan* **71**, 1, pp. 43–57, doi:10.2151/jmsj1965.71.1_43.

Ninomiya, K. (1989). Polar Comma-cloud lows over the Japan Sea and the Northwestern Pacific in Winter, *Journal of the Meteorological Society of Japan* **67**, 1, pp. 83–97, doi:10.2151/jmsj1965.67.1_83.

Nordeng, T. E. (1990). A model-based diagnostic study of the development and maintenance mechanism of two polar lows, *Tellus A* **42**, 1, pp. 92–108.

Reed, R. J. (1979). Cyclogenesis in polar air streams, *Monthly Weather Review* **107**, 1, pp. 38–52, doi:10.1175/1520-0493(1979)107<0038: Cipas>2.0.Co;2.

Rockel, B., Will, A. and Hense, A. (2008). The regional climate model COSMO-CLM(CCLM), *Meteorologische Zeitschrift* **17**, 4, pp. 347–348, doi:10.1127/0941-2948/2008/0309.

Steppeler, J., Doms, G., Schattler, U., *et al.* (2003). Meso-gamma scale forecasts using the nonhydrostatic model LM, *Meteorology and Atmospheric Physics* **82**, 1, pp. 75–96, doi:10.1007/s00703-001-0592-9.

von Storch, H. and Zwiers, F. W. (1999). *Statistical Analysis in Climate Research*, Cambridge UK: Cambridge University Press.

von Storch, H., Langenberg, H. and Feser, F. (2000). A spectral nudging technique for dynamical downscaling purposes, *Monthly Weather Review* **128**, 10, pp. 3664–3673, doi:10.1175/1520-0493(2000)128 <3664:ASNTFD>2.0.CO;2.

Wallace, J. M. and Gutzler, D. S. (1981). Teleconnections in the geopotential height field during the northern hemisphere winter, *Monthly Weather Review* **109**, 4, pp. 784–812.

Xia, X., Zahn, M., Hodges, K., Feser, F. and von Storch, H. (2012). A comparison of two identification and tracking methods for polar lows, *Tellus A: Dynamic Meteorology Oceanography* **64**, 1, p. 17196.

Xia, L., von Storch, H. and Feser, F. (2013). Quasi-stationarity of centennial Northern Hemisphere midlatitude winter storm tracks. *Climate Dynamics* **41**, pp. 901–916.

Xia, L., von Storch, H., and Feser, F. *et al.* (2016). A study of quasi-millennial extratropical winter cyclone activity over the Southern Hemisphere. *Climate Dynamics* **47**, pp. 2121–2138.

Yanase, W., Fu, G., Niino, H. and Kato, T. (2004). A polar low over the Japan Sea on 21 January 1997. Part II: A numerical study,

Monthly Weather Review **132**, 7, pp. 1552–1574, doi:10.1175/1520-0493(2004)132<1552:Aplotj>2.0.Co;2.

Yanase, W. and Niino, H. (2007). Dependence of polar low development on baroclinicity and physical processes: An idealized high-resolution numerical experiment, *Journal of the Atmospheric Sciences* **64**, 9, pp. 3044–3067, doi:10.1175/jas4001.1.

Yarnal, B. and Henderson, K. G. (1989). A climatology of polar low cyclogenetic regions over the North Pacific Ocean, *Journal of Climate* **2**, 12, pp. 1476–1491, doi:10.1175/1520-0442(1989)002 <1476:Acoplc>2.0.Co;2.

Zahn, M. and von Storch, H. (2008). Tracking polar lows in CLM. *Meteorologische Zeitschrift* **17**, 4, p. 445.

Zahn, M. and von Storch, H. (2008a). A long-term climatology of North Atlantic polar lows, *Geophysical Research Letters* **35**, 22, doi:10.1029/ 2008GL035769.

Zahn, M. and von Storch, H. (2008b). Tracking polar lows in CLM, *Meteorologische Zeitschrift* **17**, 4, pp. 445–453, doi:10.1127/0941-2948/ 2008/0317.

Zhang, Y., Wallace, J. M. and Battisti, D. S. (1997). ENSO-like interdecadal variability: 1900–93, *Journal of Climate* **10**, 5, pp. 1004–1020.

Chapter 8

Polar Low Genesis Over the North Pacific Under Different Global Warming Scenarios[*]

F. Chen[†,‡,§]**, H. von Storch**[‡]**, L. Zeng**[§]**, and Y. Du**[§]

[†]*School of Civil and Environment Engineering,*
Harbin Institute of Technology Shenzhen Graduate School,
Shenzhen, China
[‡]*Helmholtz-Zentrum Geesthacht*
Geesthacht, Germany
[§]*South China Sea Institute of Oceanology,*
Chinese Academy of Science, Guangzhou, China

Following an earlier climatological study of North Pacific Polar Lows by employing dynamical downscaling of NCEP1 reanalysis in the regional climate model COSMO-CLM, the characteristics of Polar Low genesis over the North Pacific under different global warming scenarios are investigated. Simulations based on three scenarios from the Special Report on Emissions Scenarios were conducted using a global climate model (ECHAM5) and used to examine systematic changes in the occurrence of Polar Lows over the twenty first century. The results show that with

---

[*]This chapter was originally published in *Climate Dynamics*, 43, 3449–3456, doi:10.1007/s00382-014-2117-5. Copyright (2014), with permission from Springer Nature.

more greenhouse gas emissions, global air temperature would rise, and the frequency of Polar Lows would decrease. With sea ice melting, the distribution of Polar Low genesis shows a northward shift. In the scenarios with stronger warming there is a larger reduction in the number of Polar Lows.

1.   Introduction

The North Pacific Ocean is an area where sub-synoptic Polar Lows form frequently in the cold season (Rasmussen, 2003) in association with northerly cold air outbreaks. Due to strong wind speed and heavy rainfall, Polar Lows are a significant threat to maritime human activities.

Reanalysis datasets, such as those provided by the National Centers for Atmospheric Prediction (NCEP) or the European Centre for Medium-Range Weather Forecasts (ECMWF), as well as model outputs for climate change scenarios, as those in the CMIP ensembles, do not allow determination of changing statistics of Polar Lows, due to inadequate horizontal resolutions in such data sets. Therefore, dynamical downscaling was carried out, which employs regional climate models to determine multi-decadal climatology of Polar Lows (Chen and von Storch, 2013). To do so, a regional climate model was conditioned by large-scale information from the NCEP reanalysis. It turned out that the formation and life cycle of Polar Lows during the past decades could be reconstructed this way homogeneously without exploiting sub-synoptic information in the initial fields (Chen *et al.*, 2012). This same approach is used here to determine how the frequency of Polar Lows occurrence may change in response to three SRES emission scenarios developed for the third and fourth assessment reports of the Intergovernmental Panel on Climate Change (IPCC).

In Section 2, we briefly describe the data and methodology of the simulation and tracking algorithm. Section 3 is divided into three parts in order to determine (decreasing) trends in Polar Low genesis, variations of the trend associated with different greenhouse gas emission scenarios and the northward shift of Polar Low genesis. The chapter concludes with a discussion and summary in Section 4.

2. Methodology and Data

2.1. *Downscaling and Tracking*

The basic concept is to downscale simulations from a global model to the region of the North Pacific. The Consortium for Small scale Modelling (COSMO) model in Climate Mode (COSMO-CLM; Steppeler *et al.*, 2003) was used as the regional climate model. The COSMO-CLM (CCLM) is a version of the operational weather prediction model of the Deutscher Wetterdienst and the COSMO, adapted for climate simulation purposes by the CLM-Community (http://www.clm-community.eu).

The states in the global climate change simulations are fed into the regional model at the lateral boundaries and lower boundary. Similar to a previous climatological study (Chen and von Storch, 2013), which downscaled from the NCEP1 reanalysis, spectral nudging is employed.

The CCLM model is capable of generating Polar Lows without providing seeds in the initial states (Chen *et al.*, 2012). These Polar Lows are then detected and tracked using a modified algorithm developed by Zahn and von Storch (2008). For detailed information of the model, the downscaling technology and the tracking algorithm, the readers are referred to Chen *et al.* (2012) and Chen and von Storch (2013).

2.2. *Scenario Information*

The scenarios we applied here were designed for the IPCC in 2000 (Nakicenovic *et al.*, 2000) and are known as the Special Report on Emissions Scenarios (SRESs). The SRESs consider a wide range of factors, such as population, industrialization, urbanization, and social and economic developments to estimate greenhouse gas emissions. These emission scenarios are fed into global climate models for studying the effect of increasing greenhouses gas concentrations on future state of the climate system, such as investigating the changes of storm and cyclone activities under different SRESs (Pinto *et al.*, 2007) and under increased precipitation intensity (Meehl *et al.*, 2005).

Various characteristics, which reflected the response of the climate system to different scenarios, are also investigated, such as the air temperature, precipitation, sea level rise, sea ice melting, ocean circulation change, extreme weather phenomenon, and aerosols and carbon dioxide cycle variability (Boer *et al.*, 2000; Cox *et al.*, 2000; Friedlingstein *et al.*, 2001; Stocker and Schmittner, 1997; Yonetani and Gordon, 2001).

2.3.　*Forcing Data: ECHAM5*

The global climate model used is the coupled GCM ECHAM5/MPI-OM1 (Roeckner *et al.*, 2003), which is the European Centre Hamburg Version 5/Max-Planck-Institute — Ocean Model Version 1. A recent study of multi-model weighting showed that ECHAM5/MPI-OM is one of the best models to reproduce historical observations of 21st century temperature (Raisanen *et al.*, 2010). Detailed information on the model is available in Roeckner *et al.* (2003), Marsland *et al.* (2003) and Jungclaus *et al.* (2005). The different scenario runs were initiated in the year 1948 and run until 2000 with observed forcing, and then the scenario runs were extended until 2100. Three simulations of A1B, which were initialized in different years of the pre-industrial control run, labeled A1B_1, A1B_2 and A1B_3, were run using ECHAM5/MPI-OM and then downscaled by CCLM. Similarly, one simulation forced by B1 and A2 emission scenarios was processed. The output of the global run was acquired from the World Climate Research Programme's (WCRP's) Coupled Model Intercomparison Project phase 3 (CMIP3) multi-model dataset (Meehl *et al.*, 2007).

3.　Simulation and Tracking Results

Polar Lows occur from mid autumn to mid spring; therefore "Polar Low season" (PLS) is defined as from October through April. The PLS is addressed by the first and second years; for example, the PLS 2001/2002 begins in October 2001 and ends in April 2002. For each PLS, we present the number of detected Polar Lows in the North Pacific.

3.1. *Analysis of A1B Scenarios*

To show the effect of increasing greenhouse gas concentrations on the frequency of North Pacific Polar Lows formation, we take A1B as an example. Figure 1(a) presents the time series of detected Polar Lows per PLS and the corresponding trend over the next

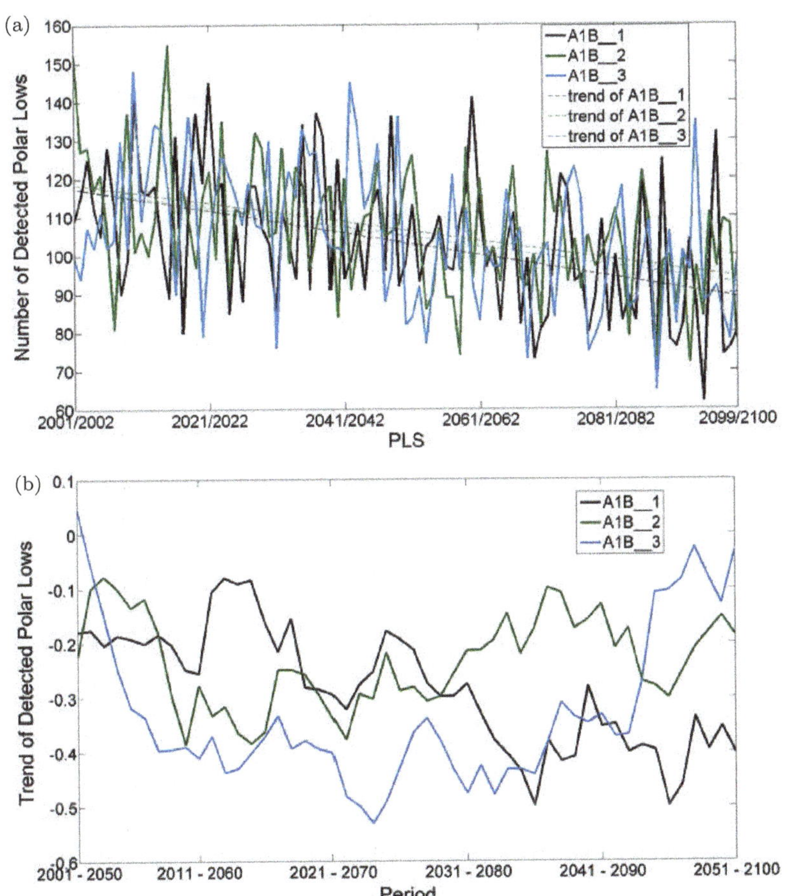

Fig. 1. (a) Number of detected Polar Lows per PLS simulation, and three realizations of A1B scenario during 2001/2002–2009/2100. The straight lines represent the trends of the three realizations. (b) Running 50-year trends of the three combined series of detected Polar Lows per PLS under A1B; the trends are labeled with the first year of the moving 50-year window © Springer-Verlag Berlin Heidelberg 2014.

Table 1.   Result of Mann-Kendall trend test for the time series of detected Polar Low number under different scenarios.

	Trend	SE	t-stat	Lower 95%	Upper 95%
A1B_1	−0.29	18.19	−4.68	−0.40	−0.21
A1B_2	−0.24	16.12	−4.12	−0.32	−0.14
A1B_3	−0.25	17.49	−4.04	−0.35	−0.16
A2_1	−0.49	21.64	−6.94	−0.57	−0.38
B1_1	−0.20	18.45	−3.40	−0.31	−0.11

century under the three realizations of A1B. They all show negative trends (decreasing number of Polar Lows), and there is no obvious difference among the three realizations. Specifically, the trend for A1B_1 is −0.29 cases/PLS; that for A1B_2 is −0.24 cases/PLS and that for A1B_3 is −0.25 cases/PLS, with corresponding standard deviations of 18.19, 16.12 and 17.49, respectively. According to the pre-whitened Mann–Kendall trend test, all three scenarios show significant decrease trends under 5% risk (Table 1).

Figure 1(b) shows 50-year running trends of the three 2000–2099 series under A1B scenario. The reduction of the number of cases varies somewhat in time, but is consistent in sign. When examining 30- and 40-year trends (not shown), a similar situation emerges, even though the variability of the series becomes much larger.

As Zahn and von Storch (2010) pointed out in their study of the Polar Low frequency over the North Atlantic in the twenty first century, both the SST of the Atlantic and the tropospheric air temperature exhibit increases. However, they found that tropospheric air temperature increases more quickly than SST, resulting in a reduction in the air-sea temperature difference and increased vertical atmospheric stability. The increased stability is less favorable for Polar Low genesis. To demonstrate that this mechanism is also operating in the North Pacific, we calculate the seasonal temperature difference between the 500 hPa and the sea surface for A1B_1. Only grid points with open water, i.e., no sea ice, are considered. For each PLS, we count the number of grid points where the mean air-sea temperature difference is equal or larger than 39 K. When this condition is met, we refer it as "critical temperature difference".

Since the number of grid points with sea ice decreases, we show in Figure 2 such grid points with a critical vertical temperature

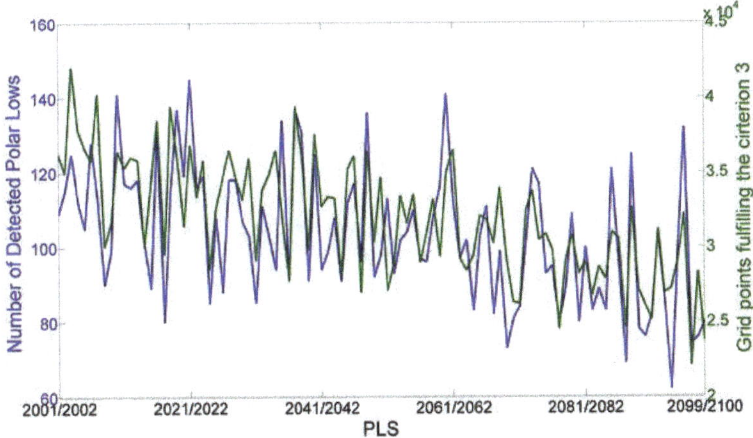

Fig. 2. Development of PLS-number of Polar Lows, derived from detecting and tracking (in blue) and PLS-number of open-sea grid points, with a vertical temperature gradient beyond the critical 39 K (green) © Springer-Verlag Berlin Heidelberg 2014.

difference compared to the overall number of eligible (ice free sea) grid points. For A1B_1, this time series shows decreases of 29% to the end of twenty first century compared to the beginning (the corresponding trend is -93.53 grid points per PLS); for A1B_2 and A1B_3, they are 24% with -71.31 grid points/PLS and 22% with -70.14 grid points/PLS, respectively. Thus, favorable conditions for Polar Low genesis decrease by nearly one third since the beginning of this century for all three realizations of A1B. The comparison with the directly detected and tracked Polar Lows, also shown in Figure 2, confirms that the proxy "vertical temperature difference" describes adequately the decreasing trend in Polar Low activity for the twenty first century. The correlation between the two curves, after removing their trends, amounts to 74.4%. For A1B_2, it is 72.8%, and for A1B_3, is 79.1% (not shown).

3.2. *Other Greenhouse Gas Emission Scenarios*

As shown in Figure 1(a), there is no obvious difference between different realizations of a single story line. Therefore, and for saving computational cost and time, we only take the first realization from the A2 and B1 scenarios (A2_1 and B1_1 respectively). In the following discussion, we refer only to A1B, A2 and B1.

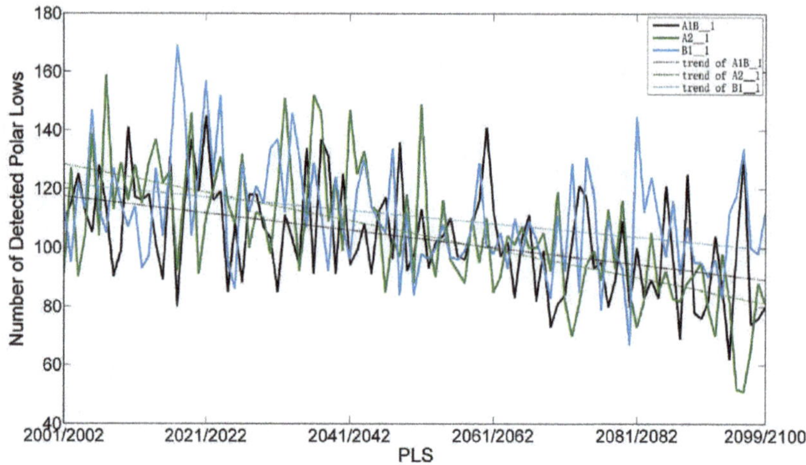

Fig. 3. Number of detected Polar Lows and corresponding trends from 2001/2002 to 2099/2100 under A1B, A2 and B1 © Springer-Verlag Berlin Heidelberg 2014.

Figure 3 presents different tracking results for the three emission scenarios. In all, the number of Polar Lows decreases: the trend of A1B is −0.29 case/PLS; for the scenario of A2, which has a much higher greenhouse gas emission, the trend is −0.49 cases/PLS. For B1_1, the trend is much smaller, with only −0.20 case/PLS. The significance of the trends is shown in Table 1. We find that with a higher greenhouse gas emission, the frequency of Polar Low occurrence has a stronger decreasing trend. This is also supported by a general decrease of grid points where the mean vertical temperature is beyond the critical 39 K. We have inferred there would be a larger decreasing trend of Polar Low occurrence frequency over the North Pacific in the next century under a global warming situation with a higher greenhouse gas emission. The higher the air temperature rises, the lower the air–sea temperature difference becomes. Consistently, the frequency of favorable conditions for Polar Low genesis declines (not shown).

3.3. *Northward Shift of Polar Low Genesis Associated with Ice Edge Melting*

For the next step, we are interested in the spatial distribution of Polar Low genesis and its variability under future global warming scenarios.

Here, we take A1B as an example. We divide the North Pacific into 12 sub-regions to compare the Polar Low information in different areas. The same sub-regions were used in Chen and von Storch (2013) to describe the spatial statistics of the formation of Polar Lows at present. The density of Polar Low genesis, the trends in number and in the change of vertical stability, as well as the correlation between the two measures of Polar Low genesis in the 12 sub-regions in the next century, are presented in Figure 4.

The region of highest Polar Low density is located to the east of Japan Island (R9), and the smallest values in R12 are in the far southwest, and in R6 and R7 in the southeast — consistent with the climatology of the last six decades. The largest trends in the 12 sub-regions are all negative, both in terms of detected and tracked Polar Lows, and in terms of the proxy "spatial percentage of critical vertical temperature difference" (in R2 the trend is small and positive). The two measures are all positively correlated (after taking out the trends), with a maximum of 60% in R4 west of the dateline and a minimum of 8% in R12 in the far southwest of the study area.

The strongest negative trend in Polar Lows genesis per open-sea grid point is found in R9 east of Japan, with -6.9×10^{-2} cases per grid point per PLS. In R4, the trend is -4.8×10^{-2} cases per grid point per PLS. For the sub-regions R1, R5, R6, and R11, the trends in detected and tracked Polar lows are about 2×10^{-2} cases per open-sea grid point per PLS; in the other sub-regions, the trends are smaller.

The correlated measure of percentage of open-sea points with a vertical difference of 39 K and beyond shows a similar pattern, with all negative trends (labeled T-VS in Figure 4) but one, albeit very small positive trend in R2. In general, the trends in the Northwest, namely, in R1, R2 and R8, exhibit small trends, while the trends in the other sub-regions are solidly negative, with largest values in R9 and R10.

The correlation of the 12 trends, derived from the tracking data and from the change in vertical temperature difference, amounts to 43%.

Another similar measure is the trend in grid points with a critical temperature difference of 39 K and more over the North Pacific in the twenty first century (Figure 5). (Before, we examined the percentage of points with a critical difference in the seasonal mean; here,

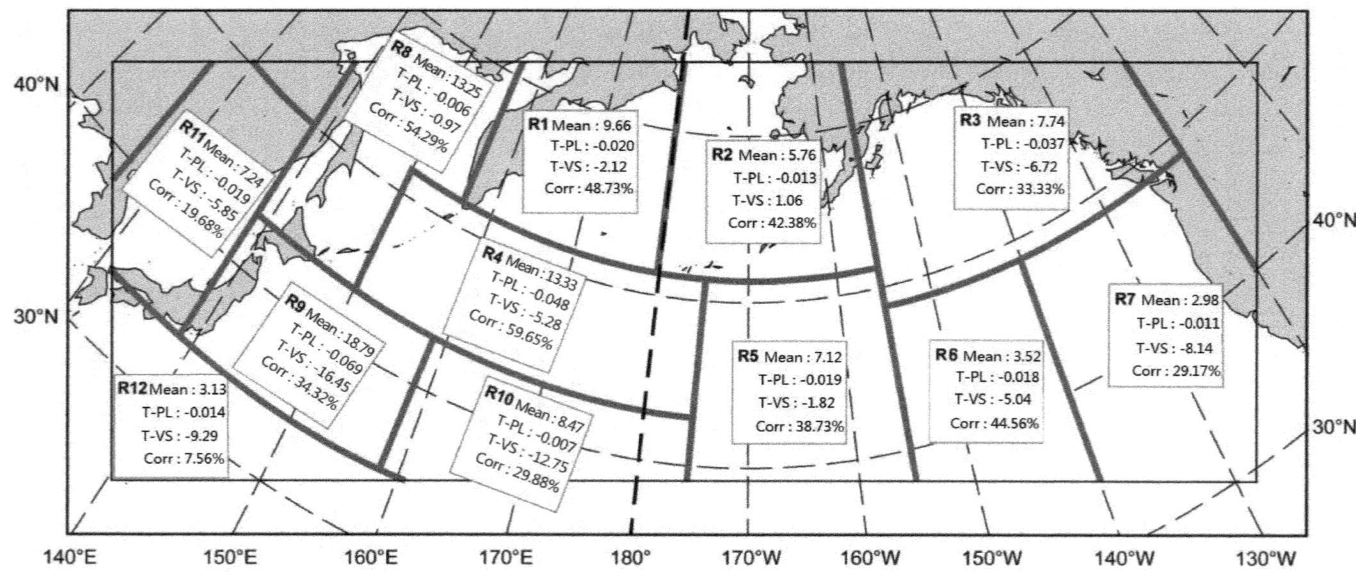

Fig. 4. Sub-regions, for which the numbers (Mean) and trends in detected (T-PL), or estimated (T-VS) Polar Lows formation were aggregated (R1–R12). Also given are the correlations (corr) between the detected and estimated Polar occurrence (after subtracting trends). The estimated number is the percentage of open-sea grid points in each sub-region, where the critical temperature difference between the sea surface and 500 hPa of 39 K or higher is met © Springer-Verlag Berlin Heidelberg 2014.

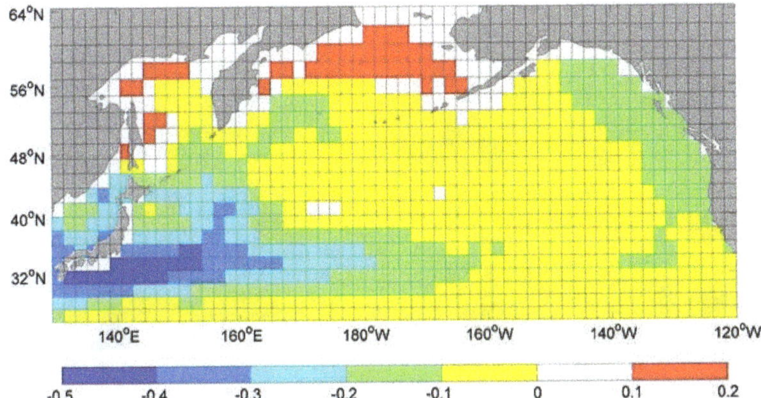

Fig. 5. Trends of the frequency of number of times (of 4-times daily values) within a PLS from 2001/2002 to 2099/2100, when the vertical temperature difference (SST — 500 hPa temperature) is 39 K or higher. Red color indicates more frequent unstable conditions, while the blue color denotes more frequent stable conditions. Units: grid points/PLS © Springer-Verlag Berlin Heidelberg 2014.

we count the number of times from four times daily data for every PLS and then determine the change.) The frequency of relatively more stable conditions, when the 39 K difference is not met, increases almost everywhere in the North Pacific Ocean, with a maximum value of −0.5 grid points/PLS east of Japan. Consistently, the largest difference in the emergence of Polar Lows is in the Northwest Pacific, and the weakest are indicated by red color in Figure 5. The tendency towards more stable condition in most of the region reflects faster warming of the troposphere, and the slower warming at the sea surface. It echoes the study of Kolstad and Bracegirdle (2008) for the possibility of cold-air outbreak change in the future. With a small decrease in cold-air outbreak over the Pacific in the twenty first century, the generation of weather systems associated with air-sea temperature difference, such as Polar Lows, would be weaken. The red areas point to the retreat of sea ice in the course of the overall warming (Figure 6).

It has to be pointed out that despite a small positive trend in this favorable condition in the northern Bering Sea and part of the Okhotsk Sea (Figures 4 and 5), there is still a negative trend in Polar Low genesis in the corresponding boxes. This may be so, because of averaging across many points, of which some show favorable

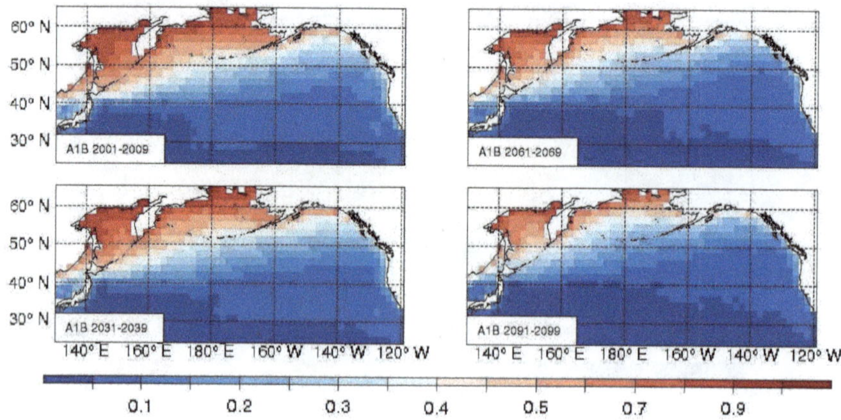

Fig. 6.   Mean proportion of ice covered time over the North Pacific Ocean during the PLS for different 10-year mean of 2001–2009, 2031–2039, 2061–2069, and 2091–2099 calculated from the ECHAM5/MPI-OM1 of A1B_1 © Springer-Verlag Berlin Heidelberg 2014.

conditions, others unfavorable. Also specifics of the land-sea distribution may play a role.

Figure 6 demonstrates this gradual decrease of sea ice in the northern part of the region. During the first decade of 2001–2009, sea ice was simulated to be present in most part of the Okhotsk Sea and the Bering Sea during 70% of PLS. In the Gulf of Alaska, nearly 50% of the time the sea is frozen along the north coast. But for the last 70 PLSs during 2031–2099, the proportion of sea ice during winter gradually decreases. In the Okhotsk Sea, the Bering Sea and in the Gulf of Alaska, the time of frozen condition is reduced by 55% at the end of twenty first century in this scenario.

As the sea ice edge retreats, the cold air-flow from the north meets the relatively warm SST earlier on its southward movement. Thus, in these regions, such as R1 and R8, the reduction of numbers of Polar Lows is the smallest. At the same time, the cold continental air has to travel a longer distance across the warmer SST before reaching the more southern regions, so the tendency for forming Polar Lows there, in R4 and R9, is further reduced.

To compare the strength of the trends between the sub-regions, we calculate area-weighted mean of their trends by dividing the overall trends by the number of sea grid points for each sub-region. The overall trend of detected Polar Lows is -0.29 cases/PLS, which amounts to -2.7×10^{-5} cases per PLS per grid point. The trends in different sub-regions are not uniform, with R9 exhibiting the strongest decreasing trend of -8.3×10^{-5} cases per PLS per grid point. For R1, R4 and R11, somewhat smaller decreasing trends prevail, with -2.0×10^{-5}, -4.8×10^{-5}, and -6.9×10^{-5}, respectively. The trend in R8 is the weakest at -0.9×10^{-5} cases per PLS per grid point. This pattern describes a northward shift of Polar Low distribution in the coming 100 years under the scenario A1B.

The reductions in sea ice extent and northward shift of Polar Low formation that are seen in the A1B simulations are qualitatively reproduced in simulations following the A2 and B1 scenarios (not shown). With more (less) greenhouse gas emission and global warming, the Polar Low formation appears to migrate more (less) northward.

4. Conclusions

We applied global climate model products under different scenarios of societal development in future, which go along with different greenhouse gas emissions and global warming developments. Corresponding spatial distribution of genesis and trend of Polar Low occurrence frequency were calculated by a model-based methodology, which we developed for a hind cast of the last 62 years from 1948 to 2010 (Chen and von Storch, 2013).

A general result is that the frequency of Polar Low occurrence has a close relationship with global warming. With larger increases in greenhouse gas emission, Polar Low frequency decreased more rapidly. The spatial distribution shows a northward shift through the twenty first century under different scenarios. The explanation for this change is mainly related to the raised atmosphere temperatures, less warming of the sea surface and a gradual retreat of the northern ice cover. The decrease in air–sea temperature difference leads to reduced vertical instability. With less frequent favorable conditions, the frequency of Polar Low occurrence is simulated to decrease. With a northward shift of the sea ice edge, Polar Lows will occur in new areas further north than in present climate.

We should point out that our regional simulation is a scenario, i.e., a possible and consistent description of future conditions related only to greenhouse gas and aerosol emissions, not a prediction of what may be considered the most probable development expected to happen in the coming 90 years.

Funding

This work is a contribution to the Helmholtz Climate Initiative REKLIM (Regional Climate Change), a joint research project of the Helmholtz Association of German Research Centers. This research was supported by the Chinese Scholarship Council (CSC), and by the Knowledge Innovation Project for Distinguished Young Scholar of The Chinese Academy of Sciences (No. KZCX2-EW-QN203).

Acknowledgments

We acknowledge the modeling groups, the Program for Climate Model Diagnosis and Intercomparison (PCMDI) and the WCRP's Working Group on Coupled Modelling (WGCM) for their roles in making available the WCRP CMIP3 multi-model datasets. Support of these datasets is provided by the Office of Science, U.S. Department of Energy. The authors thank Beate Gardeike for preparing the diagrams. The technical and scientific support, and various comments and suggestions by Dr. Beate Geyer and Dr. Matthias Zahn have greatly improved this manuscript.

References

Boer, G. J., Flato, G. and Ramsden, D. (2000). A transient climate change simulation with greenhouse gas and aerosol forcing: Projected climate to the twenty-first century, *Climate Dynamics* **16**, 6, pp. 427–450, doi:10.1007/s003820050338.

Chen, F., Geyer, B., Zahn, M. and von Storch, H. (2012). Toward a multi-decadal climatology of north pacific polar lows employing dynamical downscaling, *Terrestrial Atmospheric and Oceanic Sciences* **23**, 3, pp. 291–301, doi:10.3319/tao.2011.11.02.01(a).

Chen, F. and von Storch, H. (2013). Trends and variability of north pacific polar lows, *Advances in Meteorology* **2013**, pp. 1687–9309, doi:10. 1155/2013/170387.

Cox, P. M., Betts, R. A., Jones, C. D., *et al.* (2000). Acceleration of global warming due to carbon-cycle feedbacks in a coupled climate model, *Nature* **408**, 6809, pp. 184–187, doi:10.1038/35041539.

Friedlingstein, P., Bopp, L., Ciais, P., *et al.* (2001). Positive feedback between future climate change and the carbon cycle, *Geophysical Research Letters* **28**, 8, pp. 1543–1546, doi:10.1029/2000GL012015.

Jungclaus, J. H., Botzet, M., Haak, H., *et al.* (2005). Ocean circulation and tropical variability in the coupled model ECHAM5/MPI-OM, *Journal of Climate* **19**, 16, pp. 3952–3972, doi:10.1175/JCLI3827.1.

Kolstad, E. W. and Bracegirdle, T. J. (2008). Marine cold-air outbreaks in the future: An assessment of IPCC AR4 model results for the Northern Hemisphere, *Climate Dynamics* **30**, 7, pp. 871–885, doi: 10.1007/s00382-007-0331-0.

Marsland, S. J., Haak, H., Jungclaus, J. H., *et al.* (2003). The max-planck-institute global ocean/sea ice model with orthogonal curvilinear coordinates, *Ocean Modelling* **5**, 2, pp. 91–127, doi:10.1016/ S1463-5003(02)00015-X.

Meehl, G. A., Arblaster, J. M. and Tebaldi, C. (2005). Understanding future patterns of increased precipitation intensity in climate model simulations, *Geophysical Research Letters* **32**, 18, doi:10.1029/ 2005GL023680.

Meehl, G. A., Covey, C., Delworth, T., *et al.* (2007). The WCRP CMIP3 multimodel dataset — a new era in climate change research, *Bulletin of the American Meteorological Society* **88**, 9, pp. 1383–1394, doi: 10.1175/BAMS-88-9-1383.

Nakicenovic, N., Alcamo, J., Davis, G., *et al.* (2000). Special report on emissions scenarios, Tech. rep., The Intergovernmental Panel on Climate Change (IPCC).

Pinto, J. G., Ulbrich, U., Leckebusch, G. C., *et al.* (2007). Changes in storm track and cyclone activity in three SRES ensemble experiments with the ECHAM5/MPI-OM1 GCM, *Climate Dynamics* **29**, 2, pp. 195–210, doi:10.1007/s00382-007-0230-4.

Raisanen, J., Ruokolainen, L. and Ylhäisi, J. (2010). Weighting of model results for improving best estimates of climate change, *Climate Dynamics* **35**, 2, pp. 407–422, doi:10.1007/s00382-009-0659-8.

Rasmussen, E. A. and Turner J. (2003). Polar lows, in *A Half Century of Progress in Meteorology: A Tribute to Richard Reed*, pp. 61–78, Springer.

Roeckner, E., Bauml, G., Bonaventura, L., *et al.* (2003). The atmospheric general circulation model echam 5. Part i: Model description, Tech. rep., Max-Planck-Institut für Meteorologie.

Steppeler, J., Doms, G., Schattler, U., *et al.* (2003). Meso-gamma scale forecasts using the nonhydrostatic model LM, *Meteorology and Atmospheric Physics* **82**, 1, pp. 75–96, doi:10.1007/s00703-001-0592-9.

Stocker, T. F. and Schmittner, A. (1997). Influence of CO_2 emission rates on the stability of the thermohaline circulation, *Nature* **388**, 6645, pp. 862–865, doi:10.1038/42224.

Yonetani, T. and Gordon, H. B. (2001). Simulated changes in the frequency of extremes and regional features of seasonal/annual temperature and precipitation when atmospheric CO_2 is doubled, *Journal of Climate* **14**, 8, pp. 1765–1779, doi:10.1175/1520-0442(2001)014<1765:SCITFO>2.0.CO;2.

Zahn, M. and von Storch, H. (2008). Tracking polar lows in CLM, *Meteorologische Zeitschrift* **17**, 4, pp. 445–453, doi:10.1127/0941-2948/2008/0317.

Zahn, M. and von Storch, H. (2010). Decreased frequency of North Atlantic Polar lows associated with future climate warming, *Nature* **467**, 7313, pp. 309–312, doi:10.1038/nature09388.

https://doi.org/10.1142/9781800615816_0013

Part IV
Downscaling Regional Climatology in Coasts and Marginal Seas of China

Climatology, variability, and changes in regional climate can have significant impacts on the regional environment, society, and the economy. Climate statistics and modelling of regional climate variability involve the use of statistical methods and dynamical models to analyze and understand how climate patterns and trends vary across different regions.

Dynamical downscaling using a regional climate model (RCM) is a robust technique to provide high-resolution climate information for smaller regions, which have the potential to advance our understanding of regional climate systems by only using coarse-resolution global climate models. The technique originated in the late 1980s and has seen a tremendous growth in the past decades. Specifically, with the introduction of spectral nudging into regional climate simulations (von Storch *et al.*, 2000), the RCM has shown more added value both in the climate statistics and the mesoscale atmospheric phenomena and have been employed to reconstruct many aspects of (marine) meteorological climate in Europe and the marginal seas. Our studies of the marine climate in the region of the Bohai and Yellow Sea were inspired by these studies (mainly with respect to the North Sea and Baltic Sea) to reconstruct, to evaluate, and to apply regional climate information — in the spirit of the Four Seas framework (see Chapter 2).

The first reprinted article, namely "A model-based climatology analysis of wind power resources at 100-m height over the Bohai Sea and the Yellow Sea", is reproduced here as Chapter 9. The wind energy assessment at the 100-m wind turbine height confirmed that the modeled wind speeds are comparable to observational wind at 100-m height. Based on this, wind climatology, variability, and extreme values over the region were determined in spatial and temporal detail.

The second reprinted paper "Low-level jets over the Bohai Sea and Yellow Sea: Climatology, variability and the relationship with regional atmospheric circulations", in Chapter 10, focused on low level jets (LLJs) — a mesoscale-flow phenomenon with horizontal wind maxima within the lowest few kilometers of the troposphere. LLJs over the Bohai and Yellow Sea Region feature a strong diurnal cycle, intra-annual, and interannual variability but weak decadal variability. Pressure systems over the East Asia-Northwest Pacific region are significantly correlated with the variations of LLJ occurrence in terms of the intra-annual and interannual variability.

Another topic concerns storm surges and extreme sea level variations, which can lead to severe consequences on coastal communities and ecosystems, including coastal flooding, coastal erosion, property damage, and loss of life. By using hourly tide-gauge data along the coast of China, the third paper "Assessing changes in extreme sea levels along the coast of China", reprinted in Chapter 11, evaluated changes in extreme water levels in the past several decades along China and assessed the roles of mean sea level, astronomical tide, non-tidal component, and the tide–surge interaction in the changes of extreme sea levels. The paper concluded that changes in extreme sea levels along Chinese coasts are strongly related to the changes of mean sea levels.

Furthermore, the last paper in the list "Changes of storm surges in the Bohai Sea derived from a numerical model simulation, 1961–2006", reprinted in Chapter 12, investigated the spatial distribution, the seasonal variation, the interdecadal variability, and the long-time trend of storm surges in the Bohai Sea. To do so, simulations with the tide–surge circulation model ADCIRC was used to reconstruct the storm surges in the Bohai Sea forced by a reconstruction of wind conditions from 1961 to 2006.

A paper of this series, not reprinted here, was "High resolution wind hindcast over the Bohai and Yellow Sea in East Asia: evaluation and wind climatology analysis" (Li *et al.*, 2016b). It focused on evaluation of marine surface wind and concluded that the added value mainly lies in winds over coastal areas, especially for strong winds. The abstract reads the following:

> *A 34 year (1979–2012) high-resolution (7 km grid) atmospheric hindcast over the Bohai Sea and the Yellow Sea (BYS) has been performed using COSMO-CLM (CCLM) forced by ERA-Interim reanalysis data (ERA-I). The accuracy of CCLM in surface wind reproduction and the added value of dynamical downscaling to ERA-I have been investigated through comparisons with the QuikSCAT Level2B 12.5 km version 3 (L2B12v3) swath data and in situ observations. The results revealed that CCLM has a reliable ability to reproduce the regional wind characteristics over the BYS. Added value to ERA-I has been detected in the coastal areas with complex orography. CCLM wind quality had strong seasonal variability, with better performance in the summer relative to ERA-I, even in the offshore areas. CCLM was better able to represent light and moderate winds but had even more added value for strong winds relative to ERA-I. The spatial digital filter method was used to investigate the scale of the added value, and the results show that CCLM adds value to ERA-I mainly in medium scales of wind variability. Furthermore, wind climatology was investigated, and significant increasing trends in the south Yellow Sea especially in winter and spring were found for seasonal mean wind speeds.*

Chapter 9

A Model-Based Climatology Analysis of Wind Power Resources at 100-m Height Over the Bohai Sea and the Yellow Sea*

D. Li, B. Geyer, and P. Bisling

Helmholtz-Zentrum Geesthacht Centre for Materials and Coastal Research, Geesthacht, Germany

China has set ambitious goals for the development of offshore wind energy to meet the increasing energy demand of coastal provinces. Many studies have assessed the potential offshore wind energy in Chinese territorial waters. However, few studies have focused on the climatology, variability, and extreme climate of wind speeds and wind power, especially at hub height in this area. This type of study is important for selecting promising sites for offshore wind farms. In the present study, a 35-year (1979–2013) high-resolution (7 km) wind hindcast over the Bohai Sea and the Yellow Sea (BYS) at 100-m height was constructed using the regional climate model COSMO-CLM (CCLM) driven by the ERA-Interim reanalysis dataset. The quality of wind speeds reconstructed by CCLM was assessed by a comparison with observation data at several

---

*This chapter was originally published in *Applied Energy*, 179, 575–589, doi:10.1016/j.apenergy.2016.07.010. Copyright (2016), with permission from Elsevier.

stations. After verification, the climatology, variability, and extreme climate of winds over the BYS were spatially and temporally investigated. The results show that the 35-year mean wind speed is mostly between 7.0 and 7.5 m/s; in the coastal areas of the BYS, the mean is less than 7.0 m/s, and in the remote offshore areas, the mean is greater than 7.5 m/s. The daily mean wind speed is stronger (weaker) in winter (summer) half year, with stronger (weaker) spatial variability. Wind power density is mainly 300–500 W/m^2. The interannual variability of annual mean wind speed and the wind power are in the range of 0.1–0.3 m/s and 10–40 W/m^2, respectively. Decadal variances of the mean wind speed and the wind power are roughly within $\pm 2\%$ and $\pm 5\%$, respectively, with a stronger variability along the southwestern coasts of the Yellow Sea. The distribution patterns of extreme winds (i.e., 5, 10, 30, and 50-year return values) are generally similar, with strength increasing from the northwest to the southeast. The wind energy characteristics for water areas and potential wind farm sites are summarized.

1. Introduction

In recent years, many countries have devoted extensive resources to investigating renewable energy resources to reduce greenhouse gas emissions from the consumption of fossil fuels, mitigate climate change, and promote sustainable economic and societal development. Wind energy is a mature and rapidly growing energy source that is successfully utilized in many countries, such as Denmark and Germany (Bhattacharya *et al.*, 2016; EEA, 2009; Geyer *et al.*, 2015; Weigt, 2009). Since 2006, China has made considerable advances in wind energy development, with China's share of the annual newly installed capacity increasing from 10% in 2006 to 45% in 2014 (GWEC, 2015; IEA, 2012). The total capacity of wind power installed in China was more than 114.7 GW by the end of 2014, making China the world's leading user of wind energy (GWEC, 2015). However, wind power accounts for only 2.5% of Chinese power production; it is the third largest source, following thermal and hydro power (IRENA, 2014). Onshore wind power dominates the wind energy industry, and facilities are mainly distributed in northern and northwestern China, where transmission infrastructure is insufficient for the transmission of large-scale wind power to the eastern and southeastern coastal provinces. Offshore wind energy is an ideal source to complement thermal and hydro energy to meet energy consumption requirements in the coastal regions of China. The first major offshore wind farm, the 102 MW Donghai bridge wind farm, has been active since 2010. The total installed capacity (TIC) nationwide was 565 MW by the middle of 2013, with another 1978 MW under construction and 6712 MW planned (Wyatt *et al.*, 2014). The TIC increased to 657.9 MW by the end of 2014 (Howe, 2015). The installed capacity was 360 MW in 2015 (News, 2016), making the TIC up to 1.02 GW by the end of 2015, which falls short of the official target of 5 GW. The progress has been slow mainly due to shortage of experience for the developers in this sector, lack of proven offshore products for Chinese turbine manufactures and poor coordination between relative management departments (Bloomberg New Energy Finance, 2016). In December 2014, the National Energy Administration of China issued a "national offshore wind power development and construction plan (2014–2016)". Forty-four wind power projects, with a total of 10.53 GW in capacity, were included in the construction plan and distributed in eight provinces along the coasts. This plan together with some newly issued national favorable policies are believed to promote the development of Chinese offshore wind power in the following years.

The first stage of offshore wind energy development is to assess the wind resource and identify the most promising sites for wind farms. Many studies have estimated the offshore wind energy potential in Chinese coastal waters based on observations, reanalysis datasets, and model outputs. Observations from meteorological masts, especially at hub height, are sparse in the offshore areas, and often have limitations, such as time inconsistencies and spatial and temporal limitations. Additionally, the construction of meteorological masts is expensive and time consuming. Satellite, reanalysis, and model data, therefore, have become alternative means for assessing offshore wind energy. Qin *et al.* (2010) investigated the offshore wind energy along the Chinese coasts at 100-m hub height in the year 2005 using the MM5 model and showed that the coastal areas of southeastern China have the most abundant offshore wind resources among the Chinese coasts. Hong *et al.* (2011) evaluated available offshore wind power in China's exclusive economic zone under technical, spatial and economic constraints using a Geographical Information System. They indicated that the economic potential offshore wind power could contribute to 56%, 46% and 42% of the electricity demands of coastal regions in 2010, 2020 and 2030. Lu *et al.* (2013) suggested that about 28% of the overall potential offshore wind power could be deployed to replace thermal power in the Chinese coastal areas. Jiang *et al.* (2013) assessed the spatio-temporal variation of China's offshore wind resources based on daily QuikSCAT wind field data from 2000 to 2008 and determined that the total 10 m offshore wind energy potential is approximately 660 GW. Chang *et al.* (2014) compared wind data retrieved from ENVISAT advanced satellite synthetic aperture radar (SAR) images with in situ wind observations over the Hangzhou Bay China, and showed that SAR can capture only part of offshore wind statistics features. Chang *et al.* (2015) reconstructed wind data at 100 m over the northern South China Sea for the year 2011 by combining multiple satellite data and model simulations, and revealed that the wind power density ranges from 400 to 600 W/m^2 in the study area. Based on NCEP-CFSR reanalysis data, Wang Guosong *et al.* (2014) concluded that annual mean wind power density is greater than 200 W/m^2, and there were more than 4000 cumulative hours of wind speed greater than 6 m/s at 70-m height during 1991–2010 along the coastal areas of China.

The literature generally reveals that the offshore wind energy of China is abundant and can be a great supplement to energy production in China. However, several issues remain during the assessment of offshore wind energy. First, most assessments (using observation or model output) are based on the sea surface wind at 10-m height or at levels below hub height. Assumptions regarding the atmospheric stability effect are made when results are converted to hub height and introduce uncertainties to the assessment process. Geyer *et al.* (2015) assessed the thermal effect on the estimation of wind energy potential by comparing the annual-averaged full load hours (FLH) of wind turbines based on wind speeds at hub height (100 m) converted from 10 m and those interpolated from wind speeds of neighboring model levels. This study showed that an overestimation of approximately 200–400 FLH was calculated for the open North Sea when wind speeds were extrapolated from 10-m wind data. Second, the datasets used to assess wind energy potential are generally short term or of coarse resolution and are thus unable to capture long-term variability and fine features of offshore wind energy. Geyer *et al.* (2015) concluded that wind power estimation based on short-term observations may generate high biases, which could be in the range of

5–10% for the capacity factor (CF) on average. Therefore, realistic long-term wind data are critical and in urgent demand for accurately assessing wind power climatology and long-term variability, which are important references for selecting wind farm sites and installing wind turbines. Finally, an evaluation of long-term changes and decadal variability of offshore wind energy is not available for Chinese coastal waters. Furthermore, previous work is primarily focused on wind energy assessment; few studies have been devoted to an investigation of extreme winds over this region, which are of great importance for the construction and maintenance of wind energy facilities, as well as the sustainability assessment of already installed wind power applications.

In the present study, we investigated the climatology, variability, and extreme climate of winds at 100-m hub height over the Bohai Sea and the Yellow Sea (BYS, Figure 1) using a long-term (1979–2013) and high-resolution (7 km) atmospheric hindcast. The hindcast was constructed using a regional climate model COSMO-CLM (CCLM, Rockel *et al.* (2008), http://www.clm-community.eu/) driven by ERA-Interim (Dee *et al.*, 2011). The Bohai Sea and the Yellow Sea are abbreviated as BS and YS, respectively. The analyzed wind energy potential is not limited to Chinese water areas but also includes the western coastal areas of the Korean Peninsula. Li *et al.* (2016b) assessed the reconstructed surface wind hindcast for a comparison with in situ and satellite observations and verified that the hindcast is robust for capturing the sea surface wind climatology and variability. Added value to the driving dataset ERA-Interim was detected in the coastal areas of the BYS with complex orography. The present work is based on Li *et al.* (2016b) but with a focus on the winds at 100-m height. Additionally, the present study does not only contribute to the wind energy assessment over the BYS, the assessment routine and methods used can be easily and widely applied for other ocean areas, making high-resolution regional climate model a robust tool for wind energy evaluation in wind farm applications.

The chapter is organized as follows: Section 2 includes introductions about the regional climate model CCLM, the simulation setups, and the validation of modeled wind speeds by comparison with observation data, the wind power curve and methods to derive wind power density. Section 3 includes the results regarding climatology and long-term variability of wind and wind power. Extreme return winds at

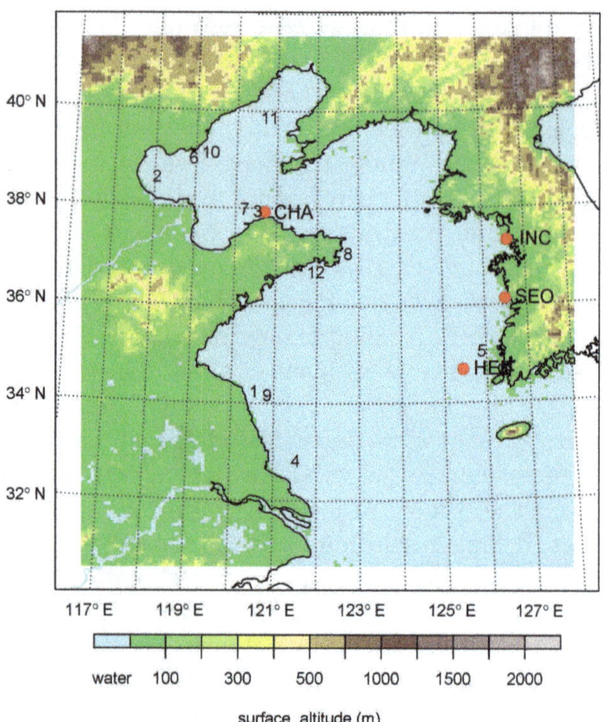

Fig. 1.   Orography of the simulation domain; red points indicate observation stations, and numbers represent the wind farm sites statistical analyzed in Table 2 that are authorized or in the planning stages. The white frame indicates the sponge zone.

100-m height over the BYS are included in Section 4. A summary and conclusions are given in Section 5.

2.   Data and Methods

2.1.   *Regional Atmospheric Hindcast*

The nonhydrostatic regional climate model CCLM version 4.14 was applied for the regional atmospheric hindcast over the BYS. This model is the climate version of the COSMO weather prediction model and was developed by Deutscher Wetterdienst. The model is widely used for simulations with temporal scales up to centuries and spatial resolutions ranging between 1 and 50 km (http://www.clm-community.eu/). The basic dynamics of the CCLM model rely on the thermo-hydrodynamic equations describing compressible flow in a moist atmosphere. These equations are formulated in a rotated geographical coordinate system with an Arakawa C-grid structure (Arakawa and Lamb, 1981) and generalized terrain-following height coordinates (Doms and Baldauf, 2015). Finite second-order differences are used for spatial discretization.

The initial and boundary conditions of the hindcast were provided by the reanalysis dataset ERA-Interim with a 6-hourly update frequency. The interior spectral nudging technique (von Storch *et al.*, 2000) was applied to the simulation at every third time-step on the horizontal wind fields with levels above 850 hPa. This technique serves to keep the resolved large-scale features in CCLM close to those of the forcing field; whereas the model can freely develop local and regional scale processes independently. The horizontal resolution is $0.0625°$ (approximately 7 km), with $168 * 190$ grid points in the longitudinal and latitudinal directions. Ten grid boxes were set as the sponge zone at the lateral boundary on each side (Figure 1). Forty levels were used in the vertical direction, with layer thicknesses increasing from the bottom to the upper levels. The heights of upper boundaries of the lowest six levels were 20, 49, 89, 143, 214, and 303 m, yielding wind speeds at heights of 10, 34.5, 69, 116, 178.5, and 258.5 m, respectively. Wind speeds at 100 m were obtained by spline interpolation between the model output at 69-m and 116-m heights and have a 3-h output interval.

2.2. *Validation of Simulated Wind Speeds*

The simulated wind speeds at 10-m height have been validated against observations in Li *et al.* (2016b), and results indicated high agreement between simulated and observed winds. To verify the model ability in reproducing horizontal wind speeds at different levels, simulated winds were vertically interpolated or extrapolated to observation heights and compared to the observed winds (Table 1).

Comparing observed and modeled wind speeds in 2006 (Figure 2, Supplement Figure S1[1] includes all observation periods for reference) reveals that CCLM has a robust performance for reproducing local

---

[1]Refer to https://ars.els-cdn.com/content/image/1-s2.0-S0306261916309515-mmc1.docx.

Table 1.   List of wind speed observation sites, their locations (lon, lat), and observation periods as well as observation heights.

Station	Longitude (°E)	Latitude (°N)	Period	Height (m)
CHA	120.717	37.933	1986/01–2012/11	40
HEU	125.451	34.687	2001/07–2013/12	68.5
INC	126.59	37.33	2004/11–2008/01	3
SEO	126.50	36.13	2005/01–2007/08	3

wind speeds. Wind speed variability and magnitude reproduced by CCLM are generally in agreement with observations, especially at station SEO. At the stations CHA and HEU, simulated wind speeds tend to overestimate observations, which may be because of frictional resistance of islands, which are not well enough resolved in the simulation. Further comparisons were made between observation and simulated wind speeds at the four stations (Figure 3) for the periods given in Table 1. Several statistical metrics, scatter plots, and quantile-quantile plots (qq-plots) were used for comparison. According to the statistical metrics, we see that mean and standard deviation of simulated winds are generally in agreement with those of observed winds: CCLM tends to overestimate the observed wind speed with mean biases less than 1.0 m/s at three out of four stations. Standard deviations of CCLM winds closely resemble those of the observed winds. These two parameters are of importance in the potential wind density estimation, which will be given in Section 4. The linear correlation coefficient is approximately 0.7, with a maximum value of 0.8 at station HEU. Root mean square errors range between 1.9 and 2.6 m/s. Based on the scatter plots, qq-plots, and Kernel density estimation contours, simulated wind distributions are generally consistent with those of observed winds, especially for weak wind speeds. Specifically, CCLM most accurately models the wind distribution at station CHA and tends to underestimate the strong wind speeds at stations INC and SEO. The model poorly models the wind distribution at station HEU, with an overestimation of by 1–2 m/s. Generally, CCLM wind speeds fit closely to the observed data.

Furthermore, a dataset of observed 10-year-mean wind speed (2000–2009) in the BYS was obtained at potential wind farm sites from the website of 4Coffshore (http://www.4coffshore.com/). The data were extrapolated from satellite data to 100 m and compared

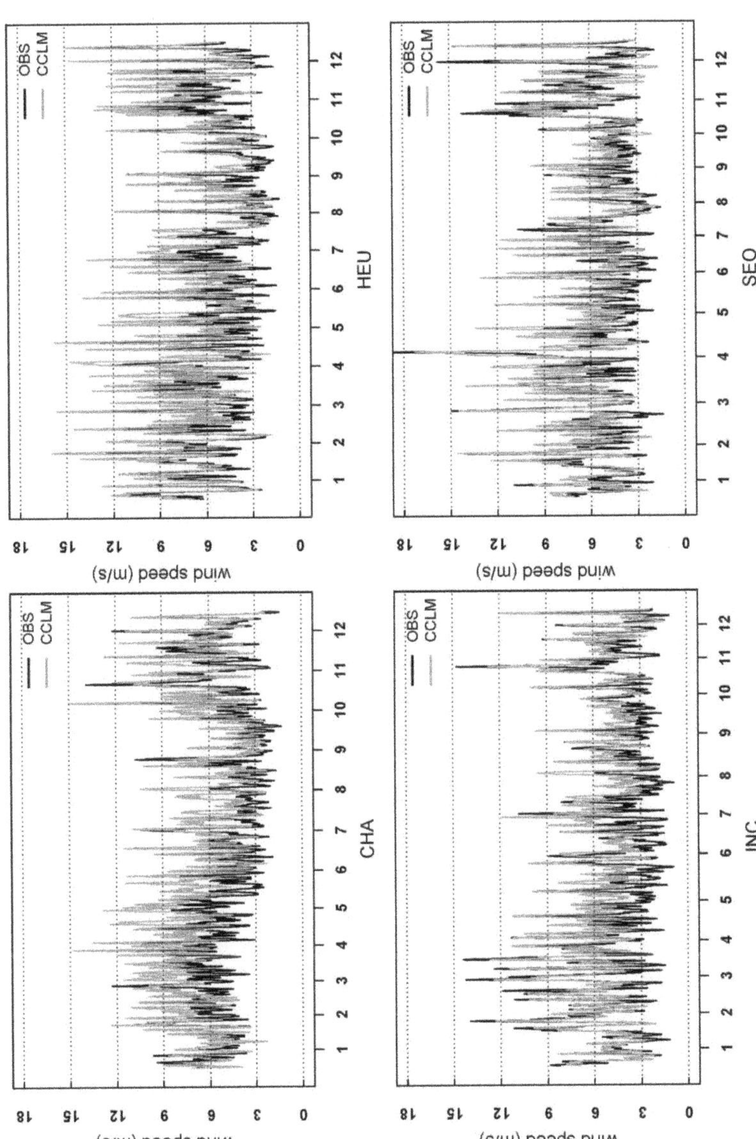

Fig. 2.   Time series of nine-point moving averages of observed (black lines) and simulated wind speeds (gray lines) during 2006 at four stations (CHA, HEU, INC, and SEO) at observation heights.

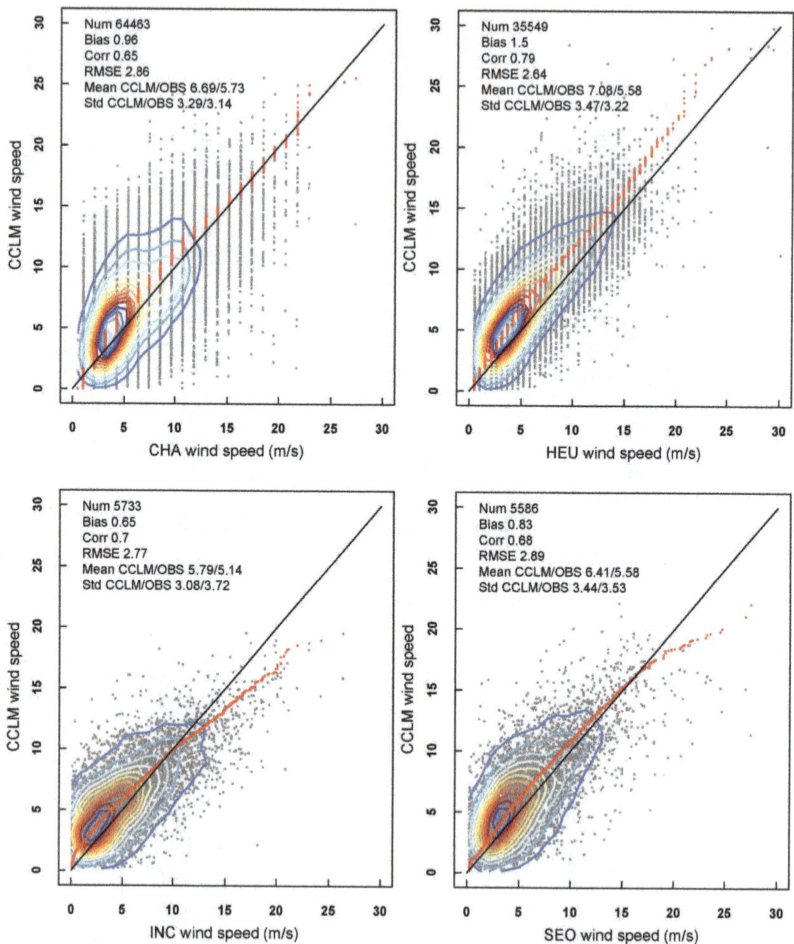

Fig. 3. Comparison of CCLM with wind observations (x-axis) at four stations (CHA, HEU, INC, and SEO): scatter plots (gray dots), qq-plots, and several statistical metrics (i.e., valid numbers of observation, bias, correlation, root mean square error (RMSE), mean, and standard deviation (std)). Kernel density estimation contours are also included. The verification periods are given in Table 1.

with modeled data (Table 2). The 10-year-mean CCLM wind speeds have minor absolute bias relative to the observed data, with values less than 0.2 m/s at 11 out of 12 sites. Based on the comparison with different observed datasets shown above, CCLM wind proves to be a reliable wind dataset and is robust for assessing potential wind energy in the BYS.

Table 2. Statistical metrics from 12 wind farm sites that are either being planned or have been authorized: station name (Name); observed mean wind speed during 2000–2009 obtained from 4Coffshore (http://www.4coffshore.com/, OBS); 10-year-mean CCLM wind speed (Mean1) from 2000 to 2009; 35-year-mean CCLM wind speed (Mean2). Standard deviation (std); Weibull parameters c and k; wind power density (WPD); capacity factor (CF); percentage of hours at rated power (HRP); downtime percentage due to calm or storm wind conditions (DT); mean duration of rated power situations (DHRP); mean duration of wind speeds below cut-in velocity (DCalm); mean duration of wind speeds above cut-out velocity (DStorm) based on 35-year CCLM wind speeds.

No.	Name	OBS (m/s)	Mean1 (m/s)	Mean2 (m/s)	Std (m/s)	C (m/s)	k	WPD (W/m^2)	CF (%)	HRP (%)	DT (%)	DHRP (h)	DCalm (h)	DStorm (h)
1	BIN	6.98	7.04	6.98	3.37	7.88	2.21	362	41	6.6	11	9	7	3
2	CAN	7.18	6.62	6.57	3.30	7.42	2.11	315	37	5.1	13.6	7	6	7
3	CHA	6.95	6.79	6.85	3.47	7.73	2.09	360	39.9	6.7	13.2	8	6	5
4	CHI	7.25	7.31	7.19	3.32	8.12	2.31	382	43.2	6.8	9.3	10	7	9
5	JEO	7.12	7.17	7.16	3.64	8.08	2.08	413	42.6	8.6	11.9	11.5	7	11
6	LAO	7.04	6.90	6.91	3.36	7.80	2.19	354	39.8	6.4	10.6	8	6	10.5
7	LON	6.95	7.03	7.08	3.63	7.99	2.07	401	41.9	8.5	12.6	9	6.5	5
8	RON	7.35	7.23	7.29	3.64	8.23	2.13	426	44	9.1	11.2	11	8	4
9	SHE	6.98	7.08	7.02	3.4	7.93	2.20	371	41.4	6.8	11.3	9	7	4.5
10	TAN	7.05	7.04	7.07	3.52	7.98	2.13	389	41.4	7.2	10.1	8	6	7.5
11	SUI	7.45	7.40	7.50	3.82	8.47	2.08	475	45.8	10.7	11.6	10	6	9
12	WEI	6.77	6.56	6.62	3.19	7.47	2.21	309	37.4	4.6	12.1	9	7	0

2.3.   *Probability Density Function and Wind Power Density*

A two-parameter Weibull distribution is widely used to characterize the probability density function of wind power (Rocha *et al.*, 2012; Pishgar-Komleh *et al.*, 2015; Seguro and Lambert, 2000):

$$p(U) = (k/c)(U/c)^{(e-1)} exp[-(U/c)^k] \tag{1}$$

where the parameters k and c indicate the non-dimensional shape parameter and scale parameter, respectively. Over the BYS region, Wang *et al.* (2015) investigated the observed wind energy potential at four marine sites in the Bohai Rim during 1990–2012, and showed that the Weibull distribution is more suitable to describe observed wind frequencies in the study area than Rayleigh and Lognormal function. Therefore, Weibull distribution is used to fit the simulated wind speed data in the BYS. Estimation of the shape parameter and scale parameter is based on an empirical model — a special case of method of moments and given as the following (Monahan, 2006):

$$c = \overline{U}/\Gamma(1 + 1/k) \tag{2}$$

$$k = (\overline{U}/\sigma)^{1.086} \tag{3}$$

where $\overline{U}$ is the mean wind speed, Γ is the gamma function, and σ is the standard deviation of wind speed. The wind power density (WPD) is then given by the following (Liu *et al.*, 2008):

$$E = \frac{1}{2}\rho c^3 \Gamma(1 + 3/k) \tag{4}$$

where ρ is the air density, which was set as a constant of 1.225 kg/m^3 in this study.

2.4.   *Synthetic Power Curve*

The power output of a wind turbine varies with wind speed and can be described using a power-velocity (i.e., synthetic power) curve. Power output can be used as an indicator of the wind energy potential of a specific wind turbine without considering its technique in detail. The curve is characterized by three important velocities: cut-in wind speed, where wind turbine starts generating electric power;

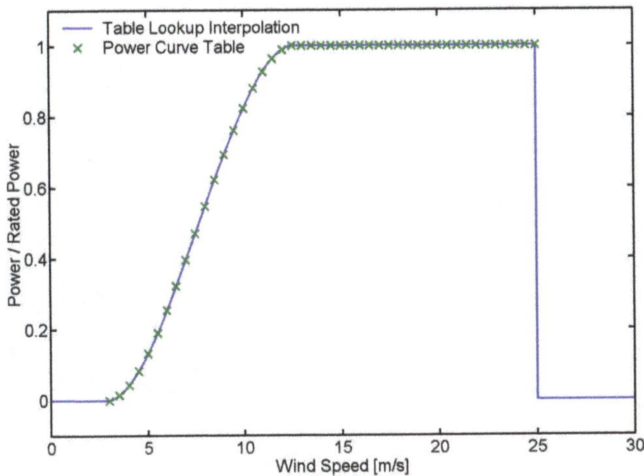

Fig. 4.   Synthetic power curve.

rated wind speed, where the wind turbine reaches its rated power, and; cut-out wind speed, where the turbine should be shut down to prevent damage. In this paper, a specific type Vestas V112-3.0 MW offshore wind turbine is used as an example. The three velocities of this kind turbine are 3 m/s, 12.5 m/s, and 25 m/s, respectively. We used a power curve normalized to the rated power (Figure 4), i.e., capacity factor, to visualize the wind energy potentials over the BYS. The spatial distribution of the climatology of the capacity factor will be obtained and shown in the following section. The calculation of some other parameters such as hours at rated power and downtime percentage of turbines are based on the 35-year CCLM wind data considering the cut-in, rated and cut-out wind speeds.

3. Climatology and Variability of Wind Speed and Wind Power

3.1. *Climatology of Wind Speed and Wind Power*

The temporal evolution of water-area-averaged daily mean wind speeds (Figure 5(a)) shows that daily mean wind speeds are higher in the winter (November to April) and lower in the summer (May to

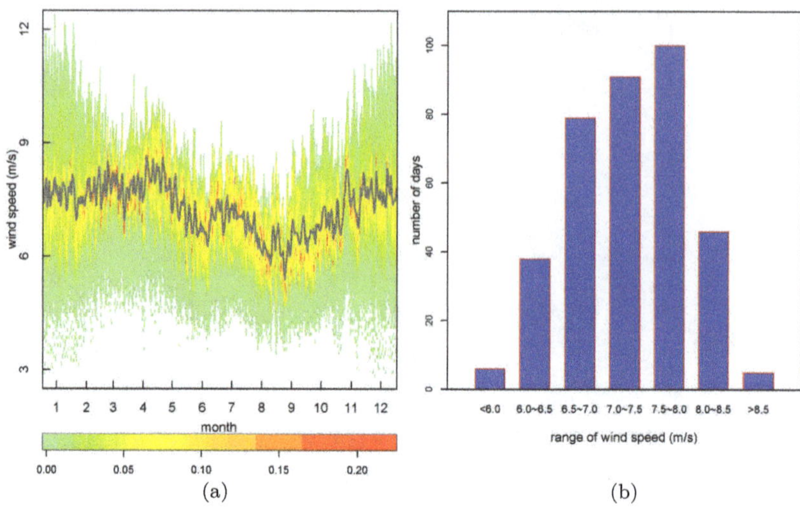

Fig. 5.   (a) Annual cycle of water-area-averaged daily mean wind speed (gray line) from 1979 to 2013 and the corresponding distribution over the domain (shaded); (b) the frequency distribution (number of days per year) of water-area-averaged daily mean wind speeds in classes of 0.5 m/s between 6.0 and 8.5 m/s from 1979 to 2013.

October). Wind speeds range from 7.0 to 8.5 m/s in the winter and 6.0–7.5 m/s in the summer. The strongest mean wind speed occurs in April, whereas the weakest mean wind is in early September. The wind speed distribution over the water areas (Figure 5(a)) indicates that the winter winds have a higher spatial variability (3.0–12.0 m/s) than summer winds (5.0–9.0 m/s). A histogram (Figure 5(b)) reveals that the water-area-averaged mean winds of most days fall in the range of 6.5–8.0 m/s, and only a few days per year have area-mean wind speeds outside this range.

In most areas of the BYS, the mean wind speeds vary from 7.0 to 7.5 m/s (Figure 6). In the northern BS and southeastern part of the BYS, wind speeds range from 7.5 to 8.0 m/s. Mean winds are less than 7.0 m/s close to the coasts of the BYS. The distribution patterns of wind climatology are similar to the annual mean wind distribution in 2005 at 100-m height in Qin *et al.* (2010), but our results are approximately 0.5 m/s lower. With reference to the validation results shown in Table 2, the results shown in Figure 6(a) provide a more accurate and robust representation of wind climatology than those in Qin *et al.* (2010). There is a significant positive

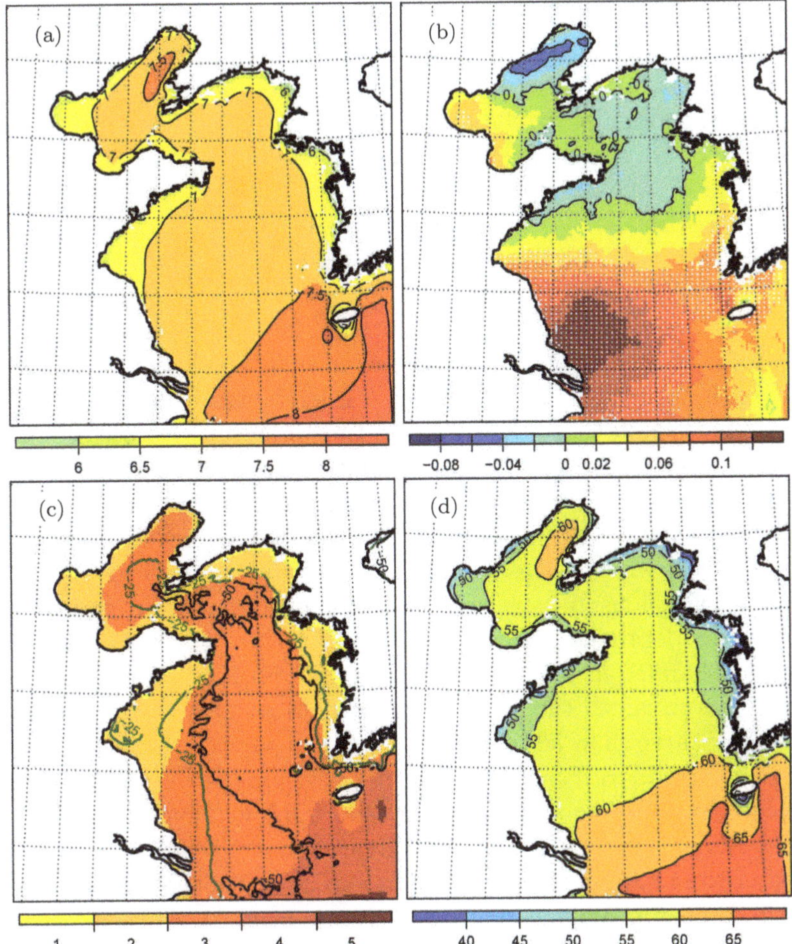

Fig. 6. Wind climatology distributions during 1979–2013: (a) annual mean wind speed (m/s); (b) linear trends of (a) with unit m/s/decade; the white dotted areas are at a 0.05 significance level; (c) wind classification (shaded) and bathymetry map (contours); (d) the mean frequency (%) per year for instantaneous wind speeds above 6.1 m/s based on 3-hourly wind dataset.

linear trend in the south YS (Figure 6(b)). Specifically, the trend estimation is greater than 0.1 m/s/decade for the southwestern coast of the YS. In most areas of the BS and the northern YS, the trends are negative. However, they are not significant ($a = 0.05$). The water-area-averaged annual mean wind speeds (not shown here) generally vary between 7.0 and 7.5 m/s; the winds in 1987, 2005, 2010, and 2013 are slightly larger than 7.5 m/s. The wind speeds in the 1990s oscillate around the 35-year mean wind speed (7.29 m/s), with smaller inter-annual variability than those in the 1980s and 2000s. The linear trend (0.05 m/s/decade) is positive and statistically significant.

Based on the 35-year mean wind speed at 100 m (Figure 6(a)), wind classification was conducted (Figure 6(c)). The classification criteria were derived from Oh *et al.* (2012) as follows: class 1 (less than 6.1 m/s), class 2 (6.1–7.1 m/s), class 3 (7.1–7.8 m/s), class 4 (7.8–8.3 m/s), and class 5 (8.3–8.9 m/s). In addition, another major factor for the evaluation of offshore wind potential is the water depth. The bathymetry map over the BYS was extracted from the ETOPO1 global bathymetry dataset (Amante and Eakins, 2009) with a spatial resolution of 1 arc-minute, and 25 and 50-m contours (Figure 6(c)). Along the coast at depths shallower than 25 m, the wind generally falls in class 2. Over parts of the BS and the southwestern coast of the YS, the wind is class 3. These areas are appropriate for wind farm applications because of abundant potential wind energy and low water depths. Over the central BYS area, the water depth is greater than 25 m and the wind resource is mostly class 3. In the southeastern area, the wind is in class 4 and the water depth is mostly greater than 50 m.

In the northern BS and southern YS, the mean frequency (%) per year for instantaneous wind speeds above 6.1 m/s based on 3-hourly wind dataset from 1979 to 2013 is greater than 60%; this frequency is greater than 65% in the southeastern YS (Figure 6(d)). In most parts of the BYS the frequency is in the range of 55–60%. In the coastal areas it is generally below 55%, except along the northern BS and southwestern YS, which are promising areas for wind farm sites.

In terms of WPD (Figure 7(a)), it is generally lower than 400 W/m^2 in the coastal areas of the BYS, is 400–500 W/m^2 in the central area of the BYS, and is greater than 500 W/m^2 in the southeastern tip of the study area. The CF (Figure 7(b)) is lower than 0.4 in the coastal areas, greater than 0.4 in the offshore areas, and greater than 0.45 in the northern central BS, some areas of the mid-YS, and the southeastern section of the study area. The spatial patterns of long-term annual mean hours at rated power (HRP) (Figure 7(c)) share similar characteristics with WPD. The annual HRP are generally in the range of 200–600 h in the coastal areas, 800–1000 h

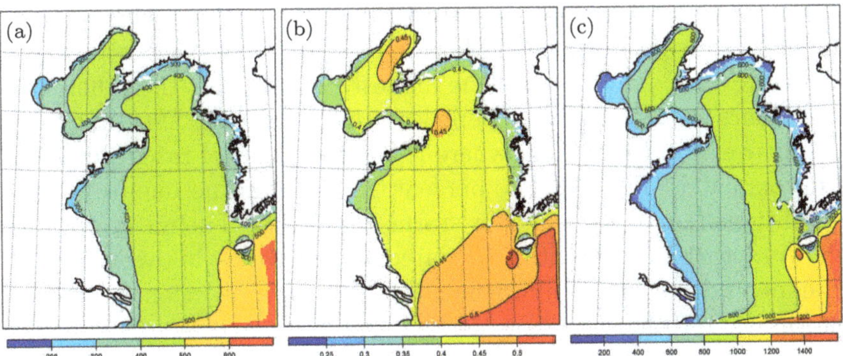

Fig. 7. Long-term average (a) wind power density (W/m^2) and (b) capacity factor derived from the wind conditions at 100 m from 1979 to 2013. Calculation of the capacity factor is based on the synthetic power curve (Figure 4). (c) Long-term annual mean of hours at rated power (h) from 1979 to 2013 for wind speeds in the range of 12.5–25 m/s.

in the central BYS, and greater than 1000 h in the southeastern tip of the study area. To summarize, the potential for wind energy is high in the offshore areas with deep water and relatively low in the coastal areas. However, the water depth in the northern part of the BS is generally shallower than 25 m, whereas the wind energy is greater than in most coastal areas (Figure 6(c)). This area is a potentially promising area for wind farm application. Another area with abundant wind energy and low water depth is the southwestern coast of the YS, where a large number of wind farm projects have been planned or are under construction according to the 4Coffshore database (http://www.4coffshore.com/).

Furthermore, the statistical metrics of 100-m wind speed and parameters related to wind power at 12 wind farm sites are shown in Table 2. As noted above, the 10-year-mean CCLM wind speed has minor differences from observation data. The 35-year-mean CCLM winds are almost identical to 10-year-mean CCLM winds (Table 2). The standard deviation is approximately 3.5 m/s. The Weibull scale parameter c is proportional to mean wind speed and is generally greater than 7.5 m/s. The shape parameter k is in the range of 2.07–2.31, which indicates that the wind distribution over the BYS is less widespread than the more widely used Raleigh distribution ($k = 2$). The wind power density is mainly greater than 300 W/m^2 and is greater than 475 W/m^2 at site SUI in the northern BS. The capacity factor is approximately 0.40 at most sites, with a maximum capacity factor at wind farm site SUI, where the percentage of hours at rated power is also the highest (10.7%). For the other sites, the hours at rated power were less than 10%, far less than wind farms over the North Sea, where the percentages of rated power are approximately 15% [3]. The percentage of downtime due to weather conditions (i.e., calm winds or storms) is 9–14%. The mean duration of wind turbines operating at rated power varies between 8–11 h; the mean duration of calm winds is 6–8 h, and the mean duration of storms is 4–10 h.

3.2. *Long-Term Variability of Wind Speed and Wind Power*

Spatial distributions of interannual variability, indicated by standard deviation calculated from annual or seasonal mean wind speed and wind power of 35 years are shown in Figure 8. We see some interannual variability of 100-m annual mean wind speed and wind power over the BYS, with values generally in the range of 0.1–0.3 m/s and 10–40 W/m^2, respectively. Lower variability is mostly over the coastal areas while it is larger over open sea areas. In terms of interannual variability of seasonal mean variables (Figure 8(b, c, e, and f)), we see stronger magnitude and spatial variability than those of annual mean variables (Figure 8(a) and (d)). During the summer season, interannual variability of mean wind speed is generally 0.2–0.5 m/s over the BYS, with larger values (greater than 0.4 m/s) in western YS and south to YS and lower values (less than 0.3 m/s) in parts of the BS and the YS coasts. The spatial distribution of the interannual variability of the summer-mean wind power is similar to one of the summer-mean wind speed. The variability in the western and southern YS (60–100 W/m^2) is larger than the one in the BS and northern YS coasts (20–40 W/m^2). For winter season, the interannual variability of wintermean wind speed and wind power is much stronger than one of summer-mean in the BS; the interannual variability for most parts of the YS is 0.3–0.5 m/s for wind speed and 50–100 W/m^2 for wind power. Still, we see a relative low variability in the coastal areas. As a summary, we see some interannual variability for annual mean wind speed and wind power, while relative strong interannual variability for seasonal mean variables.

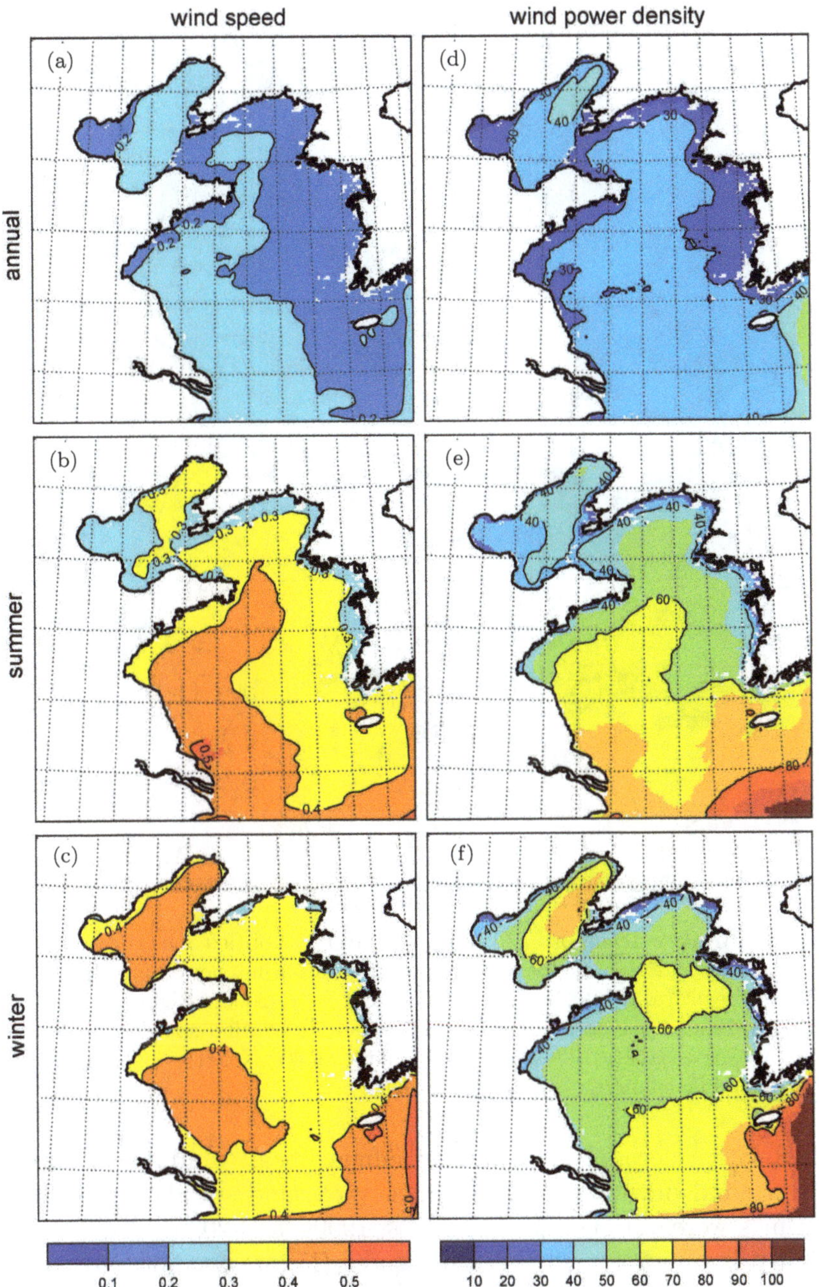

Fig. 8.   Interannual variability for (a–c) wind speed (m/s) and (d–f) wind power density (W/m^2) of: (a, d) annual mean, (b, e) summer-mean, and (c, f) winter-mean.

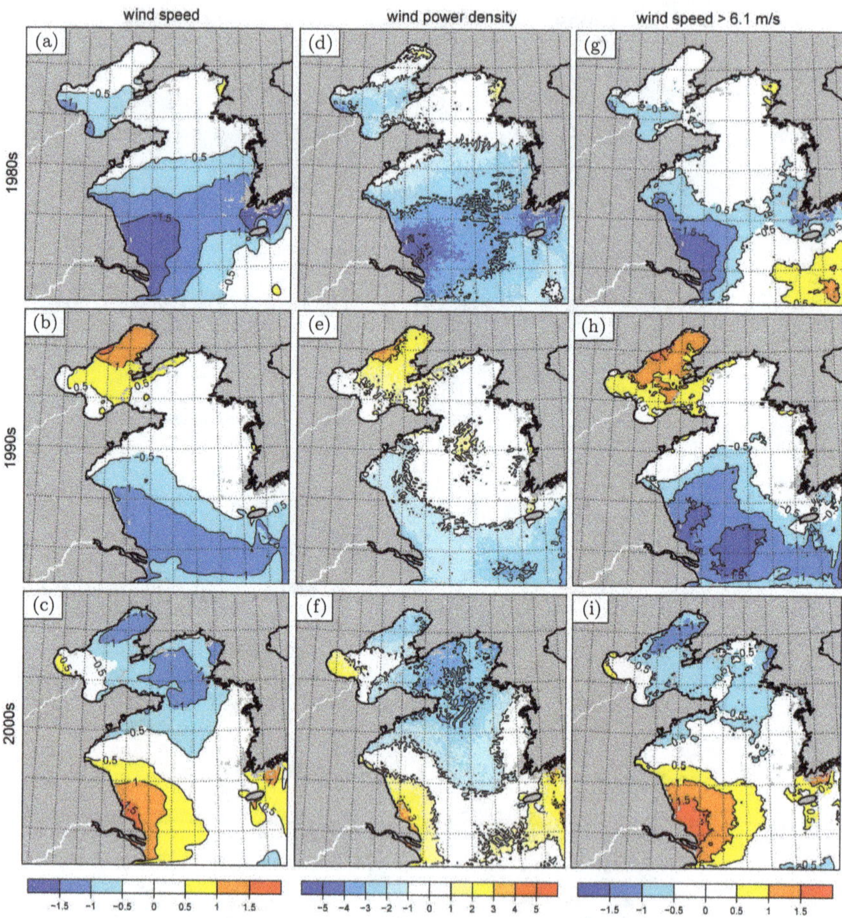

Fig. 9.   Relative deviations (%) in decadal averages of (left panel) annual mean wind speed and (middle panel) of annual mean wind power at 100 m from 35-year-mean: (a, d) 1980s, (b, e) 1990s, and (c, f) 2000s. Right panel: Deviations in decadal averages of annual frequency (%) of wind speeds over 6.1 m/s from 35-year-mean: (g) 1980s, (h) 1990s, and (i) 2000s. Note the different colorbar for the middle column.

Relative decadal anomalies in annual mean wind speed, annual mean wind power, and the frequency of wind speeds greater than 6.1 m/s were obtained for the 1980s, 1990s, and 2000s; decadal variability is minor for all three measures (Figure 9). For wide areas, the relative deviations of decadal averaged annual mean wind speeds from the 35-year-mean wind speed are within 0.5% (Figure 9(a–c)), and the frequency of change of wind speeds greater than 6.1 m/s is also less than 0.5% (Figure 9(g–i)). These two groups share similar spatial distributions and anomaly magnitudes with each other. In the 1980s, larger negative anomalies (2.0% to 0.5%) were detected in the south, especially the southwestern YS. Some negative anomalies occurred in the southern BS. In the 1990s, patterns of negative anomalies in the southern YS do not change much, whereas positive anomalies (0.5–1.5%) were detected in the BS. In the 2000s, the patterns were reversed compared to the 1990s. Positive anomalies (0.5–2%) occurred in the southwestern YS, and negative anomalies occurred in the northern YS and BS. The spatial distributions of relative deviations of decadal wind power (Figure 9(d–f)) do not change much to those of wind speeds; however, they are characterized with stronger magnitude, with a relative deviation of 1–5%.

As a summary, we see some decadal variability in mean wind speed (±2%), mean wind power (±5%), and frequency of winds greater than 6.1 m/s (±2%); additionally, it seems that in the plume and extent-plume areas of the Yangtze River, a relatively strong increase is found in the parameters investigated, which is consistent with the trends in surface wind speeds in these areas described by Li *et al.* (2016b).

Though there is no significant change on decadal climatology of wind speed, wind power density, and frequency of wind speed larger than 6.1 m/s as revealed above. Is there any significant decadal change on the seasonal variations?

Figure 10(a) shows strong variations in long-term monthly wind speed climatologies. The median wind speeds vary between 6 and 8 m/s, with largest value in April and lowest value in September. The distance between 5th and 95th, which is selected to represent spatial dispersion of wind speed, is longest in winter, and shortest in summer. It indicates that the wind speed climatologies in summer are spatially more homogeneous than those in winter. The decadal changes of monthly wind speed variability (Figure 10(b–d)) vary, but are generally within ±5%. The seasonality of wind speed deviations in 1990s and 2000s have stronger and frequenter fluctuations than those in 1980s. The spatial dispersions of monthly wind climatology deviations vary from decade to decade, and do not reveal a consistent pattern. When it comes to the wind power density (Figure 10(e–h)), the features are generally similar to those of wind speed (Figure 10(a–d)).

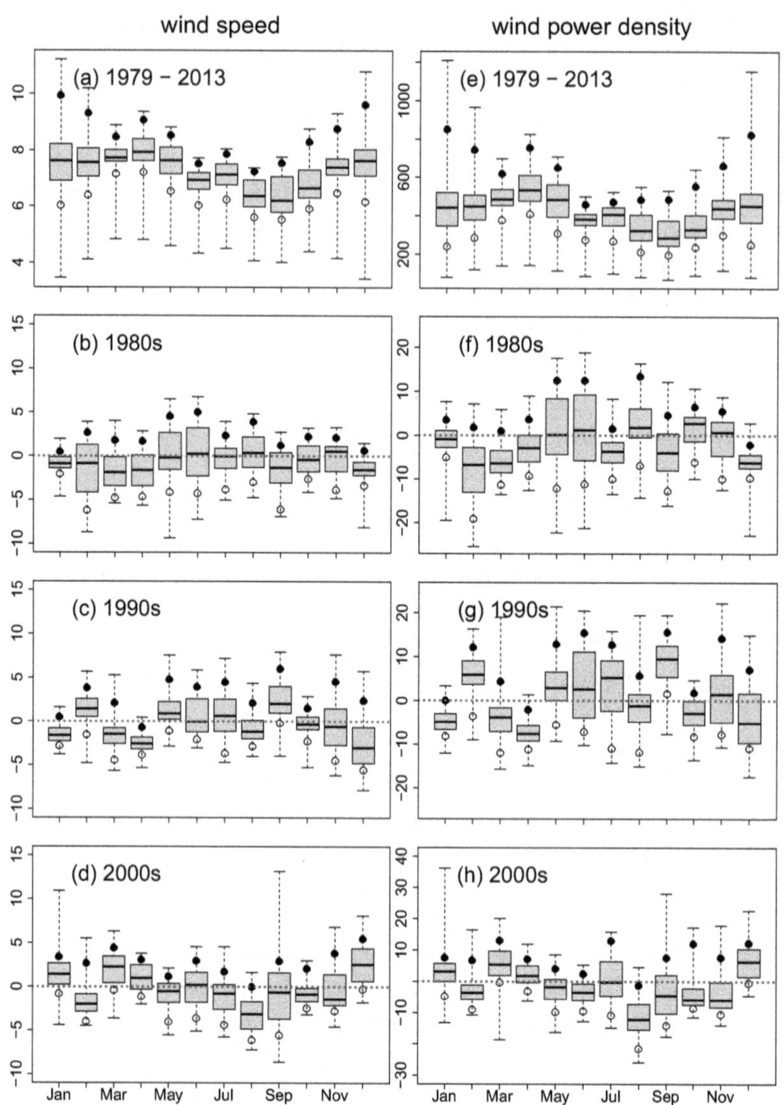

Fig. 10.   Box and whisker plots summarizing the spatial distributions of long-term (1979–2013) monthly climatologies of all sea pixels in the domain for (a) wind speed (m/s) and (e) wind power (W/m^2), the relative deviation (%) of decadal monthly climatologies for (b–d) wind speed and (g–h) wind power from those of long-term monthly climatologies. The bottom and top of the box are the 25th and 75th percentiles; the band inside the box is the median; the ends of the whiskers represent the minimum and maximum, the black circle and black point indicate 5th and 95th percentiles.

However, the fluctuation magnitudes increase. The long-term median wind power density climatologies vary in the range of 300–500 W/m^2. The decadal changes of monthly wind power density (Figure 10(f–h)) are generally within $\pm 15\%$.

To sum up, the climatologies of wind speed and wind power density over the BYS and at several potential wind farm sites are firstly given in this section. Mean wind speed and wind power density are mostly 6–8 m/s and 300–500 W/m^2, respectively. Northern parts of the BS and the southwestern coasts of the YS are characterized with abundant wind energy and shallow water depth (less than 25 m), which are appropriate areas for wind farming. In terms of long-term variability, inter-annual variability of summer- and winter-mean wind speed and wind power are generally stronger than those of annual-mean; we see some decadal variability in mean wind speed, mean wind power and frequency of winds greater than 6.1 m/s; greater variabilities are found for decadal climatologies of monthly mean variables.

4. Extreme Winds

The wind farm sites in the BYS may experience extreme wind speeds generated by extreme events such as typhoons, extratropical cyclones, and cold surges. Such events may cause damage to or failure of wind turbines. Therefore, an assessment of extreme winds is critical for identifying promising wind farm sites.

To determine extreme wind values, two types of approaches are generally used, namely the block maxima method and the peaks-over-threshold (POT) method. However, the block maxima method only considers a maximum value of each block (generally one year) and is bound miss out on some extremes, which will greatly reduce the size of data available for extreme value analysis and may generate misleading results. Additionally, each block is supposed to come from the same distribution pattern, which is not the case for annual wind data with strong seasonality. Therefore, the POT method was used to estimate return levels of extreme winds over the BYS in this study.

The extreme values, selected using POT, generally fit to the Generalized Pareto (GP) distribution (Katz *et al.*, 2005). The cumulative distribution function of GP is defined as follows:

$$F(x; \sigma, \xi) = \begin{cases} 1 - (1 + \xi(\frac{y}{\sigma}))^{-\frac{1}{\xi}}, & \text{if } 1 + \xi(\frac{y}{\sigma} > 0 \text{ and } \xi \neq 0 \\ 1 - e^{-\frac{y}{\sigma}}, & \text{if } \xi = 0 \end{cases} \tag{5}$$

where $r > 0$, represents the scale parameter, and n is the shape parameter. A suitable threshold should be chosen for the extreme value analysis using the POT method firstly. The sample mean excess (SME, Supplement Figure S2)[2] is a popular approach to determine threshold. The formula of SME is: $e_x(d) = E(X - d|X > d)$. With increasing threshold (Figure S2), SME decreases firstly, then constantly continues for a certain range and then decreases again. The lowest value in the stable region is generally used as a threshold for the POT analysis. However, eyeball detection rather than other valid methodology is available to execute a threshold selection process, making SME method hardly applied for obtaining threshold maps for extreme analysis in the model domain (cf., Outten and Esau, 2013). In the present study, a simpler method was used, which selects the lowest annual maximum as threshold. A grid-cell specific threshold rather than common threshold is used to achieve the balance between variance of parameter estimation and feasibility of GP distribution fit (cf. Van de Vyver and Delcloo, 2011). It guarantees at least 35 exceedances for each grid point. Actually, approximately 120–1200 exceedances were detected using this method for each grid. According to Figure S2, we see comparable results between the lowest annual maxima and those detected from SME plots, which prove the skill of the threshold detection method used.

In detail, the following processes (cf., Outten and Esau, 2013) were conducted to retrieve extreme winds with different return periods:

(1) Retrieve daily maximum winds at each grid point for 35 years;
(2) Select the lowest annual maxima as thresholds for each grid point;
(3) Obtain exceedances over the threshold from daily maximum wind time series;
(4) Employ a declustering method to obtain independent exceedances, with at least 48 h between any exceedances;

---

[2]Refer to https://ars.els-cdn.com/content/image/1-s2.0-S0306261916309515-mmc1.docx.

(5) Use a maximum likelihood estimation method to retrieve GP distribution fit;
(6) Calculate extreme return levels of wind speed with 5, 10, 30, and 50-year return periods.

The distribution patterns of extreme winds with return periods of 5, 10, 30, and 50 years are generally similar, with strengths increasing from northwest to southeast (Figure 11) and lower strength in the coastal areas than in open sea areas. The 5-year return winds are less than 25 m/s in the coastal areas, greater than 25 m/s in the central BYS, and greater than 27 m/s in the southeastern part of the research area. The 10-year return winds are mostly less than 27 m/s in the BS and the northern and western YS, between 27 and 31 m/s in the central and southern YS, and greater than 31 m/s in southeastern part of the research area. The 30-year return winds are between 25 and 30 m/s in most areas of the BS and the northern and western YS and are greater than 30 m/s in the other areas. The values exceed 40 m/s in some areas. The 50-year return winds are not significantly different from those in the 30-year return period; they are less than 2 m/s greater in the BS and northern and western YS and are between 2 and 4 m/s greater over the southern YS and southeastern part of the BYS.

However, extreme winds may be underestimated by model results. First, an hourly interval would be preferable for estimating extreme return values, but this interval is unavailable due to the initial setup of output frequency. According to Larsen and Mann (2006), the extreme return values estimated from a 3-h frequency output will underestimate the actual values. Second, the physiographic conditions, such as land-sea mask, are better resolved by a 7-km resolution than by a coarse resolution. However, convective parameterization was still used in the simulation. Some extreme convective processes, which are critical sources for extreme winds, cannot be captured, leading to the further underestimation of extreme return values. Third, tropical cyclone is a major source of extreme winds in this area. As shown by Li (2016), although CCLM has better performance in reproducing tropical cyclones than its forcing dataset ERA-Interim both in wind strength and variability, the modeled maximum wind speeds of tropical cyclones are still lower than the observed winds.

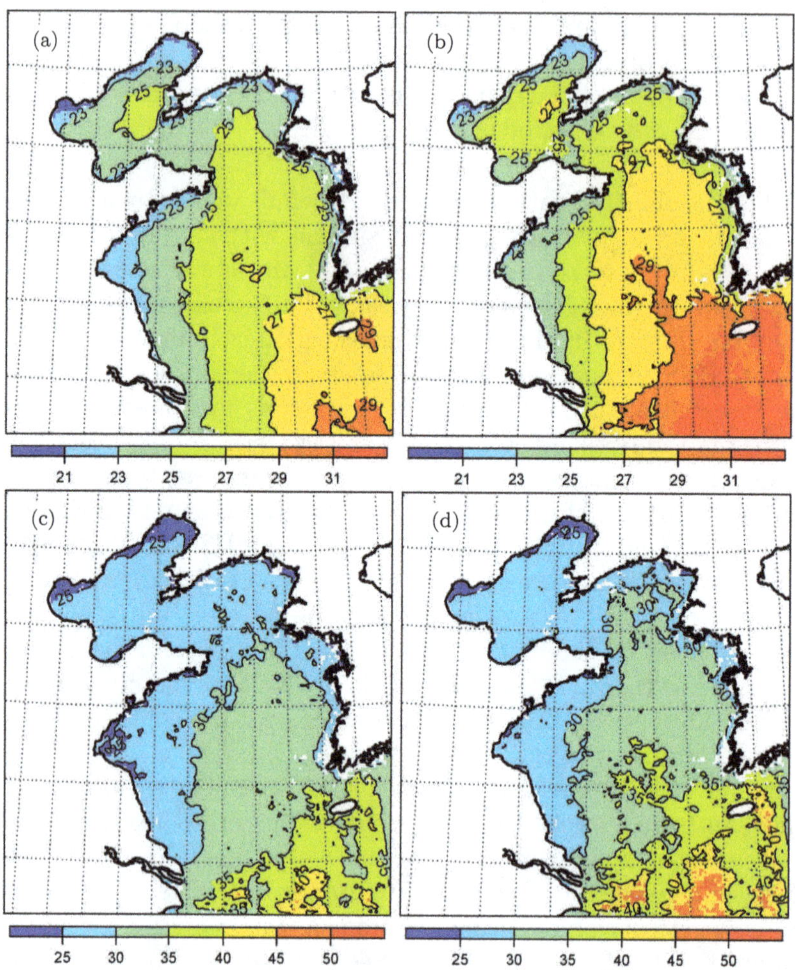

Fig. 11.   Spatial distribution of extreme wind speeds (m/s) with return periods of (a) 5 years, (b) 10 years, (c) 30 years, and (d) 50 years. Note the different colorbars in upper and lower rows.

Though there are various caveats to the assessment of extreme winds in our present work, this is the first investigation of extreme winds over the BYS at hub height and can serve as an important reference for the wind farm applications. Further simulations with high-frequency temporal output intervals and very high resolution are necessary to analyze extreme winds in this area.

5. Conclusions

With increasing energy demand as well as the requirement for the reduction of CO_2 emission in China, the development of renewable energy has grown rapidly in recent years, especially for offshore applications. In this study, we reconstructed a long-term high-resolution wind speed dataset at 100 m over the Bohai Sea and the Yellow Sea using a regional climate model, COSMO-CLM (CCLM), driven by ERA-Interim reanalysis data. This is the first investigation of offshore wind energy at hub height with high resolution for this region. The reconstructed winds were validated against observed data at several stations and proved to be a robust reflection of wind speeds. The climatology, variability, and extreme climate of winds at 100 min the Bohai Sea and the Yellow Sea were investigated based on the dataset. Statistical analysis of wind power in this area and at selected potential wind farm sites was performed. The interannual and decadal variability of wind speed and wind power were varying and were quantitatively assessed.

The main findings and conclusions drawn from this study are as follows:

(1) Statistical metrics reveal that CCLM winds are generally consistent with observations. CCLM winds are reliable for capturing the temporal evolution, variability, and wind distributions of observed winds at 100 m.

(2) The mean wind speeds are 7.0–7.5 m/s over most areas of the BYS, lower than 7.0 m/s in the coastal areas, and greater than 7.5 m/s in the central areas of the BYS. Wind power densities over most areas of the BYS are generally in the range of 300–500 W/m^2. In coastal areas with a water depth less than 25 m, the wind resource is mostly classified as class 2; wind farm applications are appropriate for implementation over northern parts of the BS and the southwestern coasts of the YS for their abundant wind resources (class 3) and shallow water depth (less than 25 m/s).

(3) The long-term standard deviation of annual mean wind speed and wind power over the BYS is in the range of 0.1–0.3 m/s and

10–40 W/m^2, respectively. A lower variability is mainly over the coastal areas, whereas the variability over open sea areas is larger. In terms of interannual variability of seasonal mean wind speed and power, we see a stronger magnitude and spatial variability than for those of the annual mean variables.

(4) The decadal variabilities in mean wind speed, mean wind power and frequency of wind speeds greater than 6.1 m/s are roughly within ±2%, ±5%, and ±2%, respectively, with a strong variability magnitude mainly in the southwestern coastal areas of the YS. The decadal variabilities of monthly mean wind speed and wind power are mostly within ±5% and ±15%.

(5) The spatial patterns of extreme winds with different return periods are generally similar: extreme wind speeds generally increase from the northwest to the southeast; coastal areas are characterized with lower extreme values than those over open sea areas.

As we can see in the present study, short-term assessment of wind energy is not advised as it may generate bias to some extent, especially at the southwestern coasts of the YS. The variations of wind speed tend to be amplified in the wind power density and have substantial influence on wind farm output, and furthermore a financial impact of the variability can be also analyzed economically (Greene *et al.*, 2010) based on the model output; however, this issue has not drawn appropriate concerns from the wind energy industry. An assumption of stationary pattern for wind resource, which is typically proposed by wind farm developers, is also not appropriate, since significant trends or non-stationary features of wind energy tend to occur in some regions. Desirable wind farm sites are generally characterized with high mean wind speed and low variability at different temporal scales (Palutikof *et al.*, 1987). The present study generated a realistic wind energy map at hub-height and quantitatively assessed the interannual and decadal variability of wind power over the BYS, which set the scientific basis and provide a reference for wind farm applications in the coastal areas of the BYS. Additionally, the climatology, variability and trend of wind speed vary from region to region. The present paradigm is also applicable for other regions in the world.

Most studies on wind energy assessments based on model data are deterministic, the uncertainty of those assessments has not been touched on before, which are probably related to the accuracy of modeled wind speed, statistical process of wind power estimation, wind turbine power curve uncertainty and so on. To shed some light on the problem, we used Taylor series approximation to quantitatively assess the uncertainty of wind power density estimation induced by artificial mean wind speed errors. The results are given in the supplementary material. Further investigation on this issue considering more uncertainty factors are encouraged to do in the future.

Additionally, two more issues about our study should be drawn attention to. First, the extreme winds may be underestimated by CCLM, and simulations with convective permitting resolution (generally less than 3 km) and more frequent output are necessary in the future for wind energy assessment. Second, we only assessed the climatology and variability characteristics of wind speed and potential wind power at 100 m; the other factors influencing wind farm applications, such as technique, economics, transportation, and politics, have not been touched on. Further investigation on these issues is required in the future.

Funding

This work was supported by China Scholarship Council (CSC201206330070) and the REKLIM project (http://www.reklim. de/).

Acknowledgments

The following institutions which provided the data are acknowledged: observation data provided by National Climate Data Center (NCDC) and by National Marine Data and Information Service of China (NMDIS), the forcing dataset ERA-I from the European Centre for Medium-Range Weather Forecasts (ECMWF). We also acknowledge the German Climate Computing Center (DKRZ) for providing the computer hardware for our simulation.

References

Amante, C. and Eakins, B. W. (2009). Etopo1 arc-minute global relief model: procedures, data sources and analysis, In: *US Department of Commerce, National Oceanic and Atmospheric Administration, National Environmental Satellite, Data, and Information Service, National Geophysical Data Center, Marine Geology and Geophysics Division.*

Arakawa, A. and Lamb, V. R. (1981). A potential enstrophy and energy conserving scheme for the shallow water equations, *Monthly Weather Review* **109**, 1, pp. 18–36, doi:10.1175/1520-0493(1981)109 <0018:APEAEC>2.0.CO;2.

Beijixing Power News. (2016). China ranks fourth globally in 2015 in offshore wind power installation, *Beijixing Power News.*

Bhattacharya, M., Paramati, S. R., Ozturk, I. and Bhattacharya, S. (2016). The effect of renewable energy consumption on economic growth: Evidence from top 38 countries, *Applied Energy* **162**, pp. 733–741, doi:10.1016/j.apenergy.2015.10.104.

Bloomberg New Energy Finance (2016). China set for a GW-scale offshore wind market in 2016, http://about.bnef.com/landing-pages/china-set-for-a-gwscale-offshore-wind-market-in-2016.

Chang, R., Zhu, R., Badger, M., Hasager, C., Xing, X. and Jiang, Y. (2015). Offshore Wind Resources Assessment from Multiple Satellite Data and WRF Modeling over South China Sea, *Remote Sensing* **7**, 1, pp. 467–487, doi:10.3390/rs70100467.

Chang, R., Zhu, R., Badger, M., Hasager, C., Zhou, R., Ye, D. *et al.* (2014). Applicability of Synthetic Aperture Radar Wind Retrievals on Offshore Wind Resources Assessment in Hangzhou Bay, China, *Energies* **7**, 5, pp. 3339–3354, doi:10.3390/en7053339.

Costa Rocha P. A., de Sousa, R. C., de Andrade, C. F. and da Silva, M. E. V. (2012). Comparison of seven numerical methods for determining Weibull parameters for wind energy generation in the northeast region of Brazil, *Applied Energy* **89**, 1, pp. 395–400, doi:10.1016/j.apenergy.2011.08.003.

Dee, D., Uppala, S., Simmons, A., Berrisford, P., Poli, P., Kobayashi, S., *et al.* (2011). The ERA-interim reanalysis: Configuration and performance of the data assimilation system, *Quarterly Journal of the Royal Meteorological Society* **137**, 656, pp. 553–597, doi:10.1002/qj.828.

Doms, G. and Baldauf, M. (2015). A description of the nonhydrostatic regional cosmo-model part i: Dynamics and numerics, *Deutscher Wetterdienst, Offenbach.*

EEA (2009). Europe's onshore and offshore wind energy potential, Tech. Rep. Technical report No 6, European Environment Agency, European Environment Agency, Office for Official Publications of the European Communities, Copenhagen.

Geyer, B., Weisse, R., Bisling, P. and Winterfeldt, J. (2015). Climatology of North Sea wind energy derived from a model hindcast for 1958–2012, *Journal of Wind Engineering and Industrial Aerodynamics* **147**, pp. 18–29, doi:10.1016/j.jweia.2015.09.005.

Greene, S., Morrissey, M. and Johnson, S. E. (2010). Wind Climatology, Climate Change, and Wind Energy, *Geography Compass* **4**, 11, pp. 1592–1605, doi:10.1111/j.1749-8198.2010.00396.x.

GWEC (2015). Global wind statistics 2014, Tech. rep., Global Wind Energy Council, Brussels, Belgium.

Hong, L. and Möller, B. (2011). Offshore wind energy potential in China: under technical, spatial and economic constraints, *Energy* **36**, pp. 4482–4491. doi: 10.1016/j.energy.2011.03.071.

Howe, M. (2015). China's offshore wind growth surges 487.9%, https://cleantechnica.com/2015/04/01/chinas-offshore-wind-growth-surges-487-9/.

IEA (2012). China wind energy development roadmap 2050, Tech. rep., International Energy Agency, Energy Research Institute, Paris.

IRENA (2014). Renewable energy prospects: China, REmap 2030 analysis, Tech. rep.,: International Renewable Energy Agency, Abu Dhabi.

Jiang, D., Zhuang, D., Huang, Y., Wang, J. and Fu, J. (2013). Evaluating the spatio-temporal variation of China's offshore wind resources based on remotely sensed wind field data, *Renewable & Sustainable Energy Reviews* **24**, pp. 142–148, doi:10.1016/j.rser.2013.03.058.

Katz, R. W., Brush, G. S. and Parlange, M. B. (2005). Statistics of extremes: Modeling ecological disturbances, *Ecology* **86**, 5, pp. 1124–1134, doi:10.1890/04-0606.

Larsen, X. G. and Mann, J. (2006). The effects of disjunct sampling and averaging time on maximum mean wind speeds, *Journal of Wind Engineering and Industrial Aerodynamics* **94**, 8, pp. 581–602, doi: 10.1016/j.jweia.2006.01.020.

Li, D. (2016a). Added value of high-resolution regional climate model: Selected Cases over the Bohai Sea and the Yellow Sea areas, *International Journal of Climatology* **37**, 1, pp. 169–179, doi:10.1002/joc.4695.

Li, D., von Storch, H. and Geyer, B. (2016b). High-resolution wind hindcast over the Bohai Sea and the Yellow Sea in East Asia: Evaluation and wind climatology analysis, *Journal of Geophysical Research: Atmospheres* **121**, 1, pp. 111–129, doi:10.1002/2015JD024177.

Liu, W. T., Tang, W. and Xie, X. (2008). Wind power distribution over the ocean, *Geophysical Research Letters* **35**, 13, doi:10.1029/2008GL034172.

Lu, X., McElroy, M. B., Nielsen, C. P., Chen, X. and Huang, J. (2013). Optimal integration of offshore wind power for a steadier, environmentally friendlier, supply of electricity in China, *Energy Policy* **62**, pp. 131–138, doi:10.1016/j.enpol.2013.05.106.

Monahan, A. H. (2006). The probability distribution of sea surface wind speeds. Part 1: Theory and SeaWinds observations, *Journal of Climate* **19**, 4, pp. 497–520, doi:10.1175/JCLI3640.1.

Oh, K.-Y., Kim, J.-Y., Lee, J.-S. and Ryu, K.-W. (2012). Wind resource assessment around Korean Peninsula for feasibility study on 100 MW class offshore wind farm, *Renewable Energy* **42**, pp. 217–226, doi:10.1016/j.renene.2011.08.012.

Outten, S. D. and Esau, I. (2013). Extreme winds over Europe in the ENSEMBLES regional climate models, *Atmospheric Chemistry and Physics* **13**, 10, pp. 5163–5172, doi:10.5194/acp-13-5163-2013.

Palutikof, J. P., Kelly, P. M., Davies, T. D. and Halliday, J. A. (1987). Impacts of Spatial and Temporal Windspeed Variability on Wind Energy Output, *Journal of Climate and Applied Meteorology* **26**, 9, pp. 1124–1133, doi:10.1175/1520-0450(1987)026<1124:IOSATW>2.0.CO;2.

Pishgar-Komleh, S. H., Keyhani, A. and Sefeedpari, P. (2015). Wind speed and power density analysis based on Weibull and Rayleigh distributions (a case study: Firouzkooh county of Iran), *Renewable & Sustainable Energy Reviews* **42**, pp. 313–322, doi:10.1016/j.rser.2014.10.028.

Qin, H., Liu, M., Wang, Y., Zhao, J. and Zeng, X. Reinvang R., *et al.* (2010). China: An emerging offshore wind development hotspot–with a new assessment of China's offshore wind potential, *WWF, CWEA and Sun-Yet-Sen University, Beijing, China*, p. 64.

Rockel, B., Will, A. and Hense, A. (2008). The regional climate model COSMO-CLM(CCLM), *Meteorologische Zeitschrift* **17**, 4, pp. 347–348, doi:10.1127/0941-2948/2008/0309.

Seguro, J. V. and Lambert, T. W. (2000). Modern estimation of the parameters of the Weibull wind speed distribution for wind energy analysis, *Journal of Wind Engineering and Industrial Aerodynamics* **85**, 1, pp. 75–84, doi:10.1016/S0167-6105(99)00122-1.

Van de Vyver, H. and Delcloo, A. W. (2011). Stable estimations for extreme wind speeds. An application to Belgium, *Theoretical and Applied Climatology* **105**, 3–4, pp. 417–429, doi:10.1007/s00704-010-0365-9.

von Storch, H., Langenberg, H. and Feser, F. A. (2000). A spectral nudging technique for dynamical downscaling purposes, *Monthly Weather Review* **128**, 10, pp. 3664–3673, doi:10.1175/1520-0493(2000)128 <3664:ASNTFD>2.0.CO;2.

Wang, J., Qin, S., Jin, S. and Wu, J. (2015). Estimation methods review and analysis of offshore extreme wind speeds and wind energy resources, *Renewable & Sustainable Energy Reviews* **42**, pp. 26–42, doi:10.1016/j.rser.2014.09.042.

Wang, G., Gao, S., Wu, B. and Xie, Y. (2014). Distribution features of wind energy resources in the offshore areas of China, *Advances in Marine Science* **32**, 1, pp. 21–29.

Weigt, H. (2009). Germany's wind energy: The potential for fossil capacity replacement and cost saving, *Applied Energy* **86**, 10, pp. 1857–1863, doi:10.1016/j.apenergy.2008.11.031.

Wiser, R., Yang, Z., Hand, M., Hohmeyer, O., Infield, D., Jensen P. H., *et al.* (2011). Wind energy. *IPCC Special Report on Renewable Energy Sources and Climate Change Mitigation*, 2011, pp. 535–608.

Wyatt, S., Govindji, A.-K., James, R., Duan, N., Xie, B., Liu, M. *et al.* (2014). Detailed appraisal of the offshore wind industry in China, *Carbon Trust*, May 2014.

Chapter 10

Low-Level Jets Over the Bohai Sea and Yellow Sea: Climatology, Variability and the Relationship with Regional Atmospheric Circulations[*]

D. Li[†,‡], H. von Storch[§,¶,‖], B. Yin[†,‡,], Z. Xu[†,‡], J. Qi[†,‡], W. Wei[††], and D. Guo[‡‡]**

[†]*Key Laboratory of Ocean Circulation and Waves,
Institute of Oceanology, Chinese Academy of Sciences,
Qingdao, China*
[‡]*Function Laboratory for Ocean Dynamics and Climate,
Qingdao National Laboratory for Marine Science and Technology,
Qingdao, China*
[§]*Institute of Coastal Research, Helmholtz-Zentrum Geesthacht,
Geesthacht, Germany*
[¶]*CliSAP, University of Hamburg, Hamburg, Germany*
[‖]*College of Oceanic and Atmospheric Sciences,
Ocean University of China, Qingdao, China*
[**]*University of Chinese Academy of Sciences, Beijing, China*

---

[*]This chapter was originally published in *Journal of Geophysical Research: Atmosphere*, 123, 5240–5260, doi:10.1029/2017JD027949. Copyright (2018), with permission from John Wiley and Sons.

[††]*State Key Laboratory of Severe Weather,*
Chinese Academy of Meteorological Sciences, Beijing, China
[‡‡]*North China Sea Marine Forecasting Center of State Oceanic*
Administration, Qingdao, China

The present study reveals climate features of low-level jets (LLJs) over the Bohai Sea and Yellow Sea (BYS) based on a 35-year (1979–2013) high-resolution (7 km) atmospheric hindcast. The regional climate model COSMO-CLM driven by the ERA-Interim reanalysis data set was used to obtain the hindcast. Through comparison with observations, the hindcast was proved to robustly reproduce the climatology, the diurnal cycle, the variability of wind profiles, and specific LLJ cases. LLJs over the BYS feature a strong diurnal cycle, intra-annual, and interannual variability but weak decadal variability. LLJs are more frequent in April, May, and June (LLJ season) and less frequent in winter over the Bohai Sea and western coastal areas of the Yellow Sea, which is due to the intra-annual variations of large-scale circulation and local land-sea thermal contrast. In the LLJ season, the heights of jet cores are generally lower than 500 m above sea level. The maximum wind speed of LLJs is mostly in the range of 10–16 m/s, and prevailing wind directions are southerly and southwesterly. The LLJs are of the nocturnal type, with the highest occurrence frequency at approximately 2300 local time. Furthermore, a low-frequency link between anomalies of LLJ occurrence and regional large-scale barotropic circulation was identified using canonical correlation analysis and associated correlation patterns. Pressure systems over the East Asia/northwest Pacific region are significantly correlated with the variations of LLJ occurrence over the BYS in terms of the intra-annual and interannual variability.

1.   Introduction

Low-level jets (LLJs) are mesoscale-flow phenomena with horizontal wind maxima within the lowest few kilometers of the troposphere (e.g., Stensrud, 1996). They are strongly linked to the deep convective activity and mesoscale convective complexes (Maddox, 1983; Means, 1954). LLJs affect transport and mixing processes in the atmospheric boundary layer by modifying the vertical wind shear and the turbulence structure, thus conditioning the formation of convective fog, clouds, and heavy rainfall events (Chen *et al.*, 2005; Muñoz & Enfield, 2011; Nuss *et al.*, 2000).

Over water regions, ocean dynamics, and the air-sea coupling processes are impacted by LLJs (Beardsley *et al.*, 1987). The coastal-parallel winds of offshore LLJs enhance the upwelling of deeper and cold waters near coasts, which results in decreases of sea surface temperature and ocean surface evaporation. This contributes to the aridity and dryness of some coastal regions such as the Peruvian coastal desert strip (Nicholson, 2010; Warner, 2004). LLJs are also significant for human activities, such as aviation safety, offshore wind energy applications, sound propagation, fishery resources, and the transport of pollutants (Arfeuille *et al.*, 2015; Nunalee & Basu, 2013).

There have been extensive studies on LLJs over regions worldwide including, but not limited to, North America (Higgins *et al.*, 1997), Europe (Soares *et al.*, 2017), South America (Marengo *et al.*, 2004), and Asia (Du *et al.*, 2014). Over the land, the most renowned LLJs that have been intensively studied are the Great Plains LLJs over North America (Blackadar, 1957; Higgins *et al.*, 1997; Parish & Oolman, 2010), which are most frequent during the warm seasons from April to October and are greatly influenced by the sloping terrain of the Rocky Mountains. They are highly ageostrophic, with wind speed maxima reached shortly after midnight.

Over water regions, typical LLJs are those found along coastal regions (Doyle & Warner, 1993). Coastal LLJs are frequently linked with large-scale atmospheric circulation, land-sea thermal contrast, and coastal terrain (Parish, 2000). They are synoptically driven but mesoscale intensified: the high-pressure system over the land and low-pressure inland are the synoptic forces of coastal-parallel flows. The local wind intensification occurs due to the sharp thermal and associated pressure gradients, with strong baroclinic structures (Burk & Thompson, 1996). Furthermore, the interaction with topography may enhance coastal LLJ winds or cause them to change in direction, when high mountain ranges exist along the coast (Chao, 1985; Jiang *et al.*, 2010). Coastal LLJs generally feature a diurnal cycle, with wind speed maximum at midafternoon (Ranjha *et al.*, 2013). The wind maxima are generally at low altitudes and are constrained by the marine atmospheric boundary layer capping the temperature inversion (Rijo *et al.*, 2018).

Ranjha *et al.* (2013) constructed a global distribution map of coastal LLJs based on ERA-Interim reanalysis for summer and winter seasons and found that they are essentially a summer phenomenon. Except for those along the southeast Arabian Peninsula (Ranjha *et al.*, 2015) and New York Bight Jet (Colle & Novak, 2010), coastal LLJs are mainly distributed along the eastern boundary current regions, including the west coasts of North America, South America, the Iberian Peninsula, and northwestern and southern Africa. Coastal LLJs over these regions have been extensively studied based on field observations (Rahn & Parish, 2007; Winant *et al.*, 1988) and/or on model and theoretical efforts (Burk & Thompson, 1996; Cardoso *et al.*, 2016; Rijo *et al.*, 2018; Soares *et al.*, 2017, 2014). However, not all LLJs in the coastal regions are the typical coastal LLJs, as defined by Ranjha *et al.* (2013). Over the Caribbean region, the wind speed exhibits a distinct jet-like profile, while the temperature shows a decreasing profile vertically. The strong meridional surface temperature gradients are thought to force Caribbean LLJs (Cook & Vizy, 2010), although this was still under debate until recently (Maldonado *et al.*, 2017).

Over the Chinese coastal areas, Wei *et al.* (2014) investigated the features and evolutions of LLJs at two sites (Tianjin and Shanghai) along the northern coast of China using wind profile radar data sets in summer. They found that nocturnal LLJs overwhelmed daytime LLJs in both strength and frequency, and distinct LLJ wind directions and heights were observed due to the different local topography and synoptic forcing at the two sites. Based on high-resolution (9 km) model data, Du *et al.* (2015) found strong LLJs off the southeastern coast of China, with jet cores at the 925-hPa level. The generation was subject to a large-scale setting enhanced by land-sea thermal contrast and coastal orographic effects. Unlike typical coastal LLJs, LLJs off the southeastern coast of China feature nocturnal wind maximum instead of midafternoon wind maximum, and the LLJ wind maxima do not reside within the sloping temperature inversion layer.

Thus, while many studies on the physics or climate of LLJs worldwide have been performed, limited studies (Du *et al.*, 2015) have examined the mechanisms behind the formation and life cycle of LLJs in Chinese coastal areas. Studies on the climatology, including the variability on diurnal and intra-annual and interannual time scales, and on decadal trends have rarely been performed for these LLJs. With our study, we present a climatology and variability of LLJs over the Bohai Sea and Yellow Sea (BYS) regions, spatially and temporally. Furthermore, the relationship of LLJs with regional large-scale forcing in low-frequency was also studied. A 35-year-long high-resolution (7 km) simulation of the regional climate model COSMO-CLM (CCLM, Rockel *et al.*, 2008) was used.

The present study is organized as follows: Section 2 describes the used data sets, including the forcing data set, observations, and CCLM data set, as well as the identification criteria of LLJs. Section 3 discusses the model evaluation using the sounding data and wind profile data from different stations. Section 4 presents the climatology and annual cycle variability of LLJs, as well as the diagnosed mechanisms for monthly variations of the LLJs. In Section 5, we describe the variability and large-scale conditioning during LLJ season. A summary and conclusions are given in the final Section 6.

2. Data and Methodology

2.1. *Regional Climate Model Simulation*

The nonhydrostatic regional atmospheric model CCLM version 4.14 was used to construct the atmospheric conditions over the BYS (see Figure 1) from 1979 to 2013. The model was developed from the Local Model of the Deutscher Wetterdienst (German Weather Service) and is now widely used in mesoscale climate studies with spatial grid resolutions in the range of 1–50 km.

The constraining conditions for CCLM used in the study were obtained from the global atmospheric reanalysis ERA-Interim (ERAI, Dee *et al.*, 2011). ERAI is produced by the European Centre for Medium-Range Weather Forecasts. It is available from 1 January 1979 to the present and is supposed to continue until the end of 2018. The horizontal resolution of ERAI is approximately 80 km (T255 spectral), and the temporal output interval is 6 h. It has shown better quality in producing low-frequency variability and stratospheric circulation than its predecessor ERA-40 (Dee & National Center for Atmospheric Research Staff, 2018).

CCLM adopts a rotated geographical coordinate with an Arakawa C-grid structure. A generalized terrain-following height coordinate system was used to keep the lowest surface of the constant vertical coordinate conformal to the orography. The horizontal resolution of

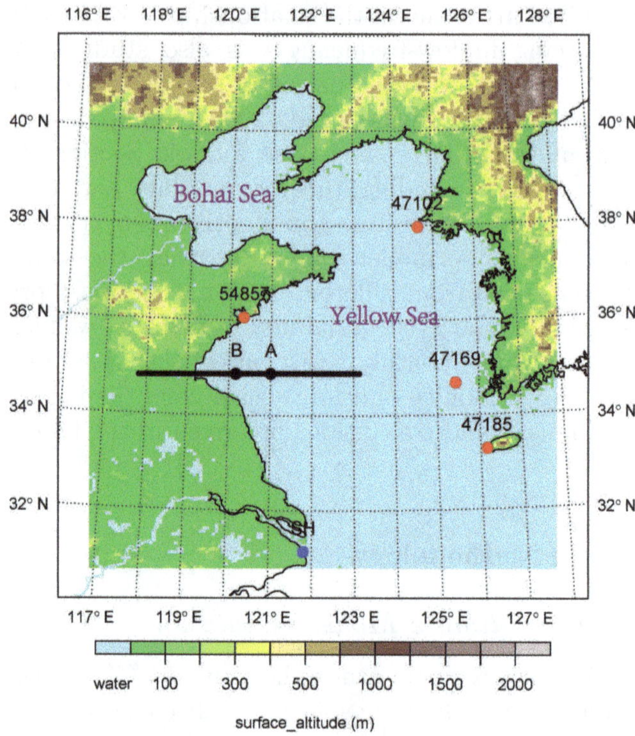

Fig. 1.   Orography of the simulation domain. Red points indicate four radiosonde observations (47102: 124.63°E, 37.97°N; 47169: 125.45°E, 34.68°N; 47185: 126.16°E, 33.28°N; 54857: 120.33°E, 36.06°N); blue point (121.81°E, 31.14°N) represents location of wind profiler radar observation. The black line marks the cross section, point A (121.00°E, 34.88°N) and point B (120.12°E, 34.87°N) are for low-level jet cases study. The white frame indicates the sponge zone.

our simulation was 0.0625° (~7 km). Ranjha *et al.* (2016) evaluated the impact of varying resolutions (from 54 to 2 km) on the model's ability to resolve features of a coastal LLJ over California coasts and concluded that 6 km is a compromise resolution that reproduces most of the features of a coastal jet. Together with another study (Du *et al.*, 2015), this indicates that such a 7-km grid resolution is reasonable for simulating the features of LLJs over the BYS.

Ten grid boxes were set as the sponge zone in the lateral boundary at each side. Forty layers were used in the vertical direction, with higher resolution in the lower troposphere. The temporal output interval for winds in vertical levels was 3 h. When identifying LLJs, we used wind data at the lowest 18 levels, that is, 10, 34.5, 69, 116, 178.5, 258.5, 357.5, 477, 618.5, 782.5, 970, 1,182.5, 1,420, 1,682.5, 1,927.5, 2,290, 2,635, and 3,007.5 m.

The simulation time step was 60 s. An interior spectral nudging technique (von Storch *et al.*, 2000) was used in the simulation every third time step on the horizontal U and V wind components with levels above 850 hPa. This aimed to keep the simulated large-scale pattern consistent with that of ERAI and to develop the local and regional physical processes on their own. The Tiedtke convective parameterization scheme (Tiedtke, 1989) was used for cumulus convection. The multilayer soil and vegetation model TERRA-ML scheme (Schrodin & Heise, 2002) was used for land surface processes. A prognostic TKE-based scheme (Mellor & Yamada, 1982) was used for vertical turbulent transport.

2.2. *Observations*

Due to the unavailability of wind observations at upper-air levels in the BYS, four radiosonde observations in the near coastal area (red points in Figure 1; 47102 [1 January 2001 to 31 December 2013], 47169 [1 January 2004 to 31 December 2013], 47185 [1 January 1997 to 31 December 2013], and 54857 [1 January 1997 to 31 December 2013]) were obtained from the atmospheric soundings data set archive of the University of Wyoming (http://weather.uwyo.edu/upperair/sounding.html) to validate the model data set. The radiosonde observations are available twice daily, at 0000 and 1200 UTC. Multiple variables, such as temperature, wind speed, wind direction, and dew point temperature, at different pressure levels (and corresponding height levels) are included in the data sets. Furthermore, more observation data were obtained from a boundary layer wind profile radar, which was operated in the Yangtze River Delta of China (blue point in Figure 1). The data are of high frequency and high vertical resolution (50 m); they have been used to study the features and evolution of LLJs along the Chinese coast (Wei *et al.*, 2013). The quality of CCLM in representing the diurnal cycle and specific LLJ cases will be assessed based on wind profile radar observations.

2.3. *Identification Criteria of LLJs*

Previous studies identified LLJs based on various criteria (Bonner, 1968; Ranjha *et al.*, 2013; Tao & Chen, 1987). Some researchers identified LLJs using the horizontal wind maxima at the 1,000-, 925-, 850-, or 700-hPa levels without requiring a vertical shear threshold of the horizontal winds (Tao & Chen, 1987; Wang *et al.*, 2013; Whyte *et al.*, 2007). Ranjha *et al.* (2013) developed an algorithm to identify the typical coastal LLJ globally based on the vertical profiles of wind speed and temperature, requiring that the wind speed maximum within a temperature inversion of the marine atmospheric boundary layer. This algorithm has been used widely in climatological studies of regional coastal LLJs (Ranjha *et al.*, 2015; Rijo *et al.*, 2018; Semedo *et al.*, 2016; Soares *et al.*, 2014). However, this method will rule out LLJs at the top of the temperature inversion layer (which has been improved by Lima *et al.*, 2018) and may rule out those are not locally generated but remotely propagating.

In the present study, we adopted the criteria defined by Bonner (1968), in which the thresholds of three parameters are defined as certain values, including the maximum wind speeds, height of maximum winds, and magnitude of vertical shear above the jets. This basic detection method defines the LLJs by examining the horizontal wind maximum vertically, without considering the associated generation mechanism. These criteria were widely used or adopted in later literature (Doubler *et al.*, 2015; Du *et al.*, 2014; Miao *et al.*, 2018; Pham *et al.*, 2008; Wei *et al.*, 2014; Whiteman *et al.*, 1997; Wu & Raman, 1998). The threshold values vary due to the strength, distribution, and background circulation of LLJs.

The following thresholds were used to identify a LLJ in a vertical column: (1) the maximum wind speed is greater than 10 m/s in the lowest 18 layers (below ~3 km); (2) the difference between the wind maximum and minimum above or the wind speed at ~3 km is greater than 5 m/s; and (3) the wind maximum does not occur at the surface (the lowest model level at 10 m). The algorithm was applied to vertical profiles of wind speeds at all model grids over the BYS every 3 h from 1979 to 2013. When a LLJ was identified, the jet location, jet height, jet speed, and direction were recorded. LLJ height is the height of the horizontal wind speed maximum, and LLJ direction refers to the wind direction at the LLJ height.

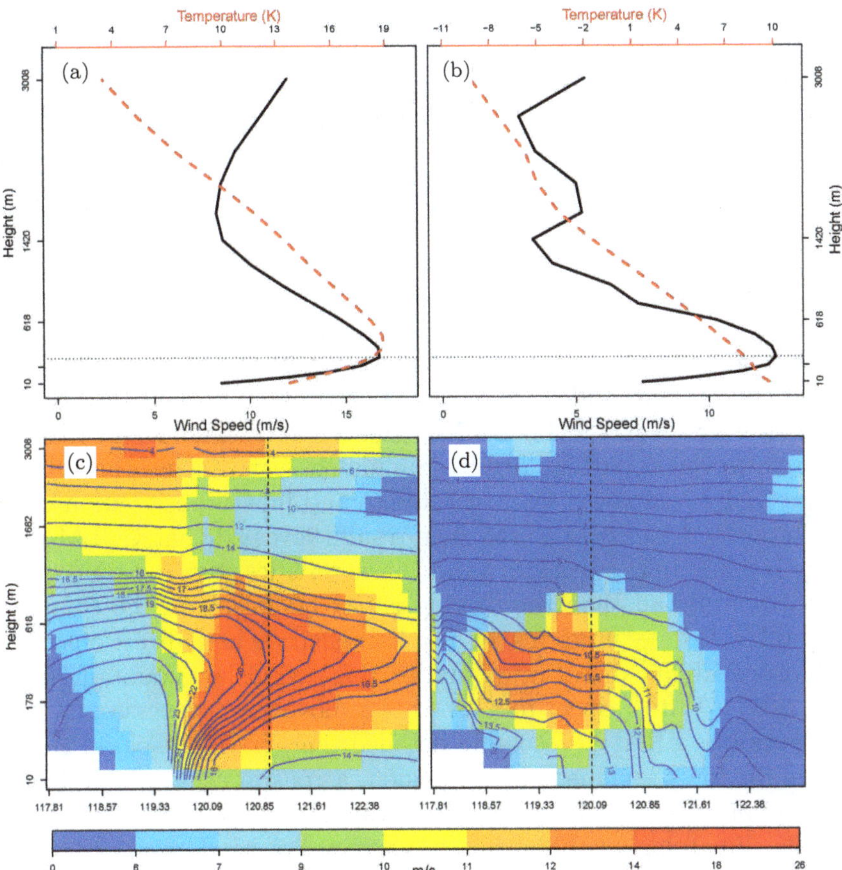

Fig. 2. The vertical profiles of wind speed and temperature for (a) point A, (b) point B, and (c and d) the black cross section in Figure 1 at 14:00 LST on 3 April 2006 (left panel) and at 20:00 LST on 17 April 2007 (right panel), respectively. The horizontal dot lines in (a) and (b) are heights of wind maxima, black lines for wind speeds, and red lines for temperature. The dashed line in (c) and (d) are profile locations of point A and point B, respectively; contours are for temperature.

Figure 2 shows the vertical profiles of two LLJ cases detected using the algorithm. The wind speed maxima are approximately 15 m/s at low levels for both cases; however, case 1 (Figure 2(c)) features a sloping temperature inversion layer, with a maximum horizontal temperature gradient. The wind speed maximum resides within the sloping layer (Figure 2(a) and (c)), and it resembles the structure of typical coastal LLJ such as Oman coastal jets (Ranjha *et al.*, 2015). Regarding case 2, the wind speed exhibits a distinct jet-like profile, since the temperature is decreasing with the height (Figure 2(b)). We can also observe a strong land-sea thermal contrast near the coasts, while there is no pronounced temperature inversion layer associated with this LLJ case (Figure 2(d)).

3.   Evaluation of the Model Data Set

The model output has been applied to investigate present surface wind climate and added value to the description of winds by downscaling over the BYS (Li, 2017; Li, Geyer, & Bisling, 2016; Li, von Storch, & Geyer, 2016a, 2016b). Simulated surface winds have been assessed by comparing against satellite and in situ observation data both on land and over water. The results revealed that CCLM reliably represents the regional wind characteristics over the BYS area, with more detail than the driving ERAI reanalysis in the complex coastal areas — in terms of wind intensities and directions, the wind probability distribution and extreme winds at mountain areas (Li, 2017). With respect to mesoscale atmospheric processes, CCLM outperforms ERAI in resolving detailed temporal and spatial structures for the phenomena of a typhoon, a coastal atmospheric front, and a vortex street. The data set has also been applied to the study of the climatology, variability, and extremes of wind energy over the BYS (Li, Geyer, & Bisling, 2016). In the following, we sought to assess how about the quality of CCLM in reproducing vertical profiles of the wind.

Four radiosonde observations (red points in Figure 1) were used to validate the reliability of CCLM in reproducing the climatology of wind profiles and its added value to ERAI. We interpolated CCLM and ERAI grid data to the radiosonde observation locations using the nearest-neighbor method and obtained temporal averaged for wind profiles. Figure 3 shows that the simulated climatology of the vertical profiles of wind speeds is generally in agreement with that of the observation data, except at station 54857, where the simulated wind speeds largely underestimated the observations at levels above 1,500 m. At station 54857, the observed wind generally shifts from southeasterly to northwesterly from the bottom level to the

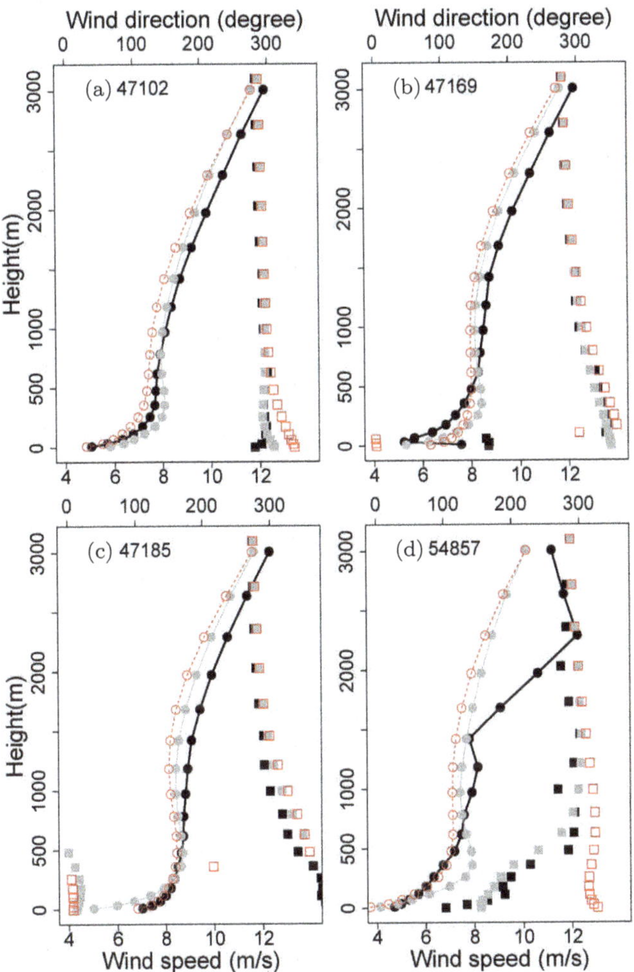

Fig. 3.   Climatological mean of vertical wind speed and wind direction of observation data (black dot line for wind speed and black squares for wind direction), CCLM data (gray dot line for wind speed and gray squares for wind direction) and ERA-Interim data (red dashed line for wind speed and red squares for wind direction).

upper level below 3,000 m, whereas at the other stations, the wind directions are generally around 300°, which was reproduced by the CCLM data set. Additionally, although CCLM does not add value to the description by ERAI in capturing wind intensities, it outperforms ERAI in capturing observed wind direction at levels below 1,000 m.

The observed strong wind at a high level for station 54857 may have been due to a wind intensification caused by the local topography or some local-scale phenomena, which were not resolved by either the coarse-resolution ERAI analysis or the CCLM analysis.

In addition to climatological features, the annual cycle was compared, showing generally consistent wind patterns and intensities between the CCLM data set and radiosonde observations at the different levels of 925, 850, and 500 hPa at the four stations (not shown here). Furthermore, based on the empirical orthogonal function (EOF) analysis method, the temporal evolution of the dominant patterns of wind speed (after subtraction of the annual cycle) has been compared between two data sets. The first two EOF modes and corresponding time series of principal coefficients, using station 47102 as an example, are given in Table 1 and Figure 4, respectively. The first EOF mode (Table 1) is dominant and explains almost 73% of the total variance; it shows that wind speed anomalies fluctuated in phase but that their intensities increased with height. The total variance explained by the second EOF mode is more than 22%, and this mode is out of phase between low (925 and 850 hPa) and upper (500 hPa) levels. All EOF patterns and temporal evolution of modeled wind speeds at different levels are consistent with those observed, which was also revealed by the results at the other three stations (data not shown), verifying the robustness of the CCLM wind data set.

Furthermore, the simulated daily climatology of wind profilers and LLJ cases at station SH (blue point in Figure 1) was compared with wind profiler radar observations for the year 2009 (Figure 5). For observations, the wind speeds are more uniform during the day than at night because of the stronger turbulent mixing in the boundary

Table 1. The first two EOFs for Radiosonde Observations (OBS) and simulation data (CCLM) at pressures levels of 925, 850, and 500 hPa at station 47102 from 2005 to 2008.

EOFs	925 hPa		850 hPa		500 hPa		Var_percentage (%)	
	OBS	CCLM	OBS	CCLM	OBS	CCLM	OBS	CCLM
EOF1	0.26	0.27	0.29	0.31	0.92	0.91	73.6	72.7
EOF2	0.68	0.70	0.62	0.59	−0.39	−0.41	22.4	24.1

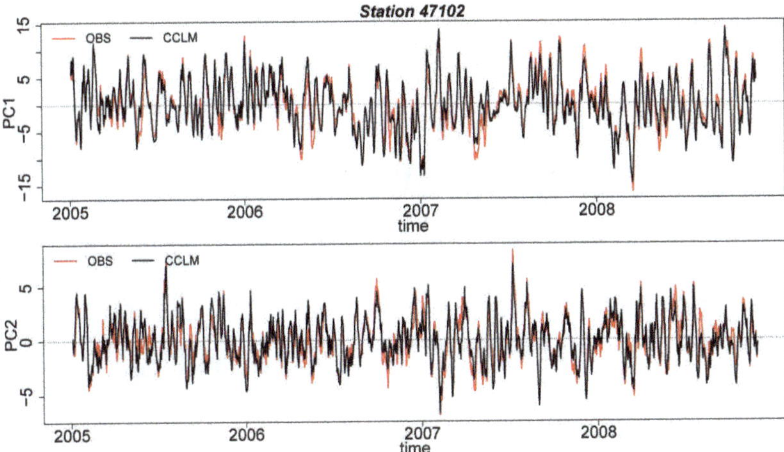

Fig. 4. Time series of the principal coefficients of the first two empirical orthogonal function modes for radiosonde observations (OBS, red line) and simulation data (CCLM, black line) at pressures levels of 925, 850, and 500 hPa, with annual cycle subtracted at station 47102 in the period 2005–2008.

layer during the day (Figure 5(a)). From the late afternoon to early morning the next day, there is a layer with larger wind speeds in the range of 300–700 m. Above 2,000 m, strong wind speeds greater than 9 m/s are persistent in the daily cycle, and the wind speed increases with height. The spatial-temporal structure of our simulated wind profile (Figure 5(b)) is consistent with the observations, with underestimations of 1–2 m/s in wind speed intensities. Strong temporal variability (data not shown) lasts for the entire diurnal cycle and generally increases from 2 m/s in the surface to more than 8 m/s above 5,000 m. The patterns are also consistent between the modeled and observed results, with some underestimation in strength by our model.

A modeled LLJ process (Figure 5(d)) from 2300 local solar time (LST) on 8 May 2009 to 0000 LST on 11 May 2009 shows a typical jet that occurred at a height of 100 to 700 m in the late afternoon and persisted until the next early morning, which is generally consistent with the diurnal cycle climatology in Figure 5(a) and (b). The strength of the LLJ from 2000 LST on 9 May to 0500 LST on 10 May 2009 is stronger than that of its predecessor and successor. The simulated onset and developing features of LLJ are generally consistent

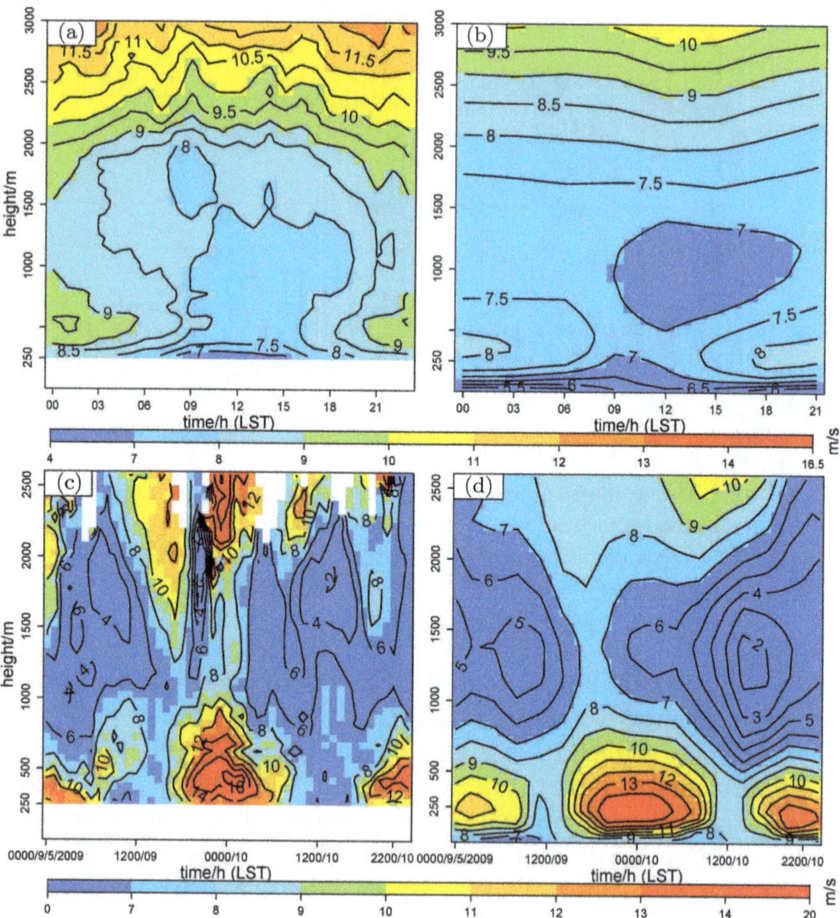

Fig. 5. (a) Observed and (b) modeled daily height-time cross section of mean wind speeds during 2009; (c) observed and (d) modeled low-level jet cases: Height-time cross section of wind speeds during 0000 LST on 9 May 2009 to 0000 LST on 11 May 2009 at station SH (blue point in Figure 1). (a) and (c) were reproduced from Wei *et al.* (2013).

with the observations (Figure 5(c)), while the simulated wind intensities of the LLJ underestimate the observations. Details of the wind structure patterns in the upper 2,000 m could not be well resolved by CCLM, which may be due to the coarse temporal resolution or some physical processes unresolved by the CCLM model.

In summary, based on the radiosonde observations and wind profiler radar observations, our model is robust in reproducing the climatology of wind profiles, the diurnal cycle, and the variability of wind speeds as well as LLJ cases.

4. Climatology and Annual Cycle of LLJs

As shown in Section 2.3, the identification criteria were applied to the high-resolution hindcast data set from 1979 to 2013. The statistical analyses in the following sections are based on the identified LLJ information data set over the BYS region.

4.1. *Intra-Annual Climatology and Variability*

The annual mean frequency of LLJ occurrence (Figure 6(a)) is in the range of 8%–20% over the BYS, being more frequent over the Bohai Sea (BS) and western coastal areas of the Yellow Sea (YS; >14%) than over the coastal areas of the Korean Peninsula. The mean wind speeds of LLJs (Figure 6(b)) are stronger over the BS and the north YS (>15 m/s) than those in the south YS and two bays of the BS (13–15 m/s). The LLJ mean heights are generally lower than 500 m in the BS and northern and northwestern YS and are mostly in the range of 500–600 m in the southeastern part of our study domain, except for in the areas around Jeju Island.

Figure 7 shows a strong intra-annual variability of LLJ occurrence over the BYS. In winter, the LLJ occurrence is very low, with values mostly <12% from December to February. In contrast, the frequency

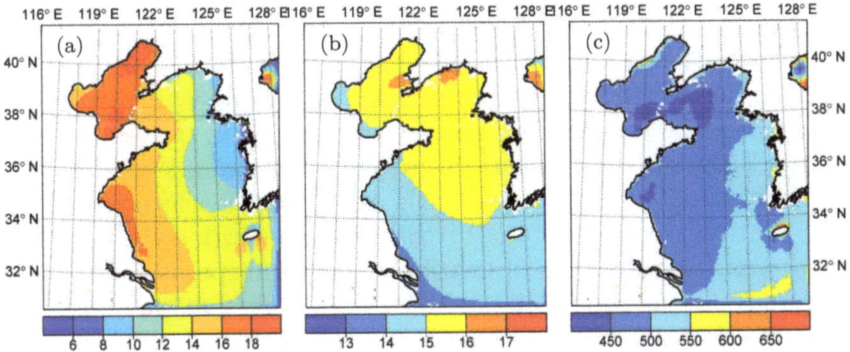

Fig. 6. (a) Annual mean frequency of occurrence (%) of low-level jet (LLJ), (b) LLJ mean wind speed (m/s), and (c) LLJ mean occurrence height (m).

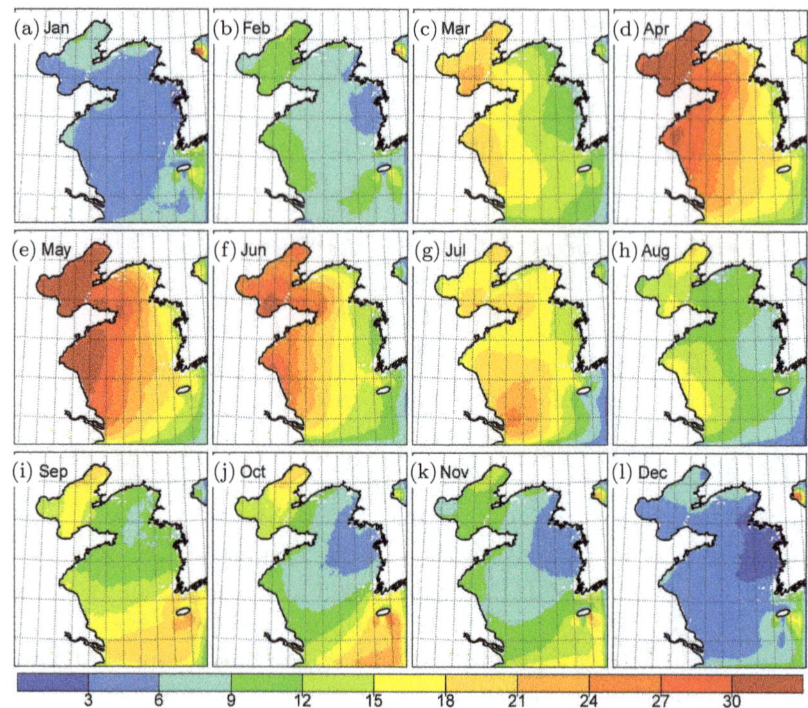

Fig. 7. Spatial distributions of monthly occurrence frequency (%) of low-level jet.

is generally greater than 21% in April, May, and June; it is even greater than 30% in April and May over the BS and part of the western YS. March, July, and August are transition periods, when it is mostly in the range of 9–24%.

The spatial patterns of LLJ generation from March to August are similar but with different intensities, and larger values are distributed in the western part of the BYS and lower values in the eastern part of the BYS. The spatial patterns of the other months are different. From September to February, LLJs are more frequent over the BS and south or southeast of the YS than over the north and middle YS. The spatial distributions of monthly mean wind speeds of LLJs (see Figure S1 in the supporting information)[1] show that the intensities over the BS are stronger from November to May (>16 m/s) and

---

[1] Refer to https://agupubs.onlinelibrary.wiley.com/action/downloadSupplement?doi=10.1029%2F2017JD027949&file=jgrd54652-sup-0001-2017JD027949-SI.pdf.

weaker in the other months, especially in July and August (<14 m/s). For the YS areas, we observe stronger LLJs from February to May, mainly in the north or middle YS areas (>16 m/s). Weaker LLJs from June to January are mainly distributed over the west YS coasts or the south YS areas (<14 m/s). In terms of the monthly mean height of LLJs (see Figure S2[2], they are mostly in the range of 450–700 m from August to December and mostly within 400–550 m from January to March. The LLJs are higher in the south YS than the north YS or the BS. The LLJ cores are generally located in the range of 400–650 m from April to June, with larger values in the south and southeast YS and west coasts of the Korean Peninsula (>500 m). LLJ cores in July are the lowest among all the months, with values mostly lower than 500 m.

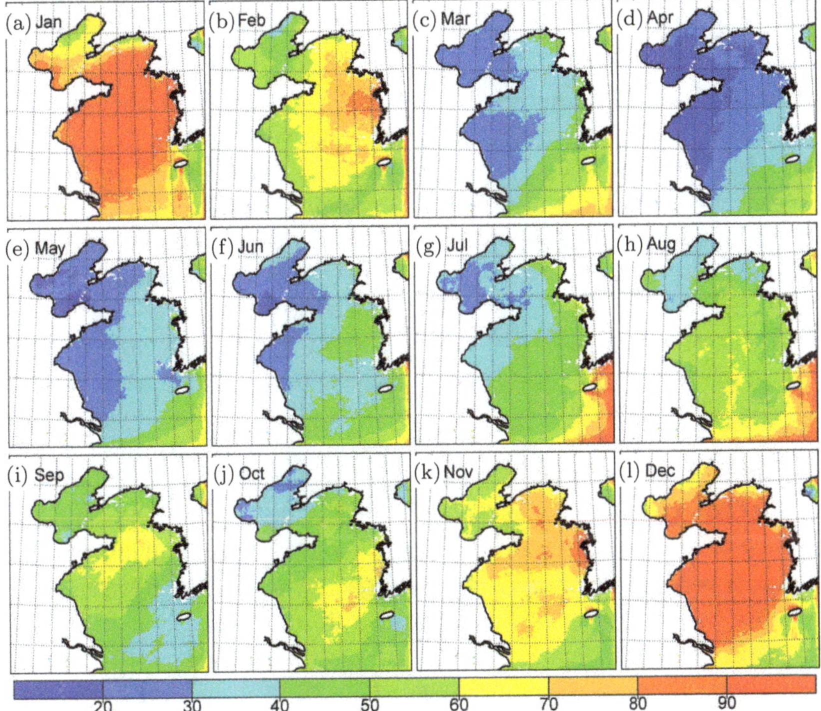

Fig. 8. Spatial distributions of relative standard deviation of low-level jet monthly frequency occurrence in percentage (%).

[2]Refer to https://agupubs.onlinelibrary.wiley.com/action/downloadSupplement?doi=10.1029%2F2017JD027949&file=jgrd54652-sup-0001-2017JD027949- SI.pdf.

However, the spatial distributions of monthly relative standard deviation (i.e., the standard deviation divided by the mean, in percentage) of LLJ occurrence (Figure 8) show different patterns from the monthly occurrence frequency of LLJs (Figure 7). They reveal that the relative interannual variability of monthly generation is generally greater than 20% in most areas and features strong variability within different months. From March to July, the relative standard deviations are mostly less than 50%, with values increasing from the northwest to southeast. In the other months, there is strong interannual variability especially in January and December, when the relative standard deviation reaches >90%. Therefore, strong interannual variability exists for LLJs over the BYS, especially during months when the LLJ occurrence is less frequent.

4.2. *Diagnosis of the Intra-Annual Variations of the Simulated LLJs*

The simulated LLJs exhibit pronounced intra-annual variability, in particular, a clear temporal mismatch between the monthly annual cycle of LLJ frequencies and wind speeds, as shown in the last section. We selected June (with high LLJ occurrence and medium LLJ wind speed) and December (with very low LLJ occurrence frequency and medium LLJ wind speed) to determine the possible reasons.

Figure 9(a) and (b) show the monthly averaged fields of the mean sea level pressure (MSLP) and the wind speed at a 404-m height from ERAI reanalysis for June and December, respectively. In June, there is a high-pressure system over the northwest Pacific Ocean and a low-pressure system over the Asian continent. Winds over the BYS at a 404-m height are predominantly southerly and southeasterly. In winter, the pressure system and wind direction reverse, with a high-pressure system over the Asian continent and a low-pressure system over the Pacific Ocean; winds are generally northerly and northwesterly. The geostrophic winds due to the large-scale pressure gradient preconditions the wind intensity associated with the LLJs over the BYS, which is consistent with Du *et al.* (2014), who found that geostrophic winds dominate actual winds near LLJ cores off the Chinese southeastern coasts. The seasonal variation of large-scale atmospheric circulations related to East Asian monsoon is thought to greatly influence the monthly variability of LLJ wind intensity over the BYS.

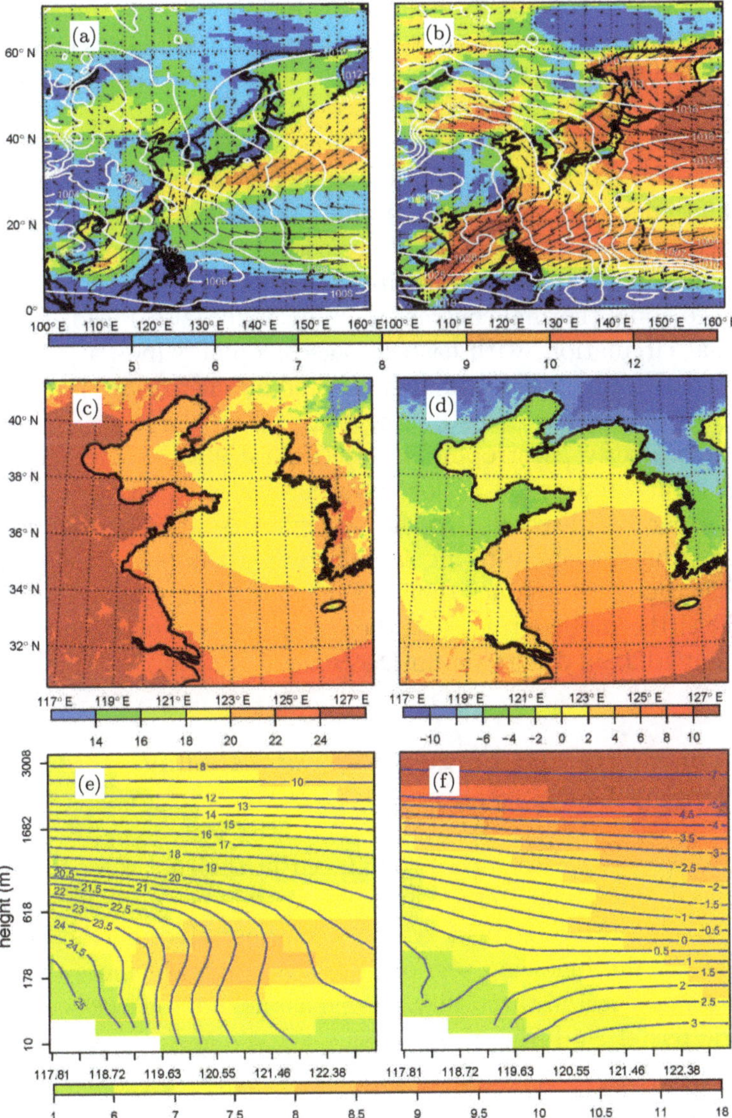

Fig. 9. Monthly climatological mean (1979–2013) for (a and b) wind speed (shading), wind vector (arrows) at 404-m height and sea level pressure (white contours) of ERA-Interim reanalysis data set; (c and d) 2-m temperature (shading and red contours) of climate model COSMO-CLM data set, (e and f) wind speed (shading) and temperature (contours) of climate model COSMO-CLM data set at black cross section in Figure 1. Left panels are for June, while right panels for December.

However, the combination of strong low-level winds and significant vertical shear of horizontal winds defines the LLJs over the BYS. The later factor contributes to the increased jet-like structure in June compared to December.

In addition to the differences in large-scale circulation, Figure 9(c) and (d) show that there are great differences in the local land-sea thermal contrast between June and December over the BYS. The temperature contours are generally coastline parallel in June, while they are mainly zonal parallel in December. The presence of a pronounced zonal thermal contrast in June (Figure 9(e)) leads to a local thermal circulation, with easterly ageostrophic winds at low levels. This ageostrophic wind is further affected by the Coriolis force, generating southerly flow and superimposing on the large-scale southerly or southeasterly geostrophic winds (Figure 9(a)). Hence, an intensified local wind speed at low level is generated, with a generally coastal-parallel direction. In the upper levels, the thermal contrast is not pronounced, resulting in relatively weak wind. The friction effect is thought to reduce the intensity of bottom winds. This strong low-level thermal contrast and the friction effect contribute to more frequent LLJs in June than in December.

5.   Variability and Large-Scale Conditioning During LLJ Season

LLJ activity is particularly pronounced in the LLJ season of April, May, and June, contributing from large-scale circulations and local land-sea thermal contrast, as was demonstrated in the previous section. In the following section, we will discuss in more detail the diurnal and decadal variability, as well as the link to large-scale atmospheric patterns, during this season.

5.1.   *Features of LLJs in the LLJ Season*

LLJ statistics in the LLJ season (Figure 10) show that more than 50% of jet core heights over the BYS area are distributed in the range of 200–400 m, more than 75% are below 500 m, and 96% are below 1500 m. In terms of the wind speed of jet cores, approximately 45% are in the range of 10–14 m/s, 96.8% are below 25 m/s. More than 3% of LLJs are characterized as having extremely strong wind speeds between 25 and 55 m/s. The jet height-wind speed distribution

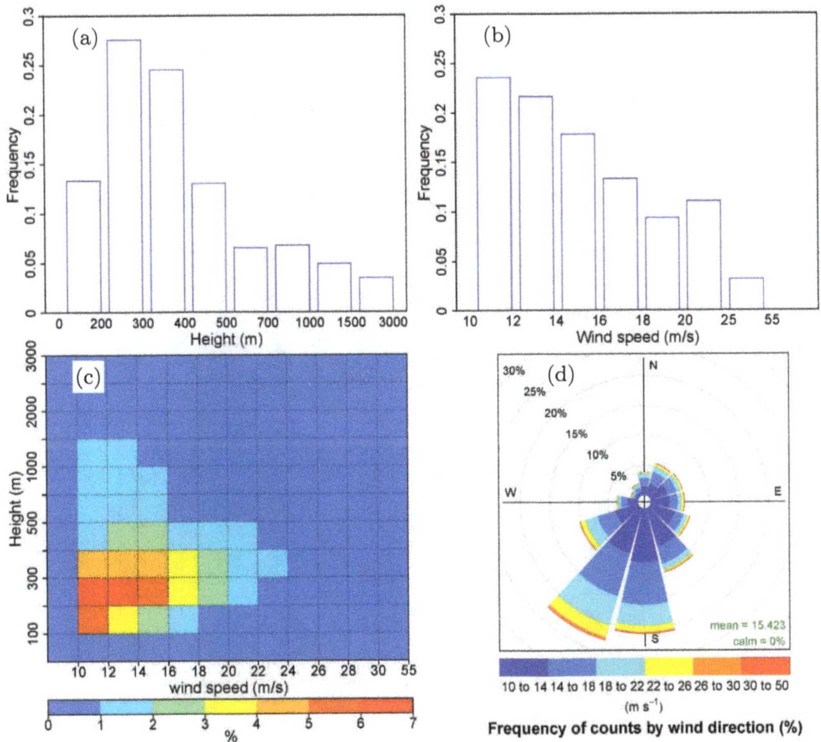

Fig. 10. Low-level jets statistics over the Bohai Sea and Yellow Sea during April, May, and June (1979–2013): (a) Jet height histogram (%), (b) jet wind speed histogram (%), (c) jet height-wind speed distribution, and (d) jet wind rose.

diagram (Figure 10(c)) indicates that the jet cores are mostly located between 200 and 400 m with speed in the range of 10–16 m/s. The prevailing wind directions of LLJs (Figure 10(d)) are southwesterly (~22.5% south-southwesterly and ~10% west-southwesterly), followed by southerly winds (more than 20%), which account for ~55% of the wind directions of LLJs; fewer than 20% of LLJs blow from the southeast, and the northwesterly LLJs are least frequent among all wind directions. The dominant directions of LLJs are coast parallel, which is a general feature of LLJs that result from the geostrophic adjustment between the pressure gradient force and Coriolis force (Soares *et al.*, 2014).

The generation of LLJs in the daytime (Figure 11(a)–(d)) is lower than in the night (Figure 11(e)–(h)) with the former generally being lower than 30%. From 1700 (LST) on, the occurrence of LLJs begins to rise in the coastal areas of the north BS and west YS;

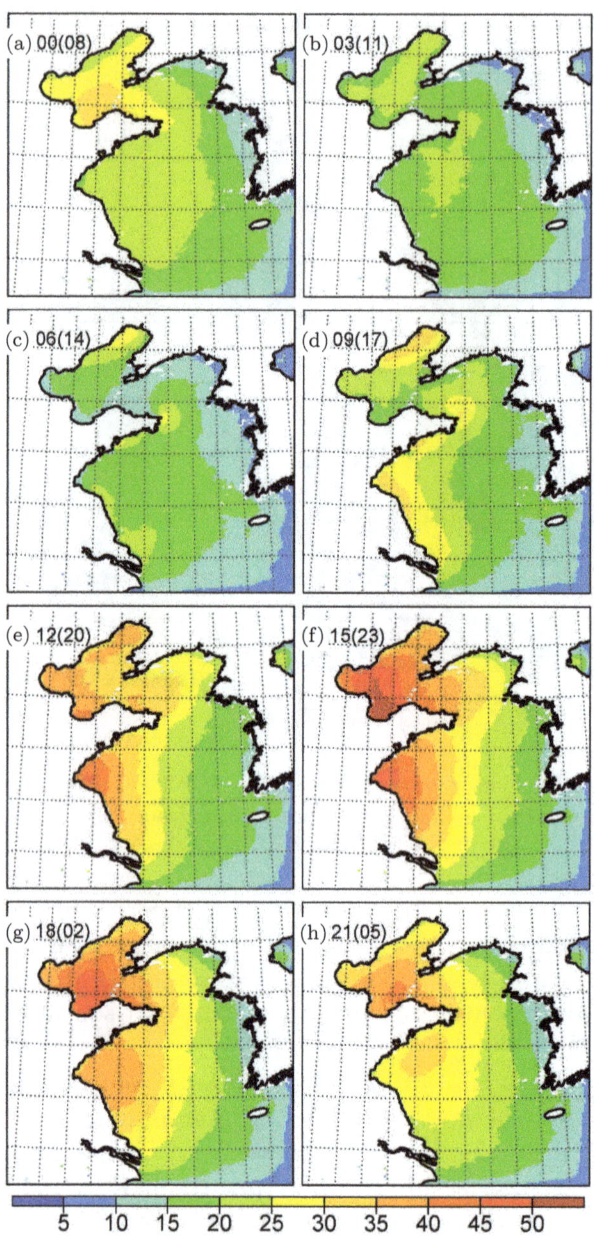

Fig. 11.   Diurnal variation of occurrence frequency (%) of low-level jet at a particular hour (UTC [LST]) in low-level jet season (1979–2013): (a–h) 00 (08), 03 (11), 06 (14), 09 (17), 12 (20), 15 (23), 18 (02), and 21 (05). UTC and LST are abbreviations of coordinated universal time and local solar time, respectively.

at 2000 (LST), the occurrence is more than 30% over the BS and west coasts of the YS and up to 45% in some coastal areas. At 2300 (LST), LLJs are the most frequent, occurring in more than 35% and 50% of the time, respectively, over the BS and in the south BS. Over parts of the north and west YS, the frequency is generally greater than 30%, while it is mostly less than 20% in the coasts of the Korean Peninsula and southeast YS. At 0200 (LST), the occurrence of LLJs drops; however, it is still larger than 35% in most parts of the BS and part of west the YS. At 0500 (LST), the areas with frequent LLJs (>35%) shrink to the middle and south BS. In most parts of the YS, the value is less than 30%.

Furthermore, the daily cycle of occurrence frequency of jet height, jet wind speed, and wind direction over the BYS areas (Figure 12) feature by strong diurnal variability, with more LLJs from 2000 (LST) in the night to 0500 (LST) in the early morning at heights between 200 and 400 m (Figure 12(a)). More LLJs feature wind speeds of 10–16 m/s in the night as well (Figure 12(b)), and the dominant LLJ directions are southerly and south-southwesterly (Figure 12(c)).

Additionally, the spatial distributions of jet occurrence frequency, mean wind speed, and mean height in the LLJ seasons of 1980s, 1990s, and 2000s were obtained (see Figure S3)[3]. The results reveal generally similar spatial patterns, with some differences in the

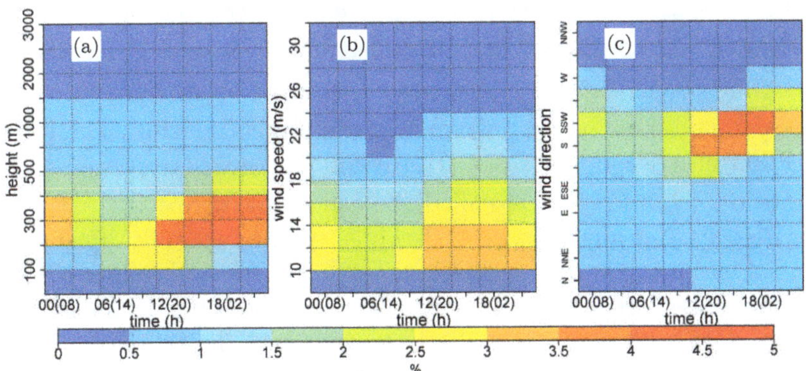

Fig. 12.   Diurnal cycle of low-level jet occurrence frequency in low-level jet season (1979–2013) for (a) jet height, (b) jet wind speed, and (c) jet wind direction.

---

[3]Refer to https://agupubs.onlinelibrary.wiley.com/action/downloadSupplement?doi=10.1029%2F2017JD027949&file=jgrd54652-sup-0001-2017JD027949-SI.pdf.

intensities for each variable among each decade. Overall, the decadal variability of LLJ features is not pronounced.

5.2. *Relationship Between LLJ Occurrence and Sea Level Pressure Patterns*

A critical driver of regional climate variability is the variation of large-scale atmospheric circulation (von Storch *et al.*, 1993), which is also applicable for regional LLJ variability. Emeis (2014) found that the occurrence of LLJs over northern Germany is correlated with the appearance of typical large-scale circulation patterns. In the present study, we found that the large-scale circulation preconditions the formation of LLJs over the BYS. Here we further investigated how the large-scale circulation distributions, averaged over the LLJ season, are related to LLJ occurrence over the BYS. The MSLP from the ERAI reanalysis data set was used, covering the northwest Pacific Ocean and East Asia (0–70°N, 100°E–160°E). Notably, we did not link the instantaneous sea level pressure field with the occurrence probability of a LLJ. Instead, we related two long-term statistics, namely, the statistic of the seasonally averaged MSLP field and the seasonal LLJ occurrence frequency.

The canonical correlation analysis (CCA) method (cf. von Storch & Zwiers, 1999) was used to study the correlation structure of a pair of random vectors $\vec{X}$ and $\vec{Y}$, that is, seasonally averaged MSLP field and seasonal LLJ occurrence frequency in the present study. The objective was to identify a pair of patterns $\vec{f}_X^1$ and $\vec{f}_Y^1$ such that the time coefficients α_{1X} and α_{1Y} in optimal approximations $\vec{X} = \alpha_{1X}\vec{f}_X^1$ and $\vec{Y} = \alpha_{1Y}\vec{f}_Y^1$ share a maximum correlation. The identification of a second pair of patterns follows the same protocol. The patterns $\vec{f}_X^1$ and $\vec{f}_Y^1$ are called the canonical correlation patterns.

Before CCA analysis, we first projected the two multidimensional sets of variables onto their EOFs to exclude noise and reduce the spatial degrees of freedom. The temporal anomalies of each multidimensional data set were used in the EOF analysis. The first five EOFs of LLJ-season-mean MSLP (explaining 84.2% of the total variance) and LLJ frequency (explaining 82.6% of the total variance) were retained for the CCA analysis.

Figure 13 shows the first two important combinations of canonical patterns of LLJ-season-mean MSLP (Figure 13(a) and (c)) and LLJ occurrence frequency anomalies (Figure 13(b) and (d)) and their

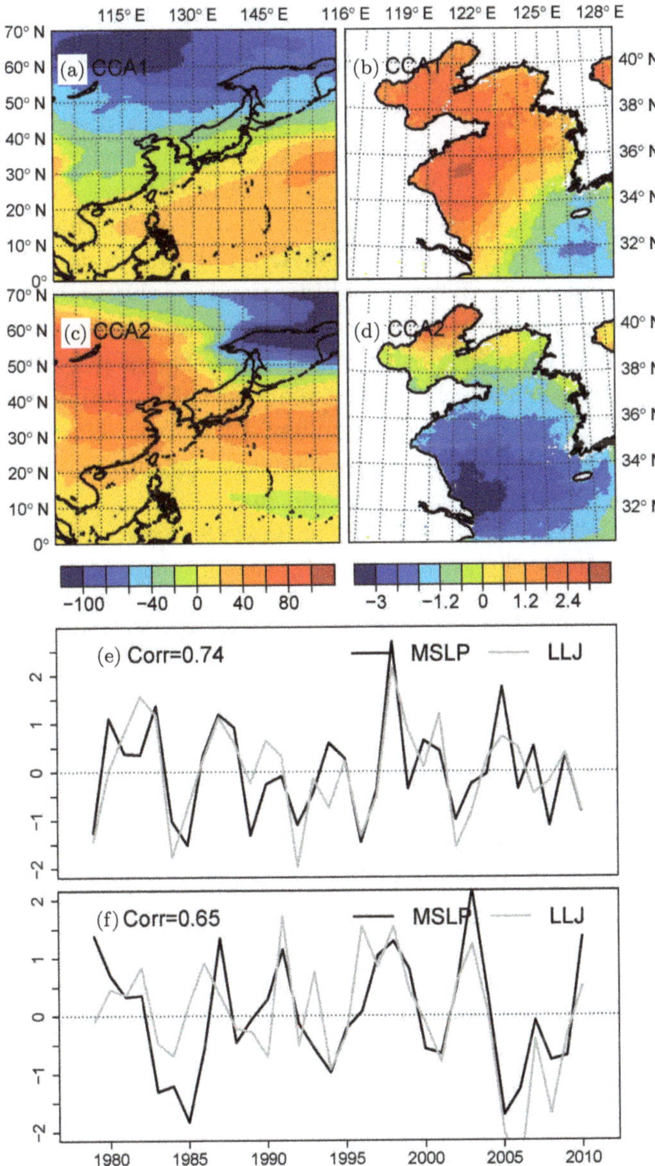

Fig. 13. First two canonical correlation patterns of MSLP (a and b, unit Pa) and LLJ (c and d, unit: %). Corresponding coefficient time series (e and f) for the first two canonical correlation analysis (CCA) patterns, respectively. The first CCA pair shares a correlation of 0.74 and second CCA pair share a correlation of 0.65. MSLP = mean sea level pressure; LLJ = low-level jet; CCA = canonical correlation analysis.

coefficient time series (Figure 13(e) and (f)). Their coefficient time series share a correlation of 0.74 and 0.65 for CCA1 and CCA2, respectively.

The first CCA pattern of LLJ-season-mean MSLP (Figure 13(a)) shows a dipolar pressure distribution, indicating reversed anomalies of the subtropical high over the northwest Pacific Ocean and the northeast cold vortex over East Asia (a cyclonic circulation with a cold core at 35°N–60°N, 115°E–145°E). The first canonical pattern of LLJ occurrence frequency highly resembles the pattern of LLJs in Figure 7(d)–(f) over the northern and western region of the BYS. When the coefficient is positive, a northeastward geostrophic flow anomaly is present, which is consistent with the dominant direction of LLJs, preconditioning more frequent LLJs over the northern and western region of BYS. A pressure contrast of approximately 1.5 hPa between the northwest Pacific Ocean and northeast Asia is related to 0.6% to 2.4% more LLJs in the most BYS region. When the coefficient is negative, the patterns of LLJ-season-mean MSLP and LLJs reverse. The coefficient time series (Figure 13(e)) reflect that CCA1 dominated in 1998 with more LLJ, and in 1979, 1984, 1992, and 1996 with fewer LLJs, in the northern and western regions of the BYS.

Another covariability is described by the second canonical pattern (Figure 13(c) and (d)). In the case of CCA2 of MSLP, there is a negative anomaly over the Sea of Okhotsk and two positive anomalies over East Asia and east to Japan. In the case of positive coefficients, the contrast between the pressure over East Asia and east to Japan, as well as the Sea of Okhotsk, induces southward or southwestward geostrophic flow anomalies, which result in fewer LLJs in the eastern and southern parts of the BYS. Based on the coefficient time series (Figure 13(f)), CCA2 dominated in the years 1991, 1998, and 2003, with fewer LLJs, and in the years 2005 and 2006, with more LLJs over the southern and eastern parts of the BYS region.

5.3. *Relationship Between LLJ Occurrence and Upper-Level Atmospheric Circulations*

To reveal the relationship between LLJ occurrence frequency and upper-level atmospheric circulations, the associated correlation pattern (ACP) approach (von Storch & Zwiers, 1999) was used in this study. ACP is a method based on a linear statistical model, which relates an index of some process with a physical field. We used the coefficient time series for the first two CCA patterns of LLJ occurrence frequency as indices to derive their relationship with geopotential heights at different pressure levels, that is, at 200, 500, and 950 hPa, in the LLJ season by means of linear correlation coefficients.

The ACP patterns at different heights (Figure 14) are rather similar among each other and to the corresponding CCA of MSLP field (Figure 13(a) and (c)), with the patterns becoming weaker with height. In the case of CCA1 (Figure 14, left panel), negative correlations prevail over northeast Asia and positive values over the northwest Pacific Ocean. In the case of CCA2 (Figure 14, right panel), there are negative values over the Sea of Okhotsk and positive values over East China and the northwest Pacific Ocean. These results indicate that the link between regional large-scale circulations and LLJ occurrence frequency is generally barotropic, with similar patterns from top to bottom. The atmospheric circulation in the bottom level has a stronger relationship with LLJ occurrence frequency than that in the upper levels.

The change of the mean state of the barotropic circulation and the associated geostrophic flow is indicative for a synoptic situation, which favor or disfavor the formation of LLJs. Indeed, we consider LLJs as short-term random events (von Storch *et al.*, 2001; similar to Polar Lows and their formation in cold air outbreaks: Kolstad *et al.* 2009) and not as deterministic features — with probabilities conditioned by the regional synoptic situation, which is well described by the barotropic and geostrophic state. The physical processes that are instrumental in actual formation of a LLJ may well not be barotropic nor geostrophic, but their presence seems to be strongly linked to changes in the low-frequency regional barotropic and geostrophic state. Since the statistics of the latter vary more slowly, we are able to construct a link of the circulation over the Asian continent and the northwest Pacific Ocean and the tendency of forming or nonforming LLJs in the coastal regions of BYS.

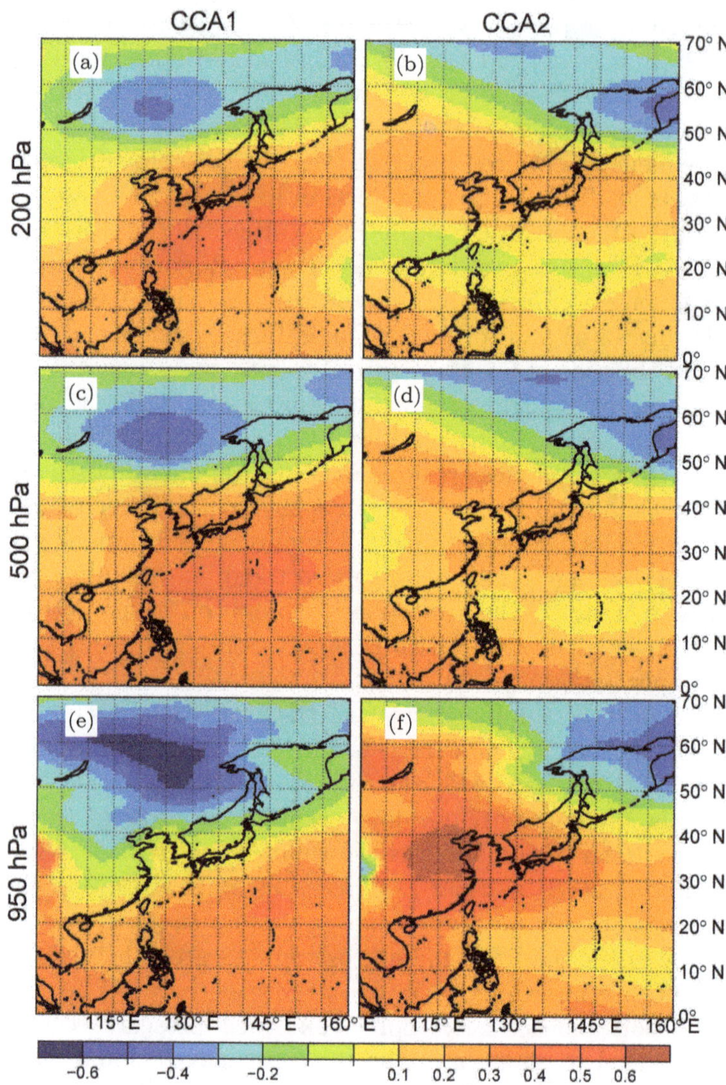

Fig. 14.   Associated correlation patterns between geopotential height anomalies (from top to bottom 200, 500, and 950 hPa) and coefficient time series for the first two canonical correlation analysis (CCA) patterns of low-level jet occurrence frequency (left panel: CCA1, right panel: CCA2).

6.   Summary and Conclusions

In the present study, the climatology and variability features of LLJs over the BYS were investigated based on a long-term (1979–2013) high-resolution (7 km) atmospheric hindcast, which was produced by a regional climate model (CCLM) constrained by ERA-Interim reanalysis. The high-resolution data set was of good quality in terms of its ability to reproduce surface wind speeds and coastal mesoscale phenomena in comparison with observations (Li, 2017; Li, von Storch, & Geyer, 2016a). In this study, we further verified the data set against several radiosonde observations and wind profiler radar observations. The CCLM data set was found to robustly capture the climatology of wind profiles, daily cycle feature, and variability of wind speeds, as well as LLJ cases.

Following the selection criteria by Bonner (1968), the occurrence, height, strength, and direction of LLJs over the BYS spanning 1979–2013 were identified. The annual occurrence of LLJs is more frequent in the BS and western part of the YS. In terms of the temporal variability of LLJs on different scales, we found that the LLJs are nocturnal type LLJs, with the highest occurrence frequency at approximately 2300 (LST). LLJs are the most frequent from April to June, with their occurrence generally exceeding 21% for much of the BYS. The frequency can be greater than 30% over the BS and part of western YS. LLJs are the least frequent in winter, with an occurrence frequency generally less than 12%. The intra-annual variations of LLJ features were found to be related to large-scale circulation and local land-sea thermal contrast. The friction effect is also important in the formation of LLJs over the BYS.

The relative interannual variability of the monthly frequency is generally greater than 20% in most areas. Strong interannual variability exists for LLJs over the BYS, especially during months when the LLJ occurrence is less frequent. In LLJ season (April, May, and June), the heights of jet cores are mostly between 200 and 400 m above sea level, with wind speed maxima mostly in the range of 10–16 m/s. The prevailing wind directions are southerly and southwesterly, which account for approximately 55% of all LLJ directions. Furthermore, we did not find strong interdecadal variability of LLJ features over the BYS in recent decades.

Furthermore, it is thought that the mean state of large-scale atmospheric barotropic circulations over the Asian continent and the northwest Pacific Ocean favors synoptic situations, which precondition LLJ occurrence over the BYS. A link between LLJ occurrence frequency and regional large-scale barotropic circulations has been shown in terms of the low-frequency variability on the interannual scale.

This is the first study to document the long-term climatology and variability of LLJs in Chinese water areas using a high-resolution model output. However, several issues should be addressed. First, the LLJ detection was based on a 3-h vertical output because of the initial model setup. A higher-frequency temporal output, that is, 1 h, may enable a more detailed description of the climatological features of LLJs over the BYS, especially for diurnal variability. Second, the detection method defines basic LLJs of jet-like wind profile, while advanced detection method (e.g., Lima *et al.*, 2018) is suggested to apply in defining typical coastal LLJs (e.g., Ranjha *et al.*, 2013) with specific generation mechanism and jet features. Third, we only investigated the link between LLJs and regional atmospheric circulations in terms of low-frequency variability; however, the influences of local baroclinicity or other mesoscale processes on LLJ features, as well as the extension of the contribution of large-scale processes versus local/mesoscale processes to LLJs in terms of long-term variability, have not been studied, and they deserve further study in the future. Finally, issues such as the impacts of LLJs on regional weather (extreme rainfall), ocean dynamics (circulation and upwelling), and human applications, such as offshore wind farms, have not been studied in the East China Sea and deserve further in-depth study.

Funding

Funding for this study was provided by the Qingdao National Laboratory for Marine Science and Technology (2016ASKJ12), the National Key Research and Development Program of China (2017YFA0604100 and 2016YFC1401404), the National Natural Science Foundation of China (41706019, 41528601, 41676006, U1606402, and 41421005), the General Financial Grant (2017M612357), and Special Financial Grant (2017T100520) from the China Postdoctoral

Science Foundation, Youth Innovation Promotion Association Chinese Academy of Sciences (CAS), CAS Interdisciplinary Innovation Team, Key Research Program of Frontier Sciences, CAS, and the Strategic Pioneering Research Program of CAS (XDA11020104 and XDA11020101).

Acknowledgments

We thank Beate Geyer and Burkhardt Rockel for their help in the regional climate modeling and statistical processing.

References

Arfeuille, G., Quintanilla-Montoya, A. L., Viesca Gonzalez, F. C. and Zizumbo Villarreal, L. (2015). Observational characteristics of low-level jets in Central Western Mexico, *Boundary-Layer Meteorology* **155**, 3, pp. 483–500, doi:10.1007/s10546-015-0005-0.

Beardsley, R., Dorman, C. E., Friehe, C. A., Rosenfeld, L. and Winant, C. (1987). Local atmospheric forcing during the coastal ocean dynamics experiment: 1. a description of the marine boundary layer and atmospheric conditions over a Northern California upwelling region, *Journal of Geophysical Research: Oceans* **92**, pp. 1467–1488.

Blackadar, A. K. (1957). Boundary layer wind maxima and their significance for the growth of nocturnal inversions, *Bulletin of the American Meteorological Society* **38**, pp. 283–290.

Bonner, W. D. (1968). Climatology of the low level jet, *Monthly Weather Review* **96**, 12, pp. 833–850.

Burk, S. D. and Thompson, W. T. (1996). The summertime low-level jet and marine boundary layer structure along the California coast, *Monthly Weather Review* **124**, 4, pp. 668–686, doi:10.1175/1520-0493(1996)124<0668:TSLLJA>2.0.CO;2.

Cardoso, R. M., Soares, P. M., Lima, D. C. and Semedo, A. (2016). The impact of climate change on the Iberian low-level wind jet: EURO-CORDEX regional climate simulation, *Tellus A: Dynamic Meteorology and Oceanography* **68**, p. 29005, doi:10.3402/tellusa.v68.29005.

Chao, S.-Y. (1985). Coastal jets in the lower atmosphere, *Journal of Physical Oceanography* **15**, 4, pp. 361–371.

Chen, G. T. J., Wang, C. C. and Lin, D. T.-W. (2005). Characteristics of low-level jets over northern Taiwan in Mei-yu season and their relationship to heavy rain events, *Monthly Weather Review* **133**, 1, pp. 20–43, doi:10.1175/MWR-2813.1.

Colle, B. A. and Novak, D. R. (2010). The New York bight jet: Climatology and dynamical evolution, *Monthly Weather Review* **138**, 6, pp. 2385–2404, doi:10.1175/2009MWR3231.1.

Cook, K. H. and Vizy, E. K. (2010). Hydrodynamics of the Caribbean low-level jet and its relationship to precipitation, *Journal of Climate* **23**, 6, pp. 1477–1494, doi:10.1175/2009JCLI3210.1.

Dee, D. P., National Center for Atmospheric Research Staff (2018). The climate data guide: ERA-Interim. Retrieved from https:// climatedataguide.ucar.edu/climate-data/era-interim, Accessed 15 March 2018.

Dee, D. P., Uppala, S. M., Simmons, A. J., Berrisford, P., Poli, P., Kobayashi, S., *et al.* (2011). The ERA-interim reanalysis: Configuration and performance of the data assimilation system, *Quarterly Journal of the Royal Meteorological Society* **137**, 656, pp. 553–597, doi:10.1002/qj.828.

Doubler, D. L., Winkler, J. A., Bian, X., Walters, C. K. and Zhong, S. (2015). An NARR-derived climatology of southerly and northerly low-level jets over North America and coastal environs, *Journal of Applied Meteorology and Climatology* **54**, 7, pp. 1596–1619, doi: 10.1175/JAMC-D-14-0311.1.

Doyle, J. D. and Warner, T. T. (1993). A three-dimensional numerical investigation of a Carolina Coastal low-level jet during gale iop 2, *Monthly Weather Review* **121**, 4, pp. 1030–1047.

Du, Y., Chen, Y.-L. and Zhang, Q. (2015). Numerical simulations of the boundary layer jet off the southeastern coast of China, *Monthly Weather Review* **143**, 4, pp. 1212–1231, doi:10.1175/ MWR-D-14-00348.1.

Du, Y., Zhang, Q., Chen, Y., Zhao, Y. and Wang, X. (2014). Numerical simulations of spatial distributions and diurnal variations of low-level jets in China during early summer, *Journal of Climate* **27**, 15, pp. 5747–5767, doi:10.1175/JCLI-D-13-00571.1.

Emeis, S. (2014). Wind speed and shear associated with low-level jets over Northern Germany, *Meteorologische Zeitschrift* **23**, 3, pp. 295–304, doi:10.1127/0941-2948/2014/0551.

Higgins, R. W., Yao, Y., Yarosh, E. S., Janowiak, J. E. and Mo, K. C. (1997). Influence of the Great Plains low-level jet on summertime precipitation and moisture transport over the central United States, *Journal of Climate* **10**, 3, pp. 481–507, doi:10.1175/1520-0442(1997)010<0481:IOTGPL>2.0.CO;2.

Jiang, Q., Wang, S. and O'Neill, L. (2010). Some insights into the characteristics and dynamics of the Chilean low-level coastal jet, *Monthly Weather Review* **138**, 8, pp. 3185–3206, doi:10.1175/2010MWR3368.1.

Kolstad, E., Bracegirdle, T. J. and Seierstad, I. A. (2009). Marine cold-air outbreaks in the North Atlantic: Temporal distribution and associations with large-scale atmospheric circulation, *Climate Dynamics* **33**, pp. 187–197, doi:10.1007/s00382-008-0431-5.

Li, D. (2017). Added value of high-resolution regional climate model: Selected cases over the Bohai Sea and the Yellow Sea areas, *International Journal of Climatology* **37**, 1, pp. 169–179, doi:10.1002/joc.4695.

Li, D., Geyer, B. and Bisling, P. (2016). A model-based climatology analysis of wind power resources at 100-m height over the Bohai Sea and the Yellow Sea, *Applied Energy* **179**, pp. 575–589, doi:10.1016/j.apenergy.2016.07.010.

Li, D., von Storch, H. and Geyer, B. (2016a). High-resolution wind hindcast over the Bohai Sea and the Yellow Sea in East Asia: Evaluation and wind climatology analysis, *Journal of Geophysical Research: Atmospheres* **121**, 1, pp. 111–129, doi:10.1002/2015JD024177.

Li, D., von Storch, H. and Geyer, B. (2016b). Testing reanalyses in constraining dynamical downscaling, *Journal of the Meteorological Society of Japan* **94A**, pp. 47–68, doi:10.2151/jmsj.2015-044.

Lima, D. C. A., Soares, P. M. M., Semedo, A. and Cardoso, R. M. (2018). A global view of coastal low-level wind jets using an ensemble of reanalyses, *Journal of Climate* **31**, 4, pp. 1525–1546, doi:10.1175/JCLI-D-17-0395.1.

Maddox, R. A. (1983). Large-scale meteorological conditions associated with midlatitude, mesoscale convective complexes, *Monthly Weather Review* **111**, 7, pp. 1475–1493, doi:10.1175/1520-0493(1983)111<1475:LSMCAW>2.0.CO;2.

Maldonado, T., Rutgersson, A., Caballero, R., Pausata, F. S. R., Alfaro, E. and Amador, J. (2017). The role of the meridional sea surface temperature gradient in controlling the Caribbean low-level jet, *Journal of Geophysical Research: Atmospheres* **122**, 11, pp. 5903–5916, doi:10.1002/2016JD026025.

Marengo, J. A., Soares, W. R., Saulo, C. and Nicolini, M. (2004). Climatology of the low-level jet east of the Andes as derived from the NCEP-NCAR reanalyses: Characteristics and temporal variability, *Journal of Climate* **17**, 12, pp. 2261–2280, doi:10.1175/1520-0442(2004)017<2261:COTLJE>2.0.CO;2.

Means, L. Y. N. N. L. (1954). A study of the mean southerly wind-maximum in low levels associated with a period of summer precipitation in the middle west, *Bulletin of the American Meteorological Society* **35**, 4, pp. 166–170.

Mellor, G. L. and Yamada, T. (1982). Development of a turbulence closure-model for geophysical fluid problems, *Reviews of Geophysics* **20**, 4, pp. 851–875, doi:10.1029/RG020i004p00851.

Miao, Y., Guo, J., Liu, S., Wei, W., Zhang, G., Lin, Y. and Zhai, P. (2018). The climatology of low-level jet in Beijing and Guangzhou, China, *Journal of Geophysical Research: Atmospheres* **123**, pp. 2816–2830, doi:10.1002/2017JD027321.

Muñoz, E. and Enfield, D. (2011). The Boreal spring variability of the Intra-Americas low-level jet and its relation with precipitation and tornadoes in the eastern United States, *Climate Dynamics* **36**, 1–2, pp. 247–259, doi:10.1007/s00382-009-0688-3.

Nicholson, S. E. (2010). A low-level jet along the Benguela coast, an integral part of the Benguela current ecosystem, *Climatic Change* **99**, 3–4, pp. 613–624, doi:10.1007/s10584-009-9678-z.

Nunalee, C. G. and Basu, S. (2013). Mesoscale modeling of coastal low-level jets: Implications for offshore wind resource estimation, *Wind Energy* **17**, 8, pp. 1199–1216, doi:10.1002/we.1628.

Nuss, W. A., Bane, J. M., Thompson, W. T., Holt, T., Dorman, C. E., Ralph, F. M., *et al.* (2000). Coastally trapped wind reversals: Progress toward understanding, *Bulletin of the American Meteorological Society* **81**, 4, pp. 719–743, doi:10.1175/1520-0477(2000)081<0719: CTWRPT>2.3.CO;2.

Parish, T. R. (2000). Forcing of the summertime low-level jet along the California coast, *Journal of Applied Meteorology* **39**, 12, pp. 2421–2433, doi:10.1175/1520-0450(2000)039<2421:FOTSLL>2.0.CO;2.

Parish, T. R. and Oolman, L. D. (2010). On the role of sloping terrain in the forcing of the great plains low-level jet, *Journal of the Atmospheric Sciences* **67**, 8, pp. 2690–2699, doi:10.1175/2010JAS3368.1.

Pham, N., Nakamura, K., Furuzawa, F. A. and Satoh, S. (2008). Characteristics of low level jets over Okinawa in the Baiu and post-Baiu seasons revealed by wind profiler observations, *Journal of the Meteorological Society of Japan* **86**, 5, pp. 699–717.

Rahn, D. A. and Parish, T. R. (2007). Diagnosis of the forcing and structure of the coastal jet near cape mendocino using in situ observations and numerical simulations, *Journal of Applied Meteorology and Climatology* **46**, 9, pp. 1455–1468, doi:10.1175/JAM2546.1.

Ranjha, R., Svensson, G., Tjernström, M. and Semedo, A. (2013). Global distribution and seasonal variability of coastal low-level jets derived from era-interim reanalysis, *Tellus A: Dynamic Meteorology and Oceanography* **65**, 1, p. 20412.

Ranjha, R., Tjernstrom, M., Svensson, G. and Semedo, A. (2016). Modelling coastal low-level wind-jets: Does horizontal resolution matter? *Meteorology and Atmospheric Physics* **128**, 2, pp. 263–278, doi: 10.1007/s00703-015-0413-1.

Ranjha, R., Tjernström, M., Semedo, A., Svensson, G. and Cardoso, R. M. (2015). Structure and variability of the Oman coastal low-level jet, *Tellus A: Dynamic Meteorology and Oceanography* **67**, p. 25285, doi: 10.3402/tellusa.v67.25285.

Rijo, N., Semedo, A., Miranda, P. M. A., Lima, D., Cardoso, R. M. and Soares, P. M. M. (2018). Spatial and temporal variability of the Iberian Peninsula coastal low-level jet, *International Journal of Climatology* **38**, 4, pp. 1605–1622, doi:10.1002/joc.5303.

Rockel, B., Will, A. and Hense, A. (2008). The regional climate model COSMO-CLM (CCLM), *Meteorologische Zeitschrift* **17**(4), pp. 347–348, doi:10.1127/0941-2948/2008/0309.

Schrodin, R. and Heise, E. (2002). A new multi-layer soil model, *COSMO newsletter* **2**, pp. 149–151.

Semedo, A., Soares, P. M. M., Lima, D. C. A., Cardoso, R. M., Bernardino, M. and Miranda, P. M. A. (2016). The impact of climate change on the global coastal low-level wind jets: EC-EARTH simulations, *Global and Planetary Change* **137**, pp. 88–106, doi:10.1016/j.gloplacha.2015.12.012.

Soares, P. M. M., Lima, D. C. A., Cardoso, R. M. and Semedo, A. (2017). High resolution projections for the Western Iberian coastal low level jet in a changing climate, *Climate Dynamics* **49**, 5–6, pp. 1547–1566, doi:10.1007/s00382-016-3397-8.

Soares, P. M. M., Cardoso, R. M., Semedo, A., Chinita, M. J. and Ranjha, R. (2014). Climatology of the Iberia coastal low-level wind jet: Weather research forecasting model high-resolution results, *Tellus A: Dynamic Meteorology and Oceanography* **66**, p. 22377, doi: 10.3402/tellusa.v66.22377.

Stensrud, D. J. (1996). Importance of low-level jets to climate: A review, *Journal of Climate* **9**, 8, pp. 1698–1711, doi:10.1175/1520-0442(1996)009<1698:IOLLJT>2.0.CO;2.

Tao, S. Y. (1987). A review of recent research on the East Asian summer monsoon in China, In *Monsoon Meteorology*, pp. 60–92.

Tiedtke, M. (1989). A comprehensive mass flux scheme for cumulus parameterization in large-scale models, *Monthly Weather Review* **117**, 8, pp. 1779–1800, doi:10.1175/1520-0493(1989)117<1779:ACMFSF>2.0.CO;2.

von Storch, H. and Zwiers, F. W. (1999). *Statistical Analysis in Climate Research*. Cambridge, UK: Cambridge University Press. doi:10.1017/CBO9780511612336.

von Storch, H., Langenberg, H. and Feser, F. (2000). A spectral nudging technique for dynamical downscaling purposes, *Monthly Weather Review* **128**, 10, pp. 3664–3673, doi:10.1175/1520-0493(2000)128<3664:ASNTFD>2.0.CO;2.

von Storch, H., von Storch, J.-S. and Müller, P. (2001). Noise in the climate system — ubiquitous, constitutive and concealing, in B. Engquist and W. Schmid. eds., *Mathematics Unlimited — 2001 and Beyond. Part II*, pp. 1179–1194, Springer Verlag. doi:10.1007/978-3-642-56478-9 6a.

von Storch, H., Zorita, E. and Cubasch, U. (1993). Downscaling of global climate change estimates to regional scales: An application to iberian rainfall in wintertime, *Journal of Climate* **6**, 6, pp. 1161–1171.

Wang, D., Zhang, Y. and Huang, A. (2013). Climatic features of the southwesterly low-level jet over Southeast China and its association with precipitation over East China, *Asia-Pacific Journal of Atmospheric Sciences* **49**, 3, pp. 259–270, doi:10.1007/s13143-013-0025-y.

Warner, T. T. (2004). *Desert meteorology*, Cambridge, UK: Cambridge University Press.

Wei, W., Zhang, H. S. and Ye, X. X. (2014). Comparison of low-level jets along the north coast of China in summer, *Journal of Geophysical Research: Atmospheres* **119**, 16, pp. 9692–9706, doi:10.1002/2014JD021476.

Wei, W., Wu, B. G., Ye, X. X., Wang, H. X. and Zhang, H. S. (2013). Characteristics and mechanisms of low-level jets in the Yangtze River Delta of China, *Boundary-Layer Meteorology* **149**, 3, pp. 403–424, doi:10.1007/s10546-013-9852-8.

Whiteman, C. D., Bian, X. and Zhong, S. Y. (1997). Low-level jet climatology from enhanced Rawinsonde observations at a site in the southern Great Plains, *Journal of Applied Meteorology* **36**, 10, pp. 1363–1376, doi:10.1175/1520-0450(1997)036<1363:LLJCFE>2.0.CO;2.

Whyte, F. S., Taylor, M. A., Stephenson, T. S. and Campbell, J. D. (2007). Features of the Caribbean low level jet, *International Journal of Climatology* **28**, 1, pp. 119–128, doi:10.1002/joc.1510.

Winant, C. D., Dorman, C. E., Friehe, C. A. and Beardsley, R. C. (1988). The marine layer off Northern California: An example of supercritical channel flow, *Journal of Atmospheric Sciences* **45**, 23, pp. 3588–3605. doi: 10.1175/1520-0469(1988)045<3588:TMLONC>2.0.CO;2.

Wu, Y. and Raman, S. (1998). The summertime great plains low level jet and the effect of its origin on moisture transport, *Boundary-Layer Meteorology* **88**, 3, pp. 445–466, doi:10.1023/A:1001518302649.

Chapter 11

Assessing Changes in Extreme Sea Levels Along the Coast of China*

J. Feng[†,‡], H. von Storch[‡], W. Jiang[†,§], and R. Weisse[‡]

[†]*Physical Oceanography Laboratory, Ocean University of China, Qingdao, People's Republic of China*
[‡]*Helmholtz-Zentrum Geesthacht, Centre for Materials and Coastal Research, Geesthacht, Germany*
[§]*Laboratory of Marine Environment and Ecology, Ocean University of China, Qingdao, People's Republic of China*

Hourly tide-gauge data along the coast of China are used to evaluate changes in extreme water levels in the past several decades. Mean sea level, astronomical tide, nontidal component and the tide-surge interaction were analyzed separately to assess their roles in the changes of extreme sea levels. Mean sea level at five tide gauges, Kanmen, Keelung, Zhapo, Xiamen and Quarrybay, show significant increasing trends during the past decades (1954–2013) with a rate of about 1.4–3.5 mm/yr. At Keelung, Kaohsiung and Quarrybay the mean high waters increased during 1954–2013 with a rate from 0.6 to 1.8 mm/yr, while the annual

---

*This chapter was originally published in *Journal of Geophysical Research Oceans*, 120(12), 8039–8051, doi:10.1002/2015JC011336. This chapter is licensed under the terms and conditions of the Creative Commons Attribution (CC BY) license (https://creativecommons.org/licenses/by/4.0/), which permits unrestricted use, distribution, and reproduction in any medium.

mean tidal range rose at the same time by 0.9 to 3.8 mm/yr. In terms of storm surge intensities, there is interannual variability and decadal variability but five tide gauges show significant decreasing trends, and three gauges, at Keelung, Xiamen and Quarrybay, exhibited significant increases of extreme sea levels with trends of 1.5–6.0 mm/yr during 1954–2013. Significant tide-surge interactions were found at all 12 tide gauges, but no obvious change was found during the past few decades. The changes in extreme sea levels in this area are strongly related to the changes of mean sea levels (MSL). At gauges, where the tide-surge interaction is large, the astronomic tides are also an important factor for the extreme sea levels, whereas tide gauges with little tide-surge interaction, the changes of wind driven storm surge component adds to the change of the extreme sea levels.

1.   Introduction

China has the largest coastal population in the world, with more than 400 million people living along the coast (National Bureau of Statistics of the People's Republic of China). The rapid economic progress of China, especially in the coastal zone, attracts more attention to the negative impacts of sea level extremes. The China Marine Disaster Bulletin (http://www.soa.gov.cn/zwgk/hygb/) shows that between 1989 and 2014, on average, extreme sea level disasters caused economic losses of 11.7 billion (RMB), 156 deaths, and affected 13.4 million people annually. As many of the Chinese coastal communities are in low-lying regions, big storm surges due to typhoons, tropical storms and winter storms frequently causes severe losses. Hallegatte *et al.* (2013) classify the China coast as one of the most vulnerable areas under climate change scenarios. It is essential to improve the understanding of sea level extremes along China coasts for coastal protection, future planning, and conservation of coastal ecosystems.

Many studies have been done in the past years about the change of mean sea level and of extreme sea levels both regionally and globally using the data from tide gauges (Marcos *et al.*, 2009; Mendez *et al.*, 2007; Menendez and Woodworth, 2010; Mudersbach *et al.*, 2013; Tsimplis and Woodworth, 1994; von Storch and Reichardt, 1997; Weisse *et al.*, 2014; Woodworth and Blackman, 2004; Woodworth *et al.*, 2007). There is indeed evidence for a general worldwide increase in extreme high-water levels and mean sea levels in the past few decades. An interesting question is, whether the increase in the tails of the distributions is mostly because of a shift or a broadening of the distribution (Bernier and Thompson, 2006; Bromirski *et al.*, 2003; Church and White, 2006; Marcos *et al.*, 2009; von Storch and Reichardt, 1997; Zhang *et al.*, 2000). Results show that there is no uniform rule worldwide: changes of extreme sea levels are sometimes mostly due to increases in mean sea level, whereas in other cases, changes in extreme and mean sea levels seem to be decoupled. It means that changes in mean sea levels may change the shape of the distribution and there are also other reasons that can change the scale of the distribution of the extreme sea levels.

Changes in the astronomical tide may also be important. Changes in the amplitudes of some diurnal and semidiurnal tidal constituents have been identified and hypothesized as changes in the propagation of the tidal due to the changes in bathymetry associated with mean sea level changing (Austin, 1991; Bolle *et al.*, 2010; Egbert *et al.*, 2004; Jay, 2009; Pickering *et al.*, 2012; Shaw *et al.*, 2010; Uehara *et al.*, 2006; von Storch and Woth, 2008; Woodworth, 2010). It should be noted that the nodal and perigean modulations on high tidal levels are important as these modulations contribute to the vulnerability of coastal areas when coupled with inter-annual sea level variation or extreme storm surges.

Another important element is the wind driven (surge) component. If the frequency, strength and tracks of weather systems alter, the storm surges may change. For the change of storm surge statistics under climate change, many works have been done in the past few years to study the past statistics of storm surges or the envisaged changes of storm surges in the future (Butler *et al.*, 2007; Karim and Mimura, 2008; Marcos *et al.*, 2011; Pascual *et al.*, 2008; Pirazzoli *et al.*, 2004; Ratsimandresy *et al.*, 2008; Weisse *et al.*, 2014; Woth, 2005; Woth *et al.*, 2006). Results show that the changes of storm surges differ for different regions.

At certain locations, in particular, in shallow seas and estuaries, significant interaction occurs between tides and surges. This interaction produces changes in amplitude and phase of the surges (Bernier and Thompson, 2007; Johns and Ali, 1980; Zhang *et al.*, 2010). For the East China Sea, Zhang *et al.* (2010) indicated that the difference in tidal residuals due to the interaction could reach up to 20 cm. For the Chinese coastal areas, few studies have been done to determine its role in changing statistics of extreme sea levels. And whether the characteristics of the changes of extreme sea levels in this area show same spatial pattern or not are still unclear.

In this work, hourly sea level data from 12 tide gauges and annual mean sea level data from 5 gauges at the China coast are used. The observed sea level are separated into their components (mean sea level (MSL), astronomical tide and surge parts), using means of a separate tidal analysis for each calendar year (Pugh, 1987). Records from 4 gauges, at least 30 years long, are used to investigate how extreme sea levels, astronomical tide and mean sea level have changed over the period 1950–2012. Changes in the surge parts and tide-surge interaction are studied at all 12 gauges. Most of the available time series are relatively short and thus of limited use for studying long term changes in sea level statistics. The reason is that most of the tide gauge data are kept confidential, so that scholars have to rely on the available data to analyze long-term changes. In our case, data from four tide gauges are available for up to 1950–2012, while for the other eight stations only limited records (1975–1997) can be used.

The outline of this chapter is as follows: A brief description and validation of the tide gauges data sets and the methodologies used in this chapter are described in Section 2. Changes in extreme sea level and every component are analyzed in Section 3. Also the contribution of each part to the extreme sea level is discussed in this part. Finally, main conclusions are drawn in Section 4.

2. Data Sets and Methodology

2.1. *Sea Level Data*

Two kinds of sea level data sets are used in this paper. The locations of the gauges are given in Figure 1.

(1) Hourly sea level data from 12 tidal gauges (Figure 1) along the Chinese coast were obtained from the University of Hawaii Sea Level Center. The time series lengths of these tide gauges are shown in Figure 1.

(2) Also annual mean sea level data of 5 gauges (2 are overlapped with the above 12), which are Yantai (1954–1994), Qinhuang-dao (1950–1994), Kanmen (1959–2013), Zhapo (1959–2013) and Macau (1925–1982), from the Permanent Service for Mean Sea Level (PSMSL) were also used to study the change of MSL. Although the annual mean sea level data can also be obtained from the hourly sea level data at Kanmen and Zhapo, the data from PSMSL are used because of its longer time series.

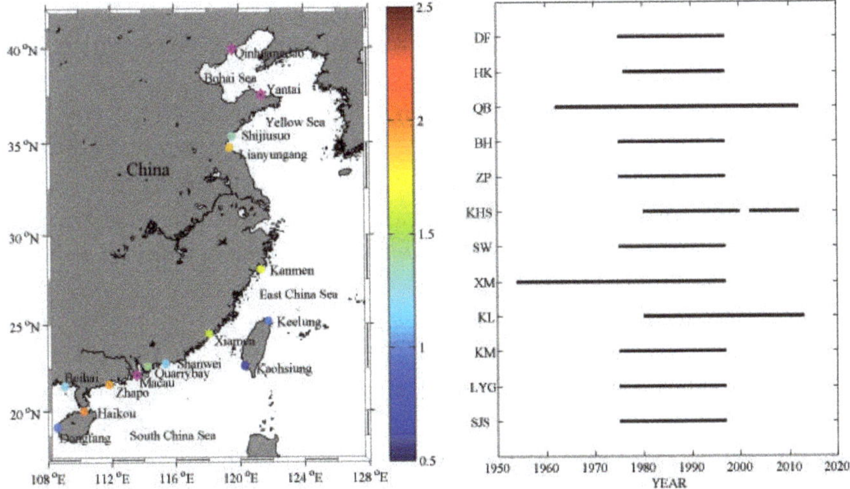

Fig. 1. (Left) Location of tide gauges used in this work, the color values (m) indicate the 99.9% of nontide residual levels of each gauges. (Right) When only annual mean sea levels are available, the gauges are marked in purple; the time lengths of hourly data at 12 tide gauges, Shijiusuo (SJS), Lianyungang (LYG), Kanmen (KM), Keelung (KL), Xiamen (XM), Shanwei (SW), Kaohsiung (KHS), Zhapo (ZP), Beihai (BH), Quarrybay (QB), Haikou (HK) and Dongfang (DF) are shown on the right.

The data have been rigorously checked for common errors such as data spikes and spurious records. Values with spurious jumps, datum shifts and time shifts were removed. In addition, 2 years (2001 and 2002) at Keelung when the data are less than 60% complete were excluded from the temporal analysis. At Hongkong sea level was recorded at North Point between 1962 and 1986 and then moved to Quarrybay. The offset between the two records is 1.02 cm. We combined these two data after shifting the earlier data by 1.02 cm.

2.2. *Methodology*

2.2.1. *Sea Level and Extreme Sea Level Characteristics*

Changes in extreme sea levels have been assessed using a percentile analysis. The hourly data are ordered in terms of height and then used to compute percentile levels (Woodworth and Blackman, 2004). Percentile values for the observed sea level have been calculated at 3 levels (99.9%, 99%, 90%), one for the total, and one after subtracting the annual medians. The trend in median supposedly represents the mean sea level rise (as we used the annual mean sea level which just equal to 50% level).

The observed sea level variation can be considered as the sum of a mean level, an astronomical tidal component and a nontidal residual (Pugh, 1987). The mean level is the average height of sea level defined over an extended period of time, usually a year. The tidal and nontidal residual components have been estimated using a harmonic analysis (Pawlowicz *et al.*, 2002).

For the tidal component, annual mean high water (MHW) and annual mean low water (MLW) tidal elevations relative to MSL have been calculated along with annual mean tidal range (MTR) (Haigh *et al.*, 2010).

Following Zhang *et al.* (2000), three indices have been used as proxies for the intensity of the storm surge climate. These are:

- *Storm surge count*: the annual number of storm surges (the nontide residuals) above a given threshold;
- *Storm surge duration*: the annual number of hours for which the storm surge levels were above a given threshold;
- *Storm surge intensity*: the annual total integral of the sea level curve above a given threshold.

As thresholds we use the hourly 99 percentile non tidal sea level variations at each of 12 tide gauges.

Several statistical methods are available to estimate tide-surge interaction. Here we follow Haigh *et al.* (2010), which determine the dependency of the nontidal sea surge peaks beyond the threshold on the different phases of tide. The timing of the surge peaks relative to the nearest high tide is noted. The tide is then divided into hour bands with respect to timing of high tide. If the surge and tide are independent processes, each surge above the threshold will have equal chance of falling in any of these bands. On the contrary, if interaction is present, then the number of surges per band would be expected to differ from one to another. A chi-square test (Haigh *et al.*, 2010) is used to quantify the level of interaction taking place at each site. The test statistic is

$$X^2 = \sum_{i=1}^{n} \frac{(N_i - e)^2}{e}$$

where N_i is the observed number of surges per tidal phase i, e is the expected number of surges per tidal phase, which is equal to peaks above the 99% threshold, divided by total number of phases (n).

When the distribution is uniform, then the expected distribution of X is $\chi^2(n)$-distributed.

Here we do not use the X-test statistic directly as other works (Antony and Unnikrishnan, 2013; Haigh *et al.*, 2010; Horsburgh and Wilson, 2007), as the X^2 increases with the length of the time series. For being able to compare the statistics, we normalized the test statistic by the number of years, so that we consider X^2/T, with T=number of years.

2.2.2. *Significance of Trends and Correlations*

Trends and correlations (between the extreme sea level and tide, surge and mean sea level) have been fitted to the time-series. Testing for the "significance" of correlations between two time series and of nonzero trends in a time series incorporates the determination of how large sample correlations and sample trends could be, when the stochastic processes, which generate the series, are not correlated or are stationary (exhibit no trends). "Significance" indicates that the actual sample correlation contradicts the assumption that there are "no correlations" between the underlying processes, or, similarly, that the detected sample trend would be unlikely to appear in limited segments of a stationary time series.

For testing these null hypotheses, that the processes X and Y share no correlation, or that segments of length L has a zero trend, standard procedures are available in the literature, namely p-value for correlations and Mann-Kendall for trends (e.g., Kulkarni and von Storch, 1995; von Storch and Zwiers, 1999).

The use of correlation depends on a normality assumption. These tests make some assumptions about the underlying processes. In the case of correlations the assumption is that the underlying processes are stationary (free of systematic trends) and serially independent, i.e., X_t and Y_t for any t are independent. In geophysical cases, not all of these assumptions are satisfied — the result is that the null hypotheses is too often falsely rejected (i.e., in cases where there are no correlations or no trends (cf. Kulkarni and von Storch, 1995)) than stipulated by the significance level (normally 5%).

Following Kulkarni and von Storch (1995), a practical remedy for avoiding such errors is to deal with normalized series (mean = 0, standard deviation $= 1$) X'_t (and Y'_t); "detrend" the time series before testing for correlations, i.e., determining the linear fit f^X_t and f^Y_t, and do the hypothesis testing with $X'_t = X_t - f^X_t$ and $Y'_t = Y_t - f^Y_t$; "prewhiten" the time series, by first determining the sample autocorrelation $\alpha = 1/L\Sigma_t X_t X_{t+1}$ of the time series X_t of length L, and forming a series $X'_t = X_t - \alpha X_t$, and then testing for the null hypothesis of no trend.

To both cases, the standard routines are applied. If the null hypothesis is rejected at the stipulated significance level of 5%, then the sample trend f^X_t, or the sample correlation $1/L\Sigma_t X_t Y_t$, is considered "significant."

2.2.3.   *Time Series Length for Meaningful Trends for Total Sea Level Variations (Including Tides)*

An important problem, before analyzing the extreme sea levels, is how many years are needed to estimate the trend of the extremes accurately. When trend is fitted to records that are not long enough, we cannot get the correct trend. As the magnitude of the trend can vary significantly depending on where the 18.6 year nodal cycle the sea level record begins and ends. Also if the sample size is too low, it may not have the power to detect an effect, or because the trend due to the climate is nonlinear. In some work (Haigh *et al.*, 2010) 36 years was used. But along the China coast, only two gauges

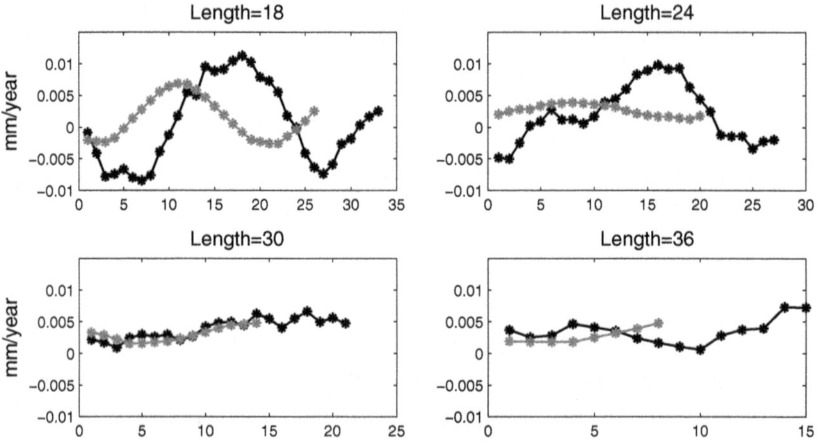

Fig. 2.   The linear trend of the extreme sea levels using different data length (18, 24, 30, and 36 years) at (black line) Quarrybay and (gray line) Xiamen, *x* axis stand for the time of the calculation for each length.

have records longer than 36 years. Here we first checked the stability of the trend when different lengths of segments L were used. Results for the two gauges with long series (Quarrybay and Xiamen) are shown in Figure 2. When only half a period (L = 18) of the nodal cycle is available, the trends vary considerably from segment to segment (Figure 2). These variations flatten out, when L = 30 year segments are used, the estimate of the segment trends become similar. The standard deviations of the trends at Quarrybay of different lengths L are 6.5 (L = 18), 4.5 (L = 24), 1.6 (L = 30) and 1.7 (L = 36) (mm/yr). At Xiamen there are 3.2 (L = 18), 1.7 (L = 24), 1.4 (L = 30) and 1.3 (L = 36) (mm/yr). Therefore in this work the length L = 30 years is used for analyzing the trends of extremes. This added another 2 gauges long enough for determining trends, namely, Keelung and Kaohsiung.

3.   Results and Discussion

In this section, sea level has been split into three component parts (MSL, tide and surge). Trends in the tide and surge component, and the interaction between surge and tide, have been separately analyzed before total extreme sea levels are analyzed.

3.1.   *Change of the MSL*

Annual MSL of the four gauges with at least 30 years of annual mean data (Keelung, Xiamen, Kaohsiung and Quarrybay) together with another five gauges (only with annual MSL data) Yantai, Qinhuangdao, Kanmen, Zhapo and Macau, were analyzed in this part. Results are shown in Figure 3, and the linear trends are listed in Table 1.

Results clearly show decadal variability at all gauges. The MSL rise rates along the China coast have a highly nonuniform distribution in space. At Macau and Quarrybay the changes are quite different even if their locations are close. Linear increasing trends are significant at 5 tide gauges: Kanmen, Keelung, Zhapo, Xiamen and Quarrybay. Rates are between about 1.4 and 3.5 mm/yr. At the other four gauges, both positive and negative trends exist, but they are all nonsignificant.

Uncertainties exist for the above results, the uncertainties may come from the land movements and the quality of tidal data.

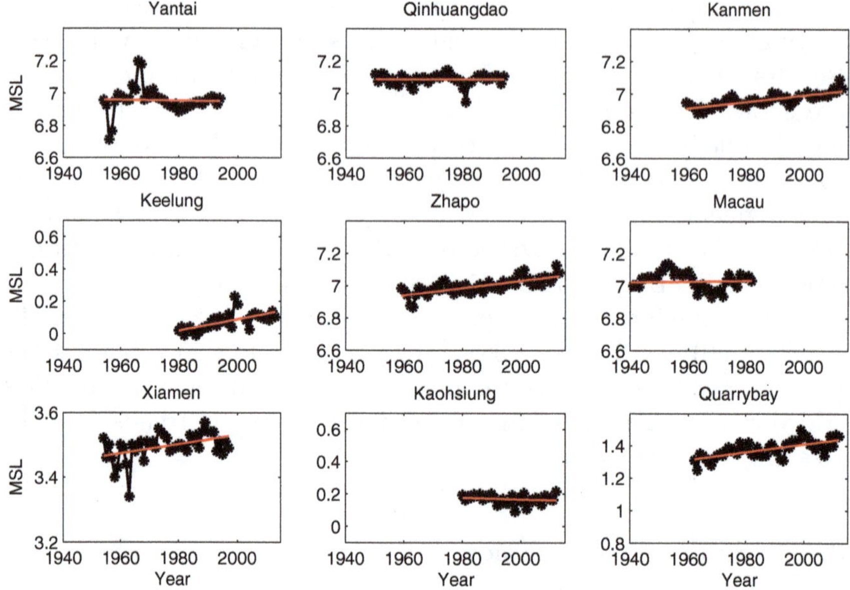

Fig. 3.   Time series of annual mean sea level (stars) with fitted linear trends at nine gauges along the Chinese coast (red straight line). Unit: m.

Table 1.   List of MSL trends at nine tide gauges.

Tide gauges	Period	Length (year)	Trend (mm/yr)
Yantai	1954–1994	41	−0.2
Qinhuangdao	1950–1994	45	0.1
Kanmen	1959–2013	54	*2.0*
Keelung	1980–2013	34	*3.5*
Zhapo	1959–2013	54	*2.3*
Macau	1925–1982	58	0.3
Xiamen	1954–1997	44	*1.4*
Kaohsiung	1980–2012	33	−0.5
Quarrybay	1962–2012	51	*2.4*

Note: Trends with significance at the 95% confidence level are italicized.

Such as in Yantai and Qinhuangdao, the annual MSL have obvious inhomogeneities in some years. Also the vertical land movements in China vary highly geographically (Hu *et al.*, 1992, 1993). Huang *et al.* (1991) summarized the rates of land movement of some tide gauges along the China coast using data from 1966 to 1988. Their results concern 5 tide gauges which we studied here: Yantai

(2.1 mm/yr), Qinhuangdao (3.9 mm/yr), Kanmen (−2.2 mm/yr), Zhapo (2.5 mm/yr) and Xiamen (1.5 mm/yr).

3.2. *Changes of the Tide*

In this part, the annual mean high water (MHW), annual mean low water (MLW) and annual mean tidal range (MTR) were used to analyze the changes of tides at four gauges, where time series were long enough. The annual results and trends are shown in Figure 4. There is considerable variability from year to year in the time series in which a significant part is related to the 18.6 year nodal cycle. This variability distorts the fitted trends, particularly over short records.

Annual values show that the tidal ranges at Xiamen are largest among the four gauges, ranging from 3.80 to 4.10 m. The tidal ranges at Kaohsiung and Keelung are very close, ranging from 0.65 to 0.85 m. At Quarrybay, the ranges are smaller than those at Xiamen while much larger than at the other two gauges, which is from 1.50 to 1.65 m.

The long time trend rates are listed in Table 2.

They are all statistically significant. At Xiamen, Kaohsiung and Keelung there are increases in MHW and decreases in MLW resulting in increases in MTR. Especially at Xiamen, the rate of the MHW-trend is 3 times of that at Kaohsiung and Keelung. At Quarrybay, there is a small decrease in MHW and an increase in MLW resulting in an overall decrease in MTR.

The increase of −0.2 to 1.8 mm/yr in MHW is small when compared with the MSL (mean sea level) trends experienced in these four gauges (−0.5 to 3.5 mm/yr; see Table 1). In relation to extreme sea levels, the increase in MHW is also important.

3.3. *Changes of the Surge*

As mentioned before, the surge is determined by subtracting the tidal harmonics from the time series of hourly sea level variations at the tide gauges. The three indices described above (frequency, duration and intensity) of storm surges at all tide gauges have been calculated and they showed quite similar characteristics. Here we only show the results of the intensity (the annual total integral (hour) of the nontidal components above the 99% (hour · m)) as it reflects more directly in the changes of the extreme sea levels.

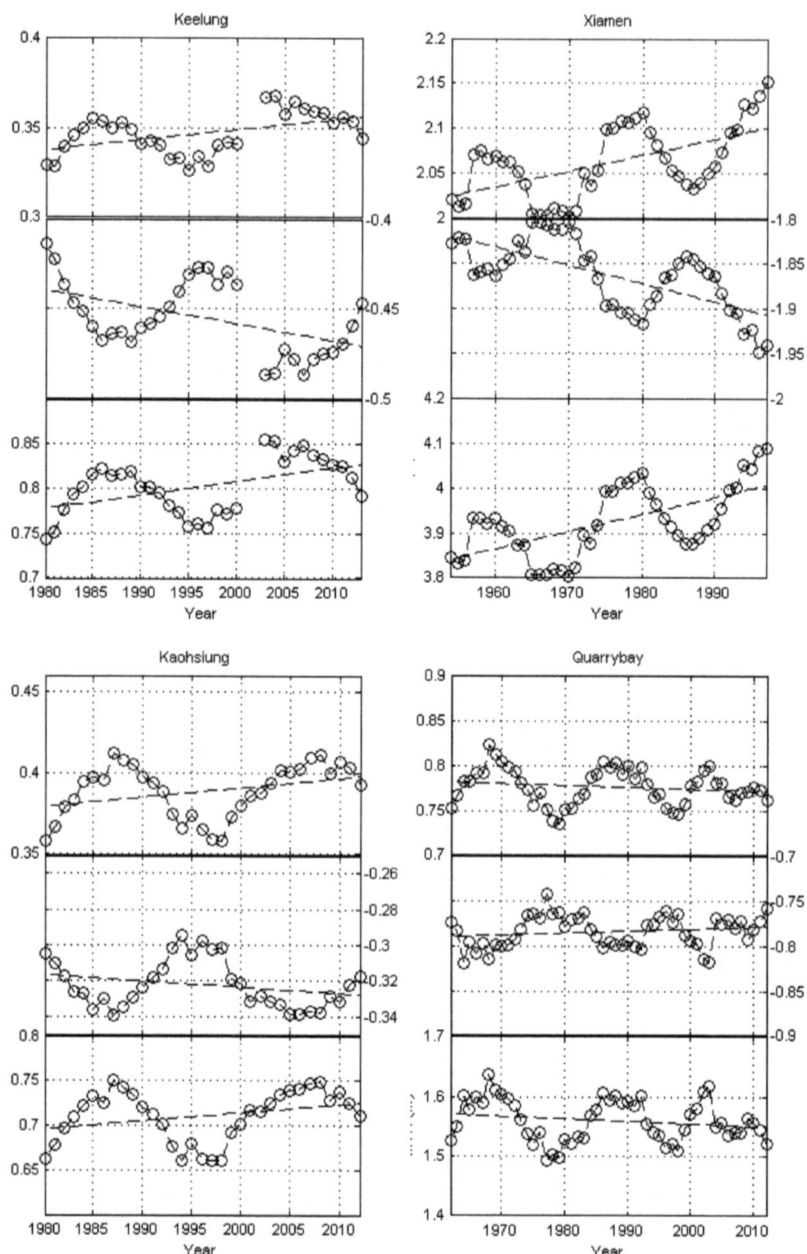

Fig. 4. (Top) Annual mean high water (MHW), (middle) annual mean low water (MLW), and (bottom) annual mean tidal range (MTR) at Keelung, Xiamen, Kaohsiung and Quarrybay. The dotted lines show the linear trends. Units: m.

Table 2.   The linear trends (mm/yr) of the MHW, MLW, and MTR at four tide gauges.

Trends at gauge	Data available	Number of years	MHW-trend (mm/yr)	MLW-trend (mm/yr)	MTR-trend (mm/yr)
Quarrybay	1962–2012	51	*−0.2*	*0.2*	*−0.4*
Xiamen	1954–1997	44	*1.8*	*−2.0*	*3.8*
Kaohsiung	1980–2012	33	*0.6*	*−0.3*	*0.9*
Keelung	1980–2013	34	*0.6*	*−0.9*	*1.5*

Note: Trends with significance at the 95% confidence level are italicized.

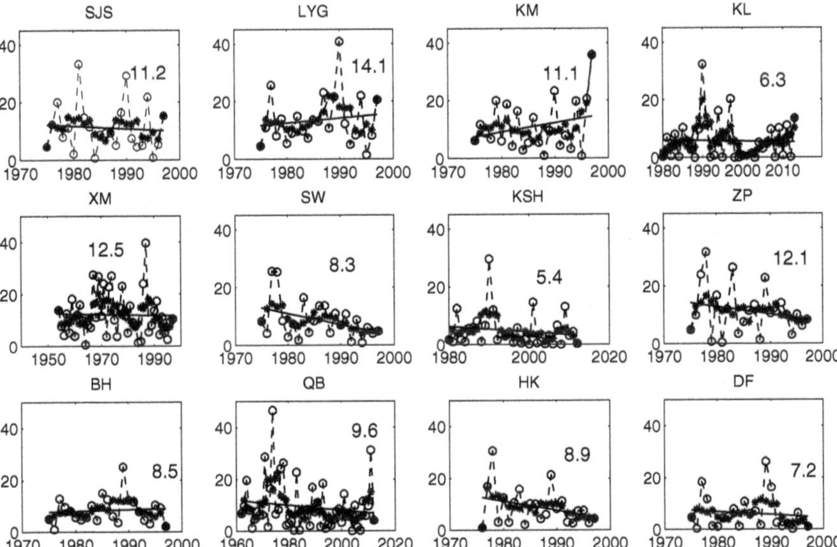

Fig. 5.   Annual surge total intensity (dashed line), and long-time trend (straight line) and the mean value of the intensity (number) at all 12 tide gauges. Unit: m × hour.

The annual average intensities at Shijiusuo, Lianyungang, Kanmen, Xiamen and Zhapo are higher and at Keelung, Kaohsiung and Dongfang the intensities are smallest of all gauges (in Figure 5). Time series of annual intensity of surges at all 12 gauges are shown in Figure 5. The interannual variability and decadal variability are obvious at most of the tide gauges, and the interannual variability is larger than the decadal variability. The intensity of surges appears to be larger in the middle of 1960s and around the 1990. There are also gauges, Zhapo and Haikou, where the decadal variability is very small. The total intensity of different gauges has maxima in

Table 3.　Long-time trend of the surge intensity at all gauges.

Tide gauge	Time	Length (years)	Trend (m * h/yr)
Shijiusuo	1975–1997	23	−0.09
Lianyungang	1975–1997	23	0.17
Kanmen	1975–1997	23	0.32
Keelung	1980–2013	34	−0.03
Xiamen	1954–1997	44	−0.02
Shanwei	1975–1997	23	*−0.39*
Kaohsiung	1980–2012	33	*−0.10*
Zhapo	1975–1997	23	*−0.26*
Beihai	1975–1997	23	0.06
Quarrybay	1962–2012	51	*−0.10*
Haikou	1976–1997	22	*−0.36*
Dongfang	1975–1997	23	−0.12

Note: Trends with significance at the 95% confidence level are italicized.

different years. The long-time trends at five tide gauges, Shanwei, Kaohsiung, Zhapo, Quarrybay and Haikou, are significantly decreasing (Table 3). At other gauges both positive and negative trends exist, but they are not significant at the 95% confidence level.

3.4.　*Changes to Tide-Surge Interaction*

The frequencies of surge above 99% level with respect to the timings of high tide for each of the 12 stations, i.e., the characteristics of tide-surge interaction (see Section 2.2), show four kinds of pattern (Figure 6). The distribution is significantly different from uniform distributions.

At Shijiusuo, Lianyungang, Kanmen, Xiamen and Haikou, the peak often occurs at a rising tide (about 4–5 h before the high tide). At Kaohsiung and Zhapo, the surge happens mostly at a falling tide (about 4–5 h after the high tide). At Shanwei, Beihai, Quarrybay and Dongfang the distribution has peaks at both, i.e., rising and falling tides. Finally, at Kanmen, Shanwei, Beihai and Dongfang storm surges tend to occur together with the high tide.

We use the test statistic X^2/T to quantify the strength of the nonuniformity of the distribution i.e., of the tide-surge interaction at all gauges (included in Figure 6). It shows that the tide-surge interactions differ among the gauges stations, also among neighboring gauges. At Lianyungang and Xiamen the interactions are much more

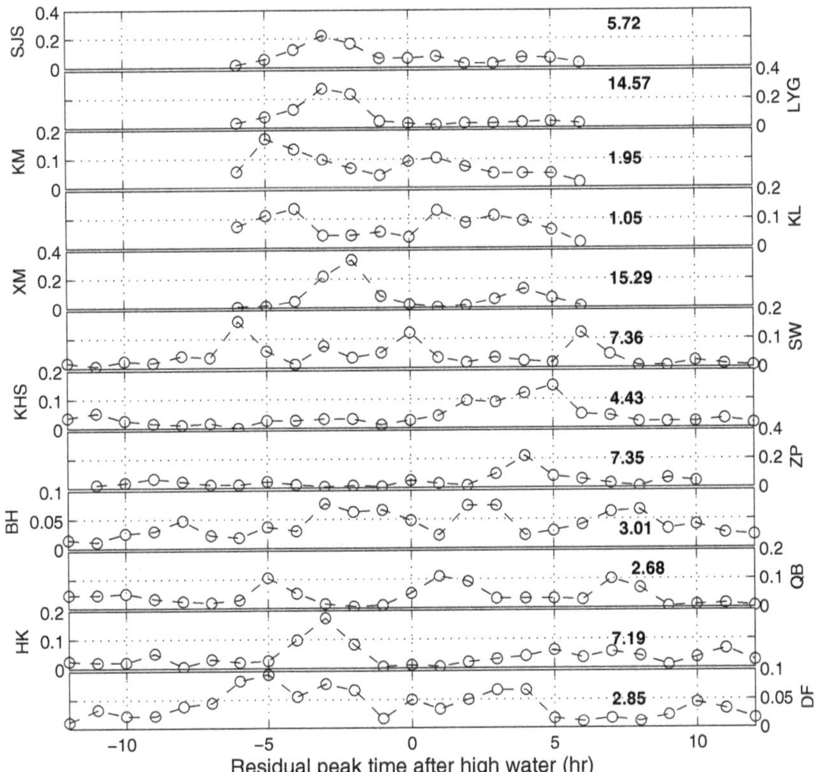

Fig. 6. Number of surge peaks above the 99% plotted with respect to timing of the high tide (in hours) for all stations, y axes is the percentage, the tide-surge interaction at all tide gauges are significant at 95% confidence levels.

pronounced than at other gauges. Interaction is weakest at Keelung and Kanmen, with $X^2/T < 2$.

In order to test whether changes in tide-surge interaction have taken place over the past decades at these 4 gauges, the phases of the surge peaks have been plotted for all 10 year overlapping periods at the four gauges with time series of at least 30 years (Figure 7). At Xiamen, Kaohsiung and Quarrybay the phases of the peaks are quite stable. Especially at Xiamen station nearly no change happened during the past decades in the distribution of the tide surge interaction. At Keelung some differences exist in the past few decades, especially for the peaks which happened on the falling tide. It seems that the tide-surge interaction become more stable as the strength of interaction increases.

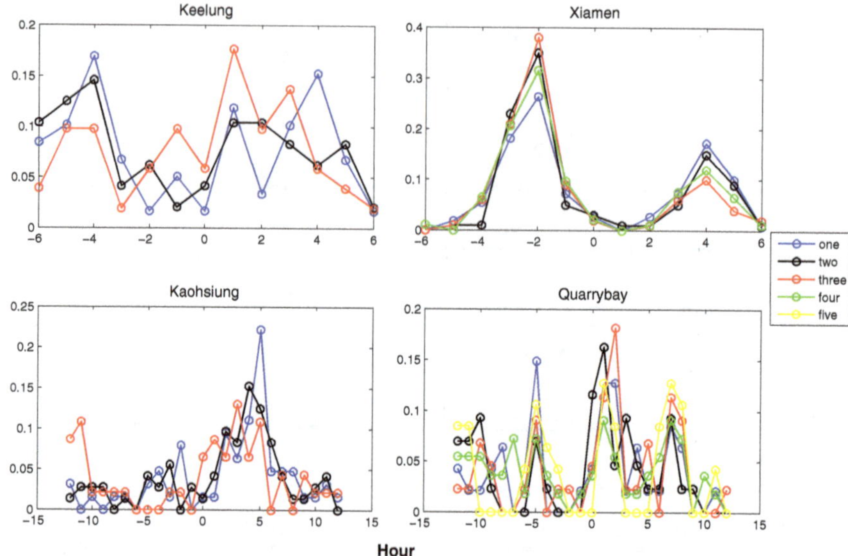

Fig. 7.   The frequencies of surge peaks plot for all 10 year overlapping periods at Keelung, Xiamen, Kaohsiung, and Quarrybay, y axes is the percentage.

In summary, results show that tide-surge interaction is important along the China coast and needs to be taken into account in the extreme sea level assessment. But as the interaction at Xiamen, Kaohsiung and Quarrybay are stable and the tide-surge interaction at Keelung is quite small, the tide-surge interaction seems to play a minor direct role in the decadal and long-time changes of extreme sea levels at those stations.

3.5. *Changes to Extreme Sea Levels*

In this section, we study the changes of extreme sea level (including the tidal components) for separating the change due to mean sea level rise and due to changing dynamical factors (tides, storms). Results are shown in Figure 8. In terms of total sea level variations, the three percentiles have statistically significant positive trends at Keelung and Xiamen. The trends are larger than that of the mean sea levels (Section 3.1) especially for the 99.9% trend. Especially at Xiamen the rate of extreme sea level is about 4 times of the mean sea level. At Kaohsiung, positive trends are also found at all percentile levels

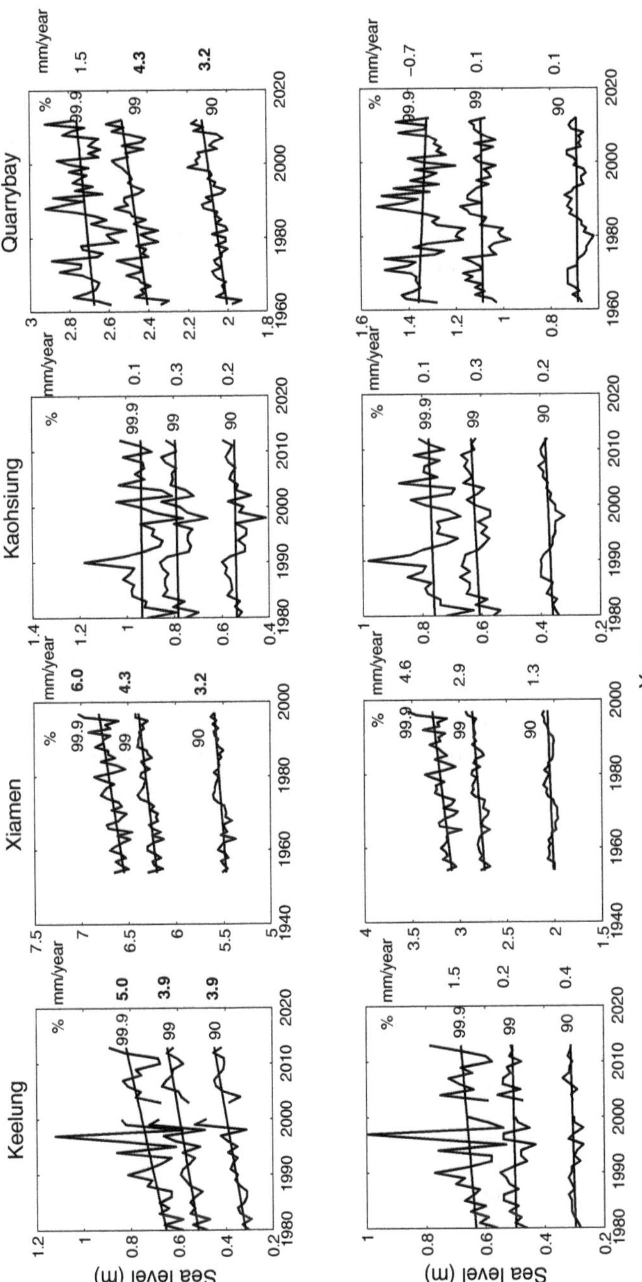

Fig. 8.   Time series of percentiles of (top) total and (bottom) reduced sea level for Keelung, Xiamen, Kaohsiung, and Quarrybay. The straight lines are the linear trends, their change rates (mm/yr) are listed at right of the figures (in bold if significant at 95%).

but they are not significant. At Quarrybay, the 99% and 90% have statistically significant positive trends, with magnitude greater than the mean sea levels. At the 99.9% level, the trend is also positive but not statistically significant.

Once the percentile time series are reduced by subtracting the annual medians mean sea levels, none of the trends are statistically significant. Positive trends remain at Keelung, Xiamen and Kaohsiung. At Quarrybay, the 99.9% percentile level has a negative trend but the 90% and 99% percentiles have positive trends.

The time series also exhibit marked decadal variability at all four tide gauges, but their characteristics of decadal variability differ due to their locations.

3.6.   *Relationship Between Extreme Sea Levels and Components*

We examine the relationship between the annual variations of the extreme sea levels (in terms of 99.9 percentiles of total sea level) and of the variations components of sea level separately. Here the annual surge intensity is considered to represent the surge part, whereas the annual mean high level can be regarded as the representative for the tide component. The correlations are shown in Table 4.

Results show that there are some characteristics in common among these four tide gauges. The extreme sea level and the MSL show significant positive correlation at all four tide gauges, especially at Keelung, Xiamen and Kaohsiung (of above 0.50). This finding is generally consistent with studies in other regional studies, e.g., English Channel (Haigh *et al.*, 2010; Pirazzoli *et al.*, 2006);

Table 4.   The correlations between the 99.9% total sea level variations and the sea level components, MSL, surge and tide.

	MSL	Surge	Tide
Keelung	*0.58*	*0.52*	−0.01
Xiamen	*0.52*	−0.05	*0.54*
Kaohsiung	*0.65*	*0.65*	*0.52*
Quarrybay	*0.37*	*0.51*	0.19

Note: Correlations with significance at the 95% confidence level are italicized.

Liverpool, UK (Woodworth and Blackman, 2002; San Francisco, USA (Bromirski *et al.*, 2003), also in a global assessment (Woodworth and Blackman, 2004).

As for the surge and tide, results differ due to their locations. The surge shows significant positive correlation (larger than 0.50) to the extreme sea level at Keelung, Kaohsiung and Quarrybay. It means that the storm surges also play important roles in the change of extreme sea levels at these three gauges. However, in Xiamen there is no obvious correlation between the extreme sea level variations and surge variations. A possible reason is that the tide surge interaction in Xiamen station is much larger than those at the other three. Nearly all surge maxima occur on the rising tide and falling tide (Section 3.4), and the tide amplitude in Xiamen is much larger than the others, which may be also an important reason (Section 3.2). Thus the sea level reaches its peak not at the time when the wind effect reached its maxima. Thus the changes of the surges were less important to the change of extreme sea levels.

The tide shows significant positive correlations, of 0.50 and more, with the variations of annual extreme sea level at Kaohsiung and Xiamen. But at Keelung and Quarrybay, no significant correlations were found. This difference may also be caused by the tide surge interaction. Figures 6 and 7, demonstrate that at Keelung and Quarrybay, the timing of many surges was close to that of the high tide time, compared to Xiamen and Kaohsiung. Thus the tide plays a more important role in Xiamen and Kaohsiung than in Keelung and Quarrybay. It seems that Kaohsiung differs from the other three gauges, as both the surge and tide show significant positive correlation to the extreme sea levels. From Figures 6 and 7, the storm surges most happen during 0 to 6 (h) after the high tide. Thus, both the surge and tide play important roles in generating the extremes. Summing up, the tide surge interaction affects the extreme sea levels through the surge part and the tide part rather than itself.

4. Conclusions

Hourly sea level data from tide gauges along the Chinese coast are used for analyzing changes of the extreme sea level in this area and to assess to what extent changes in extreme sea level over the past several decades were driven by, changes in mean sea level, tide, surge, or in tide-surge interaction. The mean sea level (MSL), tide, surge and tide surge interaction have been separately examined for trends.

Unfortunately, the data are of limited quality, both in terms of time period, spatial coverage and, possibly homogeneity. It would be favorable if more data would be made accessible for scientific analysis.

Clearly, decadal variability exists in the MSL at the selected tide gauges, while the mean sea level trends exhibit, spatially, a highly nonuniform distribution. Five of 9 tide gauges (Kanmen, Keelung, Zhapo, Xiamen and Quarrybay) show significant positive trends. The tide component was analyzed at tide gauges with data more than 30 years. Significant but spatially nonuniform trends in tide characteristics were found at all four gauges. The surge component was studied at all 12 tide gauges, and obvious interannual variability and decadal variability exists at all tide gauges. As for the long term trend, annual tide intensity at 5 tide gauges shows significant decreasing trends. Significant but spatially nonuniform tide-surge interactions were found at all 12 tide gauges. No obvious change, in particular where the tide surge interactions were strong, was found in the tide surge interaction during the past few decades.

There is evidence for an increase in extreme sea levels during the past few decades at Keelung, Xiamen and Quarrybay. Also clear decadal variability is present at these tide gauges. The variations in annual extreme sea level and in the annual MSL show significant positive correlation at all four tide gauges, which means that the change in mean sea level plays an important role in the change in extreme sea levels in this area. The surge shows significant positive correlation to the extreme sea level at Keelung, Kaohsiung and Quarrybay, thus the storm surges play important roles in the changes of extreme sea levels. The tide shows significant positive correlation to the extreme sea level at Kaohsiung and Xiamen, thus the tide plays important roles in the changes of extreme sea levels at these two gauges. Tide surge interaction plays an important role in determining which component, the surge or the tide, is more important for the change of extreme sea levels.

In conclusion, the changes in extreme sea levels along the China coast are highly affected by the changes in mean sea levels. But the changes are not totally due to the change of mean sea levels. Changes in surges and astronomic tide contribute — in a spatially nonuniform manner. The tide surge interactions are important in the changes of extreme sea levels, but not in a direct way. Thus, if we want to envisage possible changes of the extreme sea levels in one area in the future along the China coast, we need to take the characteristic of this area into consideration.

Funding

This work is supported by China Scholarship Council (201306330027) and NSFC-Shandong Joint Fund for Marine Science Research Centers (grant U1406401).

Acknowledgments

We would thank the University of Hawaii Sea Level Center and the Permanent Service for Mean Sea Level (PSMSL) for the data used in this paper. J Feng also thanks for the hospitality of the Helmholtz-Zentrum Geesthacht during a 24 month visit and for the provision of computing facilities during the time this work was carried out.

References

Antony, C. and Unnikrishnan, A. S. (2013). Observed characteristics of tide-surge interaction along the east coast of India and the head of Bay of Bengal, *Estuarine Coastal and Shelf Science* **131**, pp. 6–11, doi:10.1016/j.ecss.2013.08.004.

Austin, R. M. (1991). Modeling Holocene Tides on the NW European Continental-Shelf, *Terra Nova* **3**, pp. 276–288, doi:10.1111/j.1365-3121.1991.tb00145.x.

Bernier, N. B. and Thompson, K. R. (2006). Predicting the frequency of storm surges and extreme sea levels in the northwest Atlantic, *Journal of Geophysical Research-Oceans* **111**, C10, p. C10009, doi:10.1029/2005JC003168.

Bernier, N. B. and Thompson, K. R. (2007). Tide-surge interaction off the east coast of Canada and northeastern United States, *Journal of Geophysical Research-Oceans* **112**, C6, p. C06008, doi:10.1029/2006JC003793.

Bolle, A., Wang, Z. B., Amos, C. and De Ronde, J. (2010). The influence of changes in tidal asymmetry on residual sediment transport in the Western Scheldt, *Continental Shelf Research* **30**, 8, pp. 871–882, doi:10.1016/j.csr.2010.03.001.

Bromirski, P. D., Flick, R. E. and Cayan, D. R. (2003). Storminess variability along the California coast: 1858-2000, *Journal of Climate* **16**, 6, pp. 982–993, doi:10.1175/1520-0442(2003)016<0982: SVATCC>2.0.CO;2.

Butler, A., Heffernan, J. A., Tawn, J. A., Flather, R. A. and Horsburgh, K. J. (2007). Extreme value analysis of decadal variations in storm surge elevations, *Journal of Marine Systems* **67**, 1–2, pp. 189–200, doi:10.1016/j.jmarsys.2006.10.006.

Church, J. A. and White, N. J. (2006). A 20th century acceleration in global sea-level rise, *Geophysical Research Letters* **33**, 1, p. L01602, doi:10.1029/2005GL024826.

Egbert, G. D., Ray, R. D. and Bills, B. G. (2004). Numerical modeling of the global semidiurnal tide in the present day and in the last glacial maximum, *Journal of Geophysical Research-Oceans* **109**, C3, p. C03003, doi:10.1029/2003JC001973.

Green, J. (2010). Ocean tides and resonance, *Ocean Dynamics*, 60, 1243–1253, doi:10.1007/s10236-010-0331-1.

Haigh, I., Robert, N. and Neil, W. (2010). Assessing changes in extreme sea levels: Application to the English Channel, 1900-2006, *Continental Shelf Research* **30**, 9, pp. 1042–1055, doi:10.1016/j.csr.2010.02.002.

Hallegatte, S., Green, C., Nicholls, R. J. and Corfee-Morlot, J. (2013). Future flood losses in major coastal cities, *Nature Climate Change* **3**, 9, pp. 802–806, doi:10.1038/NCLIMATE1979.

Horsburgh, K. J. and Wilson, C. (2007). Tide-surge interaction and its role in the distribution of surge residuals in the North Sea, *Journal of Geophysical Research-Oceans* **112**, C8, p. C08003, doi:10.1029/ 2006JC004033.

Hu, H. M., Huang, L. and Yang, G. H. (1992). Recent crustal vertical movement in the Changjiang river and its adjacent area [in Chinese with English abstract], *Acta Geographica Sinica* **47**, 1, pp. 22–30.

Hu, H. M., Huang, L. and Yang, G. H. (1993). Recent vertical crustal deformation in the coastal area of Eastern China [in Chinese with English abstract], *Chinese Journal of Geology* **28**, 3, pp. 270–278.

Huang, L. R., Yang, G. H. and Hu, H. M. (1991). The isostatic datum for studying of the sea level changes along the coast of China, *Seismology Geology* **13**, pp. 8–14.

Jay, D. A. (2009). Evolution of tidal amplitudes in the eastern Pacific Ocean, *Geophysical Research Letters* **36**, p. L04603, doi:10.1029/ 2008GL036185.

Johns, B. and Ali, M. A. (1980). Numerical modeling of storm surges in the Bay of Bengal, *Quarterly Journal of the Royal Meteorological Society* **106**, 447, pp. 1–18.

Karim, M. F. and Mimura, N. (2008). Impacts of climate change and sea-level rise on cyclonic storm surge floods in Bangladesh, *Global Environmental Change-Human and Policy Dimensions* **18**, 3, pp. 490–500, doi:10.1016/j.gloenvcha.2008.05.002.

Kulkarni, A. and von Storch, H. (1995). Monte Carlo experiments on the effect of serial correlation on the Mann-Kendall test of trend, *Meteorologische Zeitschrift* **4**, 2, pp. 82–85, doi: 10.1127/metz/4/1992/82.

Marcos, M., Tsimplis, M. N. and Shaw, A. G. P. (2009). Sea level extremes in southern Europe, *Journal of Geophysical Research-Oceans* **114**, p. C01007, doi:10.1029/2008JC004912.

Marcos, M., Jorda, G. A., Gomis, D. and Perez, B. (2011). Changes in storm surges in southern Europe from a regional model under climate change scenarios, *Global and Planetary Change* **77**, 3–4, pp. 116–128, doi:10.1016/j.gloplacha.2011.04.002.

Mendez, F. J., Menendez, M., Luceno, A. and Losada, I. J. (2007). Analyzing monthly extreme sea levels with a time-dependent GEV model, *Journal of Atmospheric and Oceanic Technology* **24**, 5, pp. 894–911, doi:10.1175/JTECH2009.1.

Menendez, M. and Woodworth, P. L. (2010). Changes in extreme high water levels based on a quasi-global tide-gauge data set, *Journal of Geophysical Research-Oceans* **115**, p. C10011, doi:10.1029/2009JC005997.

Mudersbach, C., Wahl, T., Haigh, I. D. and Jensen, J. (2013). Trends in high sea levels of German North Sea gauges compared to regional mean sea level changes, *Continental Shelf Research* **65**, pp. 111–120, doi:10.1016/j.csr.2013.06.016.

Pascual, A., Marcos, M. and Gomis, D. (2008). Comparing the sea level response to pressure and wind forcing of two barotropic models: Validation with tide gauge and altimetry data, *Journal of Geophysical Research-Oceans* **113**, C7, p. C07011, doi:10.1029/2007JC004459.

Pawlowicz, R., Beardsley, B. and Lentz, S. (2002). Classical tidal harmonic analysis including error estimates in MATLAB using T-TIDE, *Computers & Geosciences* **28**, 8, pp. 929–937, doi:10.1016/S0098-3004(02)00013-4.

Pickering, M. D., Wells, N. C., Horsburgh, K. J. and Green, A. M. (2012). The impact of future sea-level rise on the European Shelf tides, *Continental Shelf Research* **35**, pp. 1–15, doi:10.1016/j.csr.2011.11.011.

Pirazzoli, P., Costa, S., Dornbusch, U. and Tomasin, A. (2006). Recent evolution of surge-related events and assessment of coastal flooding risk on the eastern coasts of the English Channel, *Ocean Dynamics* **56**, 5–6, pp. 498–512, doi:10.1007/s10236-005-0040-3.

Pirazzoli, P. A., Regnauld, H. and Lemasson, L. (2004). Changes in storminess and surges in western France during the last century, *Marine Geology* **210**, 1–4, pp. 307–323, doi:10.1016/j.margeo.2004.05.015.

Pugh, D. J. (1987). Tides, surges and mean sea level. A handbook for engineers and scientists. *Continental Shelf Research*, p. 472.

Ratsimandresy, A. W., Sotillo, M. G., Carretero Albiach, J. C., Alvarez Fanjul, E. and Hajji, H. (2008). A 44-year high-resolution ocean and atmospheric hindcast for the Mediterranean Basin developed within the HIPOCAS Project, *Coastal Engineering* **55**, 11, pp. 827–842, doi:10.1016/j.coastaleng.2008.02.025.

Shaw, J., Amos, C. L., Greenberg, D. A., O'Reilly, C. T., Parrott, D. R. and Patton, E. (2010). Catastrophic tidal expansion in the Bay of Fundy, Canada, *Canadian Journal of Earth Sciences* **47**, 8, pp. 1079–1091, doi:10.1139/E10-046.

Tsimplis, M. N. and Woodworth, P. L. (1994). The global distribution of the seasonal sea-level cycle calculated from coastal tide-gauge data, *Journal of Geophysical Research-Oceans* **99**, C8, pp. 16031–16039, doi:10.1029/94JC01115.

Uehara, K., Scourse, J. D., Horsburgh, K. J., Lambeck, K. and Purcell, A. P. (2006). Tidal evolution of the northwest European shelf seas from the Last Glacial Maximum to the present, *Journal of Geophysical Research-Oceans* **111**, C9, p. C09025, doi:10.1029/2006JC003531.

von Storch, H. and Reichardt, H. (1997). A scenario of storm surge statistics for the German bight at the expected time of doubled atmospheric carbon dioxide concentration, *Journal of Climate* **10**, 10, pp. 2653–2662, doi:10.1175/1520-0442(1997)010<2653:ASOSSS>2.0.CO;2.

von Storch, H. and Woth, K. (2008). Storm surges: Perspectives and options, *Sustainability Science* **3**, 1, pp. 33–43, doi:10.1007/s11625-008-0044-2.

von Storch, H. and Zwiers, F. W. (1999). *Statistical analysis in climate research*, Cambridge, UK: Cambridge University Press, doi:10.1017/CBO9780511612336.

Weisse, R., Bellafiore, D., Menendez, M., Mendez, F., Nicholls, R. J., Umgiesser, G. and Willems, P. (2014). Changing extreme sea levels along European coasts, *Coastal Engineering* **87**, pp. 4–14, doi:10.1016/j.coastaleng.2013.10.017.

Woodworth, P. L. (2010). A survey of recent changes in the main components of the ocean tide, *Continental Shelf Research* **30**, 15, pp. 1680–1691, doi:10.1016/j.csr.2010.07.002.

Woodworth, P. L. and Blackman, D. L. (2002). Changes in extreme high waters at Liverpool since 1768, *International Journal of Climatology* **22**, 6, pp. 697–714, doi:10.1002/joc.761.

Woodworth, P. L. and Blackman, D. L. (2003). Evidence for systematic changes in extreme high water since the mid-1970s, *Journal of Climate*, **17**, 1190–1197, doi:10.1175/1520-0442(2004)017<1190: EFSCIE>2.0.CO;2.

Woodworth, P. L. and Blackman, D. L. (2004). Evidence for systematic changes in extreme high waters since the mid-1970s, *Journal of Climate* **17**, 6, pp. 1190–1197, doi:10.1175/1520-0442(2004)017 <1190:EFSCIE>2.0.CO;2.

Woodworth, P. L., Flather, R. A., Williams, J. A., Wakelin, S. L. and Jevrejeva, S. (2007). The dependence of UK extreme sea levels and storm surges on the North Atlantic Oscillation, *Continental Shelf Research* **27**, 7, pp. 935–946, doi:10.1016/j.csr.2006.12.007.

Woth, K. (2005). North Sea storm surge statistics based on projections in a warmer climate: How important are the driving GCM and the chosen emission scenario? *Geophysical Research Letters* **32**, 22, p. L22708, doi:10.1029/2005GL023762.

Woth, K., Weisse, R. and von Storch, H. (2006). Climate change and North Sea storm surge extremes: An ensemble study of storm surge extremes expected in a changed climate projected by four different regional climate models, *Ocean Dynamics* **56**, 1, pp. 3–15, doi: 10.1007/s10236-005-0024-3.

Zhang, K. Q., Douglas, B. C. and Leatherman, S. P. (2000). Twentieth-century storm activity along the US east coast, *Journal of Climate* **13**, 10, pp. 1748–1761, doi:10.1175/1520-0442(2000)013<1748: TCSAAT>2.0.CO;2.

Zhang, W. Z., Shi, F., Hong, H. S., Shang, S. P. and Kirby, J. T. (2010). Tide-surge Interaction Intensified by the Taiwan Strait, *Journal of Geophysical Research-Oceans* **115**, p. C06012, doi:10.1029/ 2009JC005762.

Chapter 12

Changes of Storm Surges in the Bohai Sea Derived from a Numerical Model Simulation, 1961–2006[*]

J. Feng[†,‡,§], H. von Storch[§], R. Weisse[§], and W. Jiang[‡,¶]

[†]*National Marine Data & Information Service, Tianjin, China*
[‡]*Physical Oceanography Laboratory, Ocean University of China, Qingdao, China*
[§]*Helmholtz-Zentrum Geesthacht, Centre for Materials and Coastal Research, Geesthacht, Germany*
[¶]*Laboratory of Marine Environment and Ecology, Ocean University of China, Qingdao, China*

Using the tide-surge circulation model ADCIRC, the storm surges in the Bohai Sea were hindcasted from 1961 to 2006 after a regional model-based reconstruction of wind conditions. Through comparison with four storm surge cases that happened in the Bohai Sea and long-time observations at four tide gauges in the Yellow Sea, it is concluded that the model is capable of reproducing the conditions of storm surges in the past few decades in this area. The spatial distribution, the seasonal variation, the interdecadal variability, and the long-time trend were analyzed using the model results. Results show that the storm surges in

---

[*]This chapter was originally published in *Ocean Dynamics*, 66(10), 1301–1315, doi:10.1007/s10236-016-0986-3. Copyright (2016), with permission from SNCSC.

311

the three bays of the Bohai Sea are more serious than those in other areas. The storm surges exhibit obvious seasonal variations—they are more serious in spring and autumn. Obvious interdecadal variations and long-time decreasing trends take place in the Bohai Sea. Storm surge indices show statistically significant negative correlations to the Arctic Oscillation (AO) and a statistically significant positive correlation to the Siberian High (SH). Linear regression analysis was used to determine a robust link between the indices of the storm surges and the AO and SH. Using this link, conditions of the storm surges from 1900 to 2006 were estimated from the long-time AO and SH.

1.   Introduction

Storm surge is one of the major geophysical risks in coastal areas, which is often associated with significant losses of life and property (Gönnert *et al.*, 2001; von Storch *et al.*, 2015). Storm surges that happened in China can be classified into two types: tropical storm surges and extratropical storm surges (Feng, 1982; Yang, 2000). The Bohai Sea (Figure 1) is a semienclosed sea that exchanges with the Yellow Sea through the Bohai Strait. It is bounded by the Changshan Islands chain between the Liaodong and Shandong Peninsulas. It is approximately 78,000 km^2, and the average depth is about 20 m. It consists of three bays: Laizhou Bay to the south, Liaodong Bay to the north, and Bohai Bay to the west. The M2 tide is the principal tide constituent; there are two amphidromic points in the Bohai Sea, the maximum tidal range is larger than 4 m and the maximum current is 2.0 m/s. Unlike the other marginal seas of China, the Bohai Sea is less susceptible to tropical cyclones because its geographical position (37° 07–41°N) allows only a few tropical cyclones to move far northward enough to generate storm surges in this area. Cold air outbreaks and extratropical cyclones, however, usually cause significant storm surges, particularly in the winter half year.

The growing concern about recent and possible future climate changes has motivated numerous efforts to gather long time datasets to assess such change. Many studies have been done in the past years on the changes of the mean sea level and extreme sea levels using the data from tide gauges (Marcos *et al.*, 2009; Mendez *et al.*, 2007; Menendez and Woodworth, 2010; Mudersbach *et al.*, 2013; Palumbo and Mazzarella, 1982; Tsimplis and Woodworth, 1994; von Storch

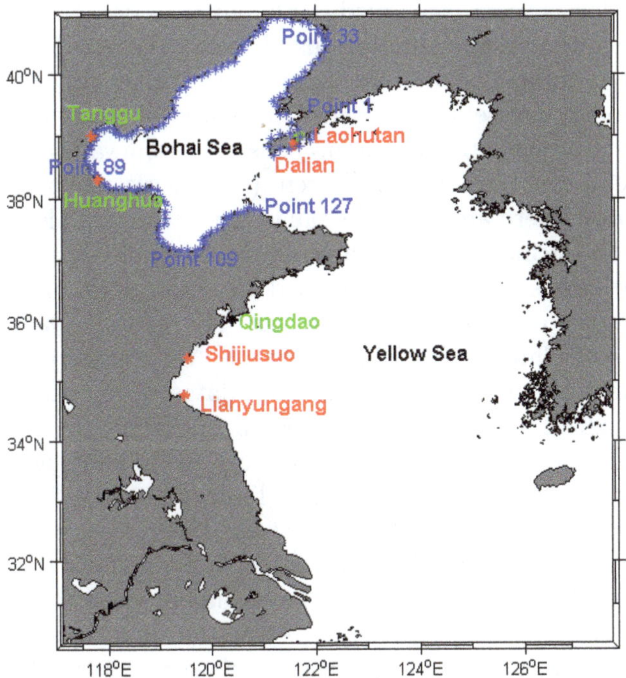

Fig. 1. Model domain of the tide-surge model and the locations of the Bohai Sea, the Yellow Sea, and the seven tide gauges: Laohutan, Dalian, Tanggu, Huanghua, Qingdao, Shijiusuo, and Lianyungang; the sea level data from Qingdao, Tanggu, and Huanghua (in green) serve as a reference to determine the quality of the simulation of three storm surge events. The data from Laohutan, Dalian, Shijiusuo, and Lianyungang (in red) are used to assess the performance of the multidecadal simulations. One hundred twenty-seven points are selected to represent the model result along the Bohai coast.

and Reichardt, 1997; Weisse *et al.*, 2014; Woodworth and Blackman, 2003, 2004; Woodworth *et al.*, 2007). There is indeed evidence for a general worldwide increase in extreme high water levels and mean sea levels in the past few decades. An interesting question is whether the increase in the tails of the distributions is mostly the result of a shift or a broadening of the distribution (Bernier and Thompson, 2006; Bromirski *et al.*, 2003; Church and White, 2006; Marcos *et al.*, 2009; von Storch and Reichardt, 1997; Zhang *et al.*, 2000).

Results show that there is no uniform rule worldwide, and the change of extreme sea levels sometimes is mostly caused by the increase in the mean sea level, but in other cases, extreme and mean sea levels seem decoupled. Unfortunately, the tide gauge data suffer from remarkable limitations. Firstly, the spatial as well as temporal coverage of the tide gauges is rather limited. According to some datasets, like the University of Hawaii Sea Level Center (data available at http://uhslc.soest.hawaii.edu) and the Global Sea Level Observation System (data accessible at www.glosssealevel.org), there are less tide gauge data available along the west coast of the Pacific, especially around the coast of China, than those along the coasts of the Atlantic and the east coast of the Pacific. Moreover, the duration of most of the tide gauges available along the coasts of China is less than 30 years. Very unfortunate for our study, there are no publicly accessible datasets for the Bohai Sea, even though such gauges have been in operation for many years. Secondly, these historical data exhibit various inhomogeneities, such as an ever-increasing accuracy, which limit the utility of the data for climatic studies (Jones, 1995; Karl *et al.*, 1993).

The limitations of the in situ data for building long-time datasets have been partly overcome by the availability of regional ocean models (Langenberg *et al.*, 1999). Over the past decades, numerical models have become one of the most useful tools for generating consistent and realistic long-time environment offshore data (Weisse *et al.*, 2009). Such "hindcasts" make use of long, high-resolution "analyses" of atmospheric conditions.

For the change of storm surge in the past and the possible future storm surge trends, many researches using numerical models have been done in the past few years (Feng *et al.*, 2014; Pascual *et al.*, 2008; Ratsimandresy *et al.*, 2008; Weisse *et al.*, 2014; Woth *et al.*, 2006). Such numerical model results have the advantage of generating information at locations and for times without observed data. Also, the model simulations have high temporal sampling rate, every hour or even less.

These extratropical storm surge hazards in the Bohai Sea have been studied by many researches in recent years. Yin *et al.* (2001), using a tide-surge model involving wave effect to simulate a typical storm surge case that happen in 1964, found that the model result was improved when taking the wave effect into consideration. Zhao and Jiang (2011a,b) studied the effects of tracks of cold air outbreaks in the Bohai Sea using a numerical model. Zhao and Jiang (2011a,b) studied the effects of coastal geometry on storm surges in the Bohai Sea. Feng *et al.* (2012) studied the effects of typhoon paths on the storm surge in Tianjin, located in the Bohai Bay. Zhang and Sheng (2015) used the Climate Forecast System Reanalysis (CFSR) data and POM hindcasted the extreme sea levels in the northwest Pacific area, and they estimated the 50-year return levels in this area. But, no works using the high-resolution atmospheric data have been done to study the temporal trends of storm surges in the Bohai Sea.

In the present study, storm surge conditions were obtained using the tide-surge circulation model ADCIRC, forced with a regional model-generated wind dataset from 1961 to 2006 (Wu *et al.*, 2011). The simulated surge results were validated by examining four storm surge cases in detail and by comparing the simulated and observed long-time observation data at four tide gauges in the Yellow Sea. Furthermore, the spatial distribution, the seasonal variations, the interdecadal variability, and the long-time trend were analyzed. The relationship between the variability of the storm count, duration, and intensity and the Arctic Oscillation (AO) and Siberian High (SH) were analyzed. The linear regression analysis was used for relating the storm surges to the states of the AO and the SH. It should be noticed that this study deals only with the impact of changing regional wind conditions of the Bohai Sea and the Yellow Sea. As a result, the rise in the mean sea level, which we believe is minor for the change of the wind driven part of the extremes (Feng *et al.*, 2014), has been ignored.

The outline of this chapter is as follows: a brief description and verification of the long-time regional wind datasets, the setup of the tide-surge circulation model, and all the data used in this study are described in Section 2. Our approach for taking into account serial correlations in time series before testing correlations of pairs of time series and the significance of trends is also sketched in Section 2. Section 3 is devoted to verifying the quality of the hindcast data with observations from different sources. The model results and discussion are analyzed in Section 4, and the main conclusions are drawn in Section 5.

2. Data and Model Setup

2.1. *The Long-Time Wind Dataset*

A 46-year (from January 1961 to December 2006) high resolution 10-m hindcast wind was supplied by Wu *et al.* (2011) for the whole Bohai Sea and Yellow Sea. This dataset is a hindcast dataset using the Fifth-Generation Penn State/NCAR Mesoscale Model (MM5) forced by the ERA-40 reanalysis (before 2000) and NCEP-FNL reanalysis data (since 2000). Some observational data have been assimilated, using a Newton nudging scheme: GTS transmitted local observations (1960–2007), Typhoon best track data from UNIsys (1947–2007, http://weather.unisys.com/hurricane/). and Digital Typhoon data (1978–2007). For inserting typhoons, the model wind was nudged toward a given vertical/radial wind profile (for details, refer to Wu *et al.* 2011). Obviously, the simulation does not return a homogeneous description of the weather stream across the entire time span 1961–2006. However, since the corrections introduced by the local data may be considered mostly homogeneous (statistically uniform across time), this may not be a too troublesome limitation for extratropical storms; for typhoons, the situation may be more severe, as the quality of the best track data has undergone some improvements through the advent of satellites and other changes (Barcikowska *et al.*, 2012; Hallegatte, 2007). A homogeneity test had been done in Wu *et al.* (2011) by correlating Wu winds and Climate Forecast System Reanalysis (CFSR) from 1980 to 2006. The strong correlation (about 74%) and the lack of obvious phase difference throughout the whole period indicate that the inhomogeneities are limited.

In order to drive the tide-surge model, 3 hourly, instantaneous values of the horizontal wind components at 10-m height with the resolution of $0.1° \times 0.1°$ were extracted. In the following sections, we name this dataset as "the Wu dataset."

Here, we test whether the dynamically downscaled Wu dataset provides added value over the driving ERA-40 especially near the coastline, which plays a more important role in the storm surges. Following Winterfeldt *et al.* (2011), the Brier skill score S is defined by

$$S = 1 - \sigma^2_{WO}/\sigma^2_{EO} \tag{1}$$

where σ^2_{WO} denotes the error variance between the Wu data W and verifying satellite observations O, and σ^2_{EO} is the corresponding error variance between the ERA-40 reanalysis wind fields E and the satellite observations. S measures the skill of the Wu wind data W relative to the ERA-40 data E in describing the reference as given by O. When $S > 0$, the Wu wind fits the observation (satellite data) better than the ERA-40 reanalysis wind. In this case, added value or additional knowledge in comparison with the reanalysis was achieved.

The satellite observations used here are the Blended Sea Winds from the National Climate Data Center (NOAA), which contain globally gridded, high-resolution ocean surface vector winds on a global $0.25°$ grid. They were generated by blending observations from multiple satellites (Zhang *et al.*, 2006). Prior to 1990, there were fewer satellites used. Therefore, we use the period 1990–1999.

Figure 2 shows the result of this exercise. When all wind speeds are taken into consideration, in most part of the Bohai Sea and Yellow Sea, the skill score is above 0, but there are still some areas with skill scores below 0, such as in the middle of the Liaodong Bay and north of the Yellow Sea. However, when satellite wind speeds of above 15 m/s are considered, the skill score in almost the entire domain is above 0, especially in the coastal area, which means that there is an added value for strong near-surface wind speeds near the coast.

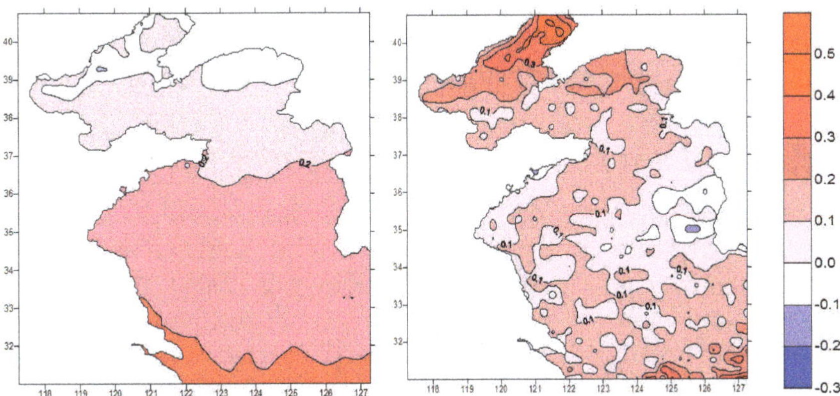

Fig. 2. Brier skill score S of the Wu wind data compared with ERA-40 wind, using Blended Sea Winds of NOAA as a reference; time period 1990–1999. Left: scores for all wind speeds. Right: scores for times when the satellite wind speeds are above 15 m/s.

We conclude that the Wu data are indeed more realistic than the ERA-40 data in the coastal areas of the Yellow Sea and in the Bohai Sea, at least between 1990 and 1999.

2.2.   *Tide-Surge Model Setup*

The dynamical downscaling of storm surges was carried out by driving the numerical tide-surge model ADCIRC. This model is an unstructured grid finite-element hydrodynamic model which allows both high grid resolution where solution gradients are large and low grid resolution where solution gradients are small (Bacopoulos *et al.*, 2012; Blain *et al.*, 1998; Ebersole *et al.*, 2010; Jr and Westerink, 1991; Mattocks and Forbes, 2008). The 2D ADCIRC model has been successfully used to study storm surge conditions in the Bohai Sea (Feng *et al.*, 2014, 2012).

The model domain covers the entire Bohai Sea and Yellow Sea (Figure 1), which is considered large enough to simulate storm surges accurately in the Bohai area. The unstructured triangular grids have a horizontal resolution varying from 10 km at the open boundary of the Yellow Sea to 1 km along the coast of the Bohai Sea. The model is forced by tides and winds. The amplitudes and phases of the tide constituents (M2, S2, K1, O1, K2, P1, N1, and O1) are taken from the OSU Tidal Prediction Software (OTPS) (Egbert *et al.*, 1994). The tide-surge model has been run twice with and without the wind to get the storm surge conditions in the Bohai Sea, and all the model results have been stored hourly at every grid point.

2.3.   *Local Sea Level Data*

Hourly observed storm surge data are available for three storm surge cases at three tide gauges, namely Qingdao, Huanghua, and Tanggu (marked in green in Figure 1). The data are provided by the National Marine Environmental Forecasting Center, but the data only last for a few days. We use these data to validate the performance of our modeling system in simulating storm surge events in detail (Section 3.1).

Another hourly sea level dataset comprises observations at four tide gauges (Dalian, Laohutan, Lianyungang, and Shijiusuo) in the Yellow Sea (marked in red in Figure 1). The data are accessible on http://uhslc.soest.hawaii.edu. For Dalian, the data are available for 1975–1990, for Laohutan from 1991 to 1997, for Lianyungang from 1975 to 1997, and for Shijiusuo from 1975 to 1997. After the removal of all the tidal components, by subtracting the astronomical tide components by means of a harmonic analysis (Pawlowicz *et al.*, 2002), two annual "storm surge indices" (SSIs) are derived for each year (Conte and Lionello, 2013). The positive SSI is the mean of the three highest, independent surge levels, while the negative SSI is the mean of the three lowest independent surge levels. Only the storm events separated by at least 120 h are considered as independent. The SSIs are used to determine the skill of our model system (Section 3.2).

2.4. *Testing for Significance of Correlations and Trends*

Testing for the significance of correlations between two time series and the significance of the presence of trends in a time series incorporates the determination how large sample correlations and sample trends could be, even if the stochastic processes, which generate the series, are not correlated at all and are stationary (exhibit no trends). The question is how large "sample correlations," derived from limited times series, may be like this so that they do not contradict the assumption that there are "no correlations" between the underlying processes. Similarly, the significance of trends indicates that such trends can hardly appear in limited segments of an infinite stationary time series.

For testing these null hypotheses, namely the processes X and Y share no correlation, or segments of length L of the process have no trend, standard procedures are available in the literature, namely p value for correlations and Mann-Kendall for trends (e.g., Kulkarni and von Storch, 1995; von Storch and Zwiers, 1999).

These tests make some assumptions about the underlying processes. In the case of correlations, the assumption is that the underlying processes are stationary (free of systematic trends) and serially independent, i.e., X_t and X_{t+1} for any t are independent. For most practical purposes, it is sufficient to examine the autocorrelation at lag 1: $Corr(X_t, X_{t+1}) = 0$. In the case of trends, the assumption is the independence of X_t'. However, in geophysical cases, these assumptions are not satisfied — the result is that the null hypotheses are more often falsely rejected (i.e., in cases where there are no correlations or no trends (cf. Kulkarni and von Storch 1995) than stipulated by the significance level (normally 5 %).

A practical remedy for avoiding such errors is to deal with normalized series (mean = 0, standard deviation = 1) X_t' (and Y_t'), and "detrend" the time series before testing for correlations, i.e., determining the linear fit f_t^X and f_t^Y, and do the hypothesis testing with $X_t' = X_t - f_t^X$ and $Y_t' = Y_t - f_t^Y$; "prewhiten" the times series, by first determining the sample autocorrelation $\alpha = 1/L\Sigma_t X_t X_{t+1}$ of the time series X_t of length L, and forming a series $X_t' = X_t - \alpha X_t$, and then testing for the null hypothesis of no trend.

To both cases, the standard routines are applied. If the null hypothesis is rejected at the stipulated significance level of 5 %, then the sample trend f_t^X, or the sample correlation $1/L\Sigma_t X_t Y_t$, is "significant."

3.   Model Validation

3.1. *Storm Surge Case Validation*

In order to examine the feasibility of simulating storm surges with the ADCIRC model, four storm surge events were simulated. One was caused by Typhoon 8509 (Miame) in August 1985, and the other three were caused by cold air outbreaks, 0310 in October 2003, 0703 in March 2007, and 0904 in April 2009. Comparisons of observed and modeled surge levels at different tide gauges are shown in Figure 3 for the four events.

For Typhoon 8509 (Miame), the simulated storm surge in Qingdao is in good agreement with the observation, with the simulated error of the maximum storm surge within 10% and the phase error within 1 h. In Tanggu, the simulated result is as good as that in Qingdao; the simulated error of the maximum storm surge is about 10%, and the phase error is about 1 h.

For the cold air outbreak event 0310, the simulated storm surges at the two tide gauges coincide with the observations. The simulated error of the maximum storm surge is within 10%, and the phase error is within 3 h. Similar results are also found for the other cold air outbreak event 0703 and 0904, with the simulated error of the maximum storm surge within 10% and the phase error close to zero,

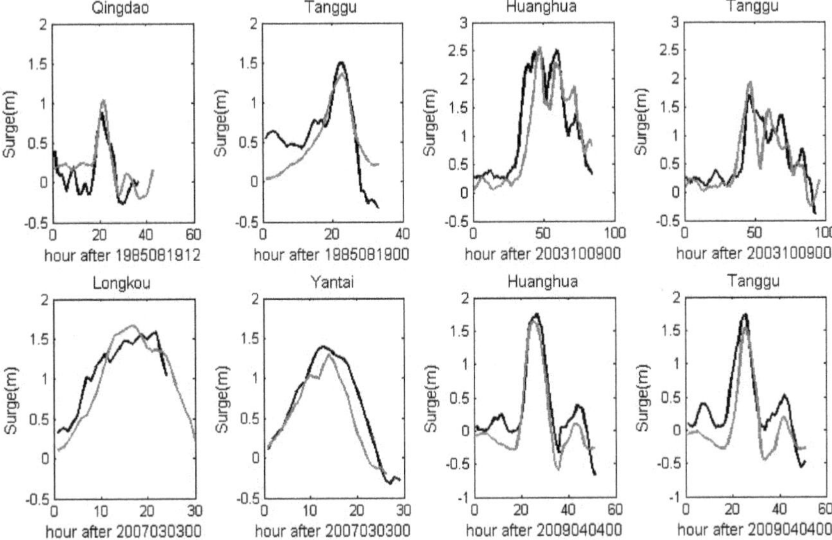

Fig. 3. Simulated (gray line) and measured (black line) storm surges at different locations for Typhoon 8509 (Miame), and the cold air outbreaks 0310, 0703, and 0904.

suggesting a good agreement between the simulated storm surges and the observations at the two stations.

We conclude that this setup of the tide-surge model and the tidal and wind forcing used in this study are sufficient to simulate the storm surge cases in the model domain.

3.2. *Long-Time Validation*

The time series of positive and negative SSIs (see Section 2.3) for the four tide gauges Dalian, Laohutan, Lianyungang, and Shijiusuo (red in Figure 1) are shown in Figure 4. It should be noted that there are no publicly accessible datasets for the tide gauges of the Bohai Sea, so all the tide gauges in this study are located in the Yellow Sea, except for Dalian which is just at the margin of the Bohai Sea.

In Dalian, the model underestimates the positive SSI in all the years except for the years of 1982 and 1983, with the differences mostly less than 15 cm except in 1982, while the negative SSIs are generally well reproduced.

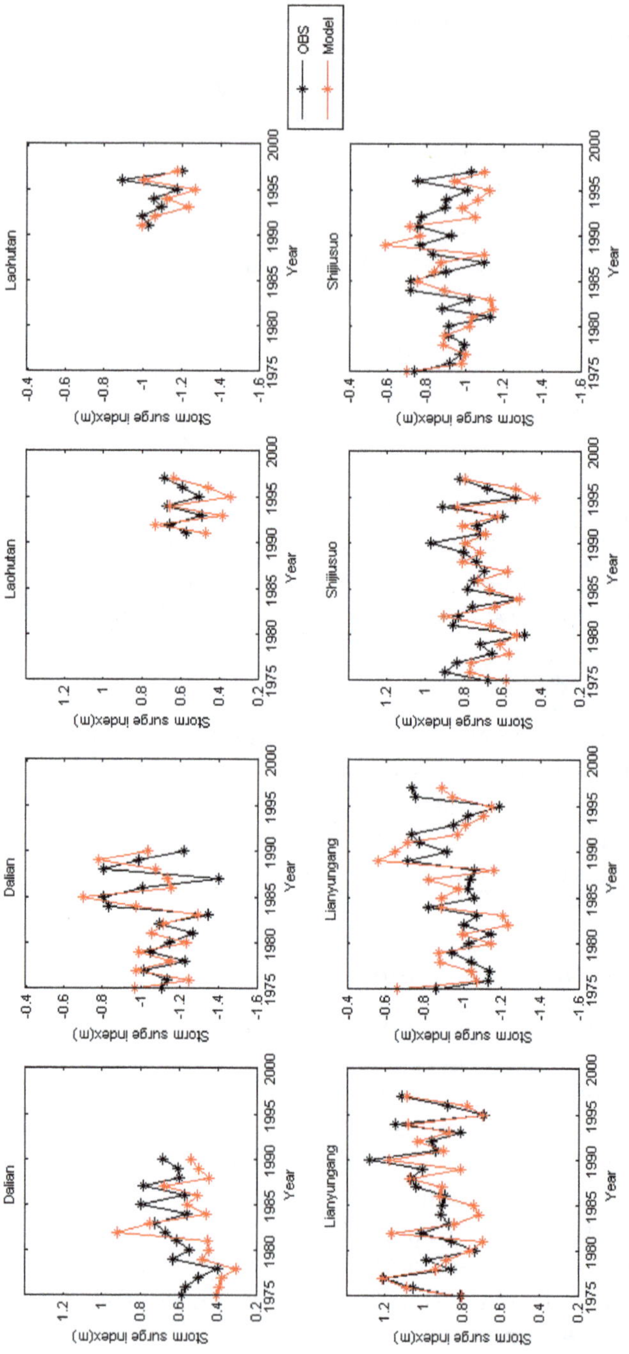

Fig. 4. Positive (left) and negative (right) storm surge indices (SSIs) at the four tide gauges Dalian, Laohutan, Lianyungang, and Shijiusuo between the observation (black line) and model simulation (red line). For locations, refer to red marks in Figure 1.

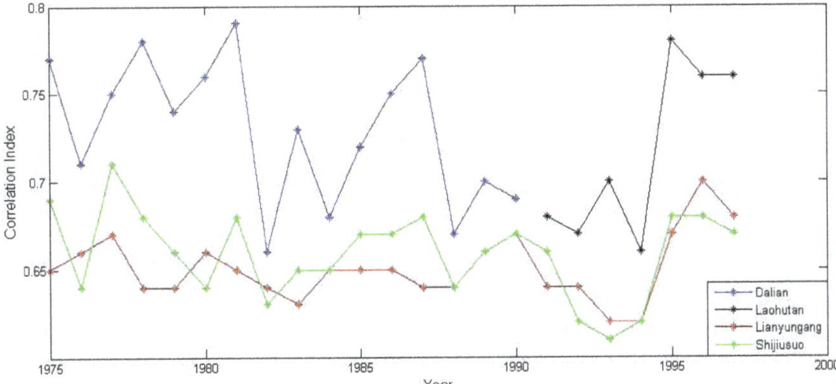

Fig. 5.   Annual correlations between the hourly modeled results and the observations at the four Yellow Sea tide gauges Dalian, Laohutan, Lianyungang, and Shijiusuo.

In Laohutan, Lianyungang, and Shijiusuo, the modeled positive SSIs are in good agreement with the observations. The errors amount to 15 cm and less at Laohutan, 10 cm in Lianyungang with exceptions in 1984, 1985, and 1989, and also 10 cm in Shijiusuo with exceptions in 1976 and 1991. The negative SSIs are also mostly well reproduced, with the exception of Shijiusuo, where in about 5 years, the simulated differences range between 15 and 20 cm.

The annual correlations between the hourly modeled results and the observations (after removal of tidal components) at the four tide gauges are also determined (Figure 5). In Dalian and Laohutan, the correlations are larger range from 0.68 to 0.78. In Lianyungang and Shijiusuo, the correlations are smaller from 0.62 to 0.70.

In summary, comparisons between model and in situ measurements of the storm surges show a good agreement, thus enhancing the confidence in the quality of the data produced through the present hindcast.

4.   Results and Discussion

In this section, the modeled results are used to analyze the seasonal, the interdecadal variability, and the long-time trend of the storm surges in the Bohai Sea. The relationship between the variations in regional climate and the storm surges in the Bohai Sea are also analysed in this section.

4.1. *Annual Cycle*

The annual cycle of mean sea levels and extreme sea levels have been widely studied (Kang *et al.*, 2008; Singh, 2001; Tsimplis and Woodworth, 1994). In order to describe the annual cycles of the storm surges along the Bohai Sea coast, 127 points were selected along the Bohai Sea coastline (blue dots in Figure 1). Forty-six-year means of monthly maximum storm surge levels of all the points are plotted for the whole period in Figure 6. As shown in this figure, storm surges in three areas, which represent the Liaodong Bay (grid points 25–35), the Bohai Bay (74–94), and the Laizhou Bay (100–115), are more serious than the others.

In all the three bays, the 46-year mean of monthly maxima have two annual peaks, with one in spring and the other in fall. In the other areas around the Bohai Sea, the monthly mean maxima are quite small, less than 0.6 m all year round. In the Liaodong Bay, the peaks are in April and August, and in the Bohai Bay, the peaks can be found in April and October, while in the Laizhou Bay, April and December are the months when the peaks appear.

Figure 6 displays the seasonal characteristics of maximum water levels in the Bohai Sea, which correspond well with empirical studies (Feng, 1982; Yang, 2000). These characteristics can be well related to the two main causes of storm surges in the Bohai Sea, namely cold air outbreaks and extratropical cyclones.

Storm surges caused by cold air outbreaks happen mostly from late autumn to spring. In winter, the cold-core high from Siberia and Mongolia passes the Bohai Sea to South China, and in spring and autumn, the Bohai area is the place where cold fronts most often sweep. These weather systems often lead to cold air outbreaks in northern China.

Previous studies (Ding and Krishnamurti, 1987; Zhu *et al.*, 2000) show that cold air outbreaks hit China along four major tracks, that is, from the northwest (44%), from the west (33%), from the north (18%), and from the east (9%). The wind directions of cold air outbreaks along the first track vary from north to northeast, while cold air outbreaks from the second track usually bring strong northwest

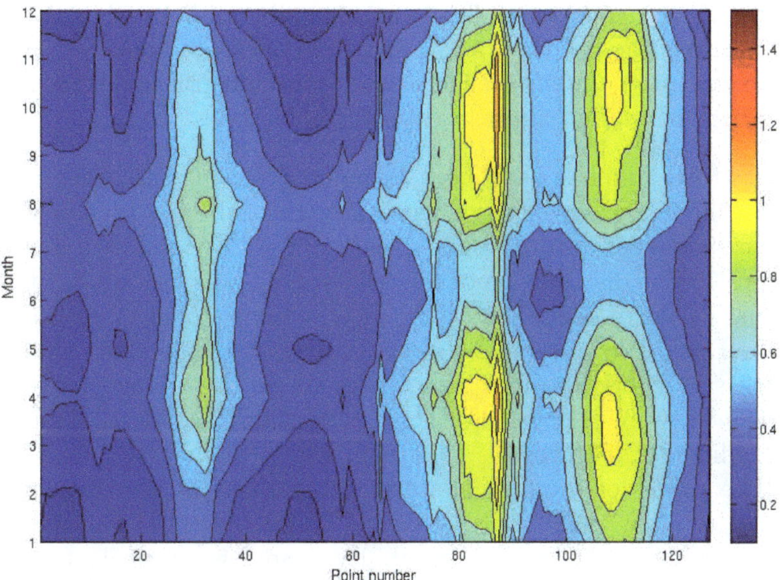

Fig. 6.   Forty-six-year means of simulated monthly maximum storm surge levels (m) at the selected 127 points along the Bohai Sea (blue dots in Figure 1). The Liaodong Bay is represented by grid points 25–35, the Bohai Bay by 74–94, and the Laizhou Bay by 100–115.

winds. On the other hand, cold anticyclones from the third track commonly induce strong northeast winds, and those of the fourth track mainly result in high easterly winds over the Bohai Sea. These four major tracks make the storm surges more serious in the Bohai Bay and the Laizhou Bay because of the wind directions. The storm surges caused by extratropical cyclones mostly happen in winter and spring; this can also cause the storm surges in the Liaodong Bay but are not as serious as those caused by cold air outbreaks.

4.2.   *Variations of the Storm Surge Climate*

Following Zhang *et al.* (2000), three indices have been used as proxies for storminess. They are:

(1) *Storm surge count*: The annual number of storm surges above a given threshold.

(2) *Storm surge duration*: The annual number of hours during which storm surges above a given threshold affect an area.
(3) *Storm surge intensity*: The annual total integral under the storm surge curve and above a given threshold.

These three indices have been calculated for all the surge events above the 99 percentile surge levels at each of the 127 points. Values of the 99 percentile surge levels at each of the 127 points are shown in Figure 7. The distribution of the 99% surge levels was similar to the annual cycles in Section 4.1. The most serious storm surges happen in the Bohai Bay and Laizhou Bay. Indices of points 33, 89, and 109, which stand for the three bays where the most serious surges have happened, were selected as examples to show the characteristics of the indices (cf. Figure 8).

The means of the annual storm surge count at the three points are quite close, about 13 per year. The interannual variability is large, whereas the decadal variability is somewhat weak. On average, the indices reached high values in the 1980s and low values prevailed in the 1990s. Figure 8 shows that at point 33, the annual storm surge count was smaller in the 1960s and in the years after

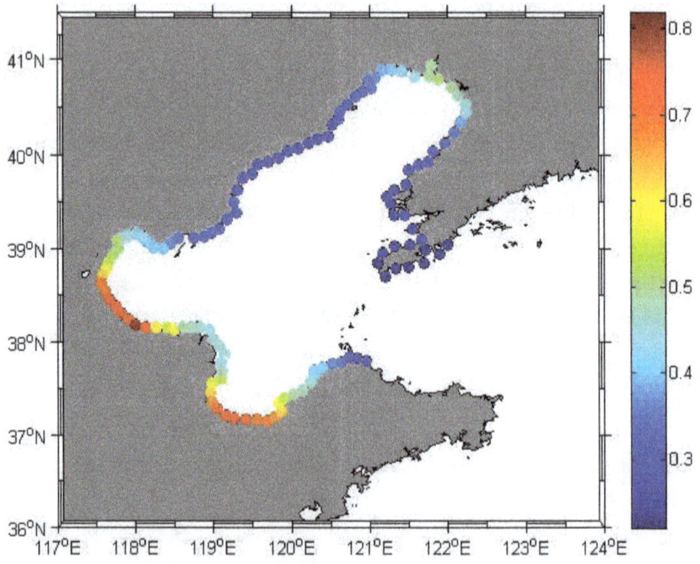

Fig. 7.   Values of the 99 percentile surge levels at each of the 127 points.

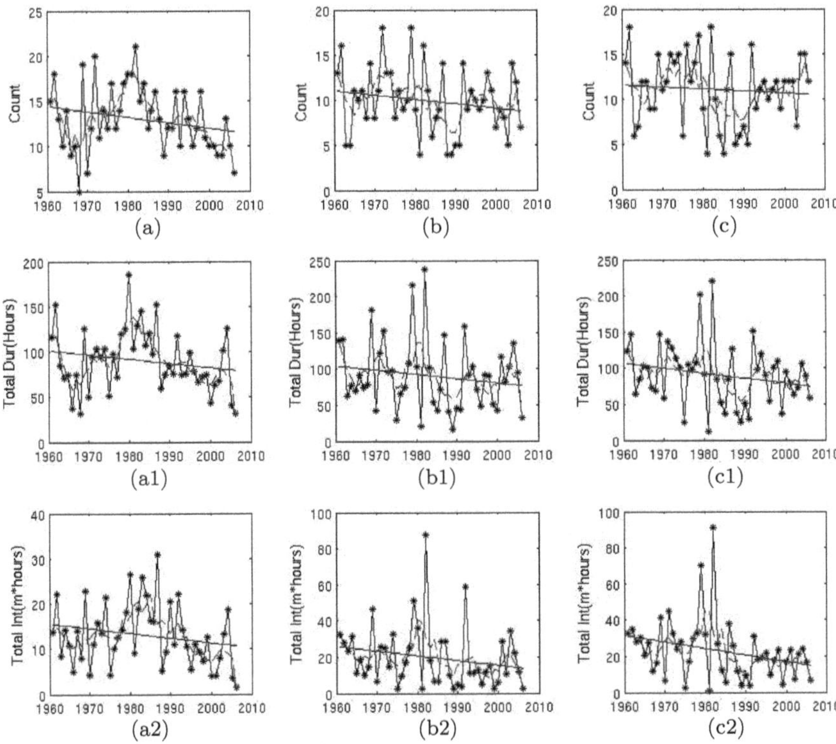

Fig. 8.　Annual total count (black dot line) (a, b, c), total duration (a1, b1, c1), and total intensity (a2, b2, c2), at point 33 (a, a1, a2), point 89 (b, b1, b2), and point 109 (c, c1, c2), with linear trend (blue line), and 5-year running mean (red line).

1990, but became larger in the middle of the whole period. The time series of annual storm surge counts at point 89 and point 109 are similar, with smaller values in the 1980s and from 1965 to 1970. A downward linear trend is obtained from the annual storm surge counts of these three points (significant at 95% confidence intervals tested by the Mann-Kendall test). This negative trend is the most obvious at point 33, which decreases by 0.07 (times per year) during the whole period. At point 109, the linear trend is weak with mere 0.03 (time per year). The annual total durations at the three points show characteristics similar to the counts, including a downward linear trend (significant at 95% confidence intervals), which is stronger at points 33, 89, and 109, where the decrease amounts to about 0.97, 1.20, and 1.40 h/year, respectively, during the whole period.

The means of the annual storm surge intensity at points 89 and 109 are larger than at point 33. Largest intensities are found in 1980 (points 89 and 109) and 1985 (point 33). Moreover, the downward trend is true for intensity (significant at 95% confidence intervals), with decreases of 9.6 m·hour (point 33), 24 m·hour (point 89), and 33 m·hour (point 109).

Therefore, it can be concluded that similarities exist among the three points, in particular, a linear decreasing trend in all of the three indices, which means that the storm surges caused by the wind forcing in the Bohai Sea are getting alleviated. There are also some differences among the three points. At point 33, the three indices show quite similar distributions: the annual total duration and intensity of the surges are higher when the annual count is larger, whereas at points 89 and 109, the distributions of the annual total duration and intensity of the surges do not covary with the annual total count.

When calculating the linear trends at all of the 127 points around the Bohai Sea (Figure 9), it is found that all the three indices at the 127 points have the downward trends, with only some of them significant at 95% significance intervals. The smallest downward trends of the surge count are shown to be at the Bohai Bay and the Laizhou Bay, while the strongest decreasing rates of the annual total duration and of intensity are at the Liaodong Bay. Researches have showed that the cold waves that can affect north China in recent 50 years have decrease trend (Gregory *et al.*, 2001; Meng *et al.*, 2013; Wang and Ding, 2006; Zou *et al.*, 2006). In the Bohai Sea, the storm surges are mostly caused by the cold waves; this may be the reason why storm surges at all points show same downward trends.

4.3. *Large-Scale Conditioning of Storm Surge Genesis*

An important issue in assessing changes in storminess is to determine whether the variability in extreme surges is related to variations in regional climate, which can be represented by different indices (Woodworth and Blackman, 2004). The cold air outbreaks and extratropical cyclones are two main weather systems that cause significant

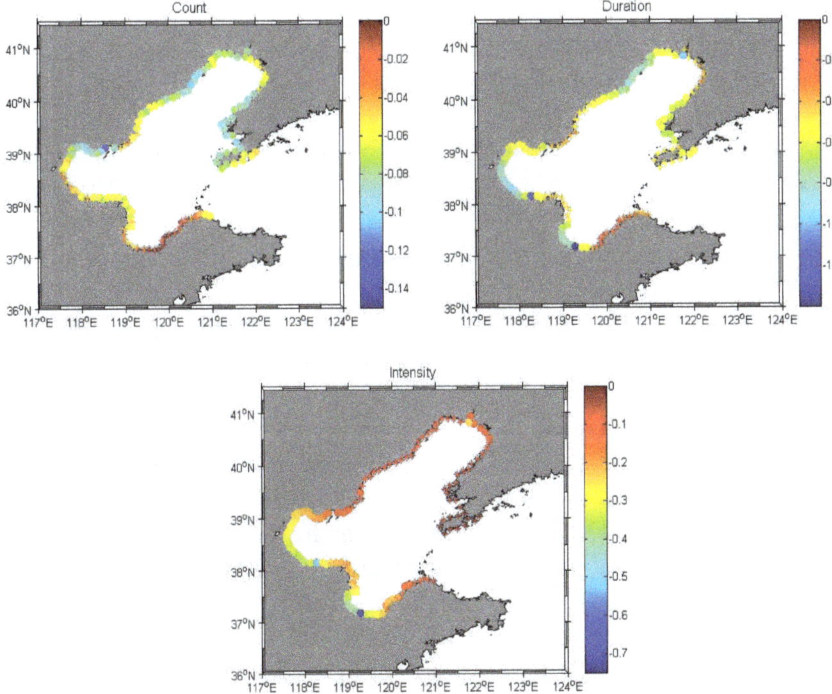

Fig. 9. Decreasing rate of the total count, total duration, and total intensity of the storm surges around the Bohai Sea. The dots mark the points that have a significant nonzero trend at the 95% confidence level when using the Mann-Kendall test, while the stars mark those not significant at the 95% confidence level.

storm surges in the Bohai Sea. Much research shows that cold air outbreaks and extratropical cyclones in China areas are strongly affected by the AO and the SH (Ding, 1990; Francis and Vavrus, 2012; Gong, 1999; Gregory *et al.*, 2001; Jeong *et al.*, 2011; Li, 1955; Wang and Ding, 2006; Zhang *et al.*, 2004, 2012; Zou *et al.*, 2006). Results show the changes of AO and SH are important factors of the frequency and intensity change of the cold air outbreak and extratropical cyclone, and the change of AO can affect the track change of the extratropical cyclone in China. In this way, the AO and SH can affect the change of storm surges in the Bohai Sea. Thus, the relationships between the storm surges and AO and the SH are studied to gain insights into the dynamical causes of the storm surge changes in the Bohai Sea.

The AO is an index of the dominant pattern of nonseasonal sea-level pressure variations north of 20°N latitude, and it is characterized by pressure anomalies of one sign in the Arctic with the opposite anomalies centered about 37°–45°N latitude (Thompson and Wallace, 1998). When AO is positive, surface pressure is low in the polar region. This helps cold Arctic air locked in the polar area. When the index is negative, there tends to be high pressure in the polar region and greater movement of frigid polar air into middle latitudes (Hansen *et al.*, 2010). The AO index used in the present study is accessed from the University of Washington (http://jisao.washington.edu/ao/). The SH is an accumulation of cold or very cold dry air in the northeastern part of Eurasia (Gong and Ho, 2002). A recent study (Gong and Ho, 2002) shows that weakening of SH is the prime driver of warmer winters in almost all of inland extratropical Asia. The SH intensity is calculated based on the data supplied by the National Center for Atmospheric Research (NCAR) (https://climatedataguide.ucar.edu/climate-data/ncar-sea-level-pressure) by using the index defined in Appendix.

The AO index (September–May) and the SH (September–May) are shown in Figure 10. The AO index undergoes interdecadal variations and has a linear upward trend between 1950 and 2010. Studies (Wang and Ding, 2006; Zhang *et al.*, 2004, 2012; Zou *et al.*, 2006) show that when the AO increase, then the synoptic-scale baroclinic waves in the westerlies will decrease in the number and intensity; in this way, the cold waves occurring in north China will decrease. Also, when the AO index was in the positive phase (the intensity of the polar vortex increase), the extratropical cyclones in 30°–60° decreased (Zhang *et al.*, 2012). At the same time, the SH decreased on average from 1980 to 2010; research (Wang and Ding, 2006) shows that when the SH decrease, then the temperature of the low-level cold heap will increase; thus, the intensity of the cold wave will decrease. The AO and the SH vary mostly independent of each other, with a (detrended 5-year running mean) correlation of merely 0.16.

Thus, in the linear regression model,

$$SS_t(x) = a_A(x)AO_t + a_S(x)SH_t + \epsilon(x) \tag{2}$$

With the storm surge index $SS_t(x)$ in year t and at location x as predicted, and the indices of the AO and SH as predictors, the regression

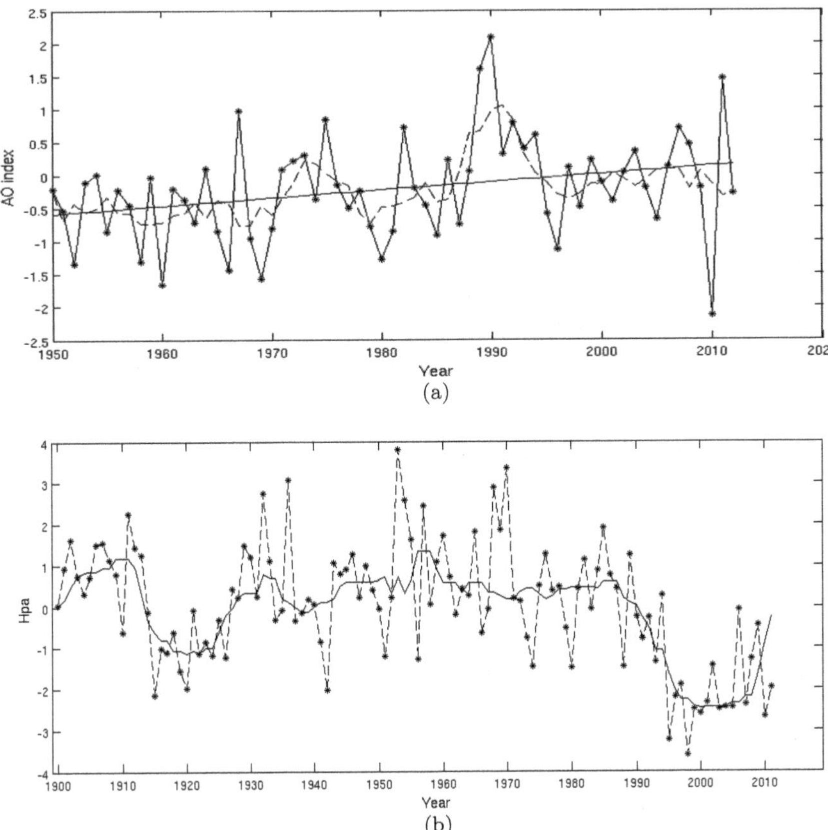

Fig. 10. (a) The monthly mean AO index (1950–2012) from September to May (black dot line), the linear trend (black line), and the 5-year running mean (dotted line). (b) The anomalies of the Siberian High (September–May) intensity from 1900 to 2011 (dotted line) and the 5-year running mean (black line).

coefficients are of first order proportional to the correlations $c_A(x)$ and $c_S(x)$

$$a_A(x) = c_A(x)s_A/s_{SS}(x) \quad \text{and} \quad a_S(x) = c_S(x)s_S/s_{SS}(x) \qquad (3)$$

with s_A, s_S, and $s_{SS}(x)$ representing the sample standard deviations of the AO index, the SH index and the considered storm surge index at location x.

Using the procedure of detrending and pre-whitening described in Section 2.4, we have determined at which location x these correlations are significantly nonzero. The correlations between the 5-year running means of the storm surge indices (as the interannual variability is too large, here, we used 5-year running mean to study decadal and long-time trend) and the 5-year running means of the AO and the SH at all of the 127 points are shown in Figure 11.

Results show that the storm surge indices and the AO have a significantly negative correlation along the southwestern coast of the Bohai Sea, including the Bohai Bay and most of the Laizhou Bay, but no significant correlations exist at the Liaodong Bay. The correlation between the storm surge indices and the SH shows different characteristics. Significantly positive correlations can be found at most of the north and northeast of the Bohai Sea, which is around the Liaodong Bay. The correlations of the three indices at the Bohai Bay and the Laizhou Bay show different characteristics. As to the storm count, some points in the Bohai Bay and the Laizhou Bay show a significantly positive correlation to the SH. On the other hand, although the correlations between the storm duration and the storm intensity at most points are larger than 0.3, significant correlations are only found at only a small number of these points. The differences between Liaodong Bay, Bohai Bay, and Laizhou Bay, can also be explained by the different weather systems, that can cause storm surges in the Bohai Sea, as we talked in Section 4.1.

Results from the linear regression analysis of the indices at points 33, 89, and 109 are shown in Figure 12. The coefficient of determination (R^2) and p (if $p \leq 0.05$, then it is significant at the 95% confidence level) values are shown in Table 1. Results from the linear regression analysis show characteristics similar to the numerical model results. The linear decreasing trends of these two results are quite similar. At points 89 and 109, the linear regression analysis results also reproduce the interdecadal variations well. Results show that the AO and the SH can explain about 10–40 % of the changes of the storm surges at the three points. Using the linear regression analysis method and the long-time AO index and SH index, the indices of the storm surges from 1900 to 2006 at the three points were estimated, with results shown in Figure 12.

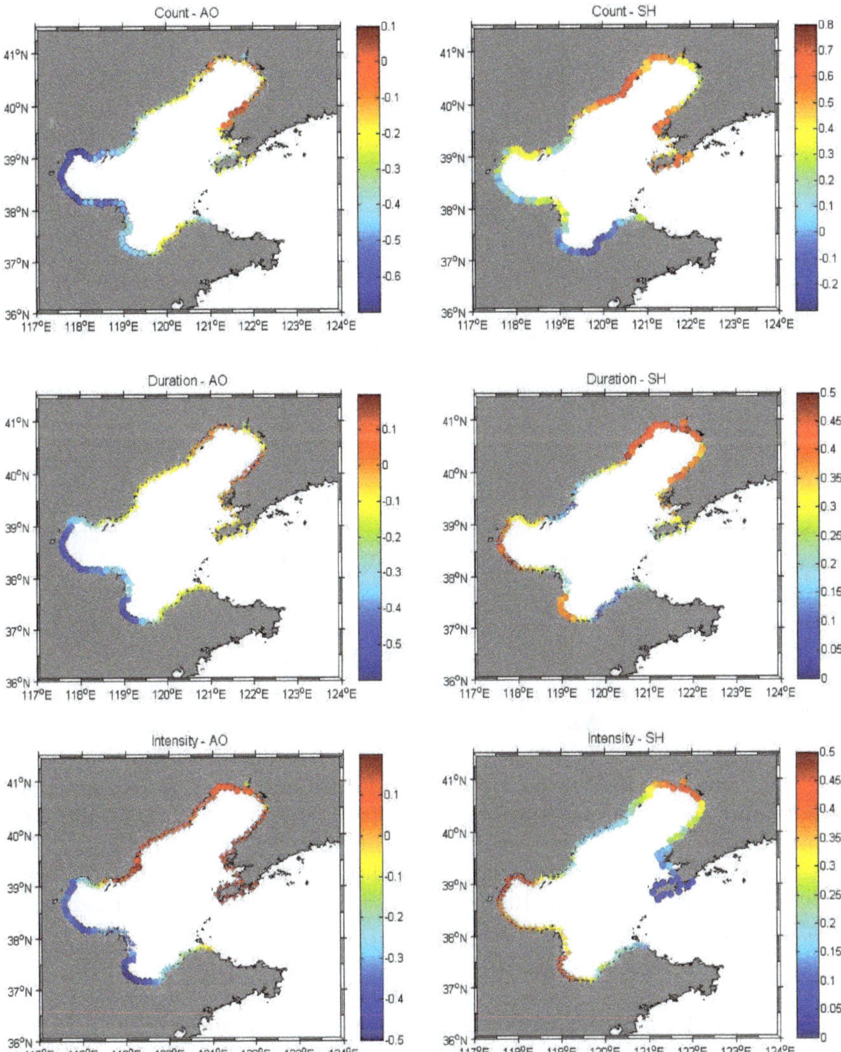

Fig. 11. Correlations between storm surge indices and the AO (left) and the SH (right) colored values indicate the value of correlations without prewhitening and detrending, and the dots mark the points where the correlations are found to be significantly nonzero at the 95% confidence level after prewhitening and detrending.

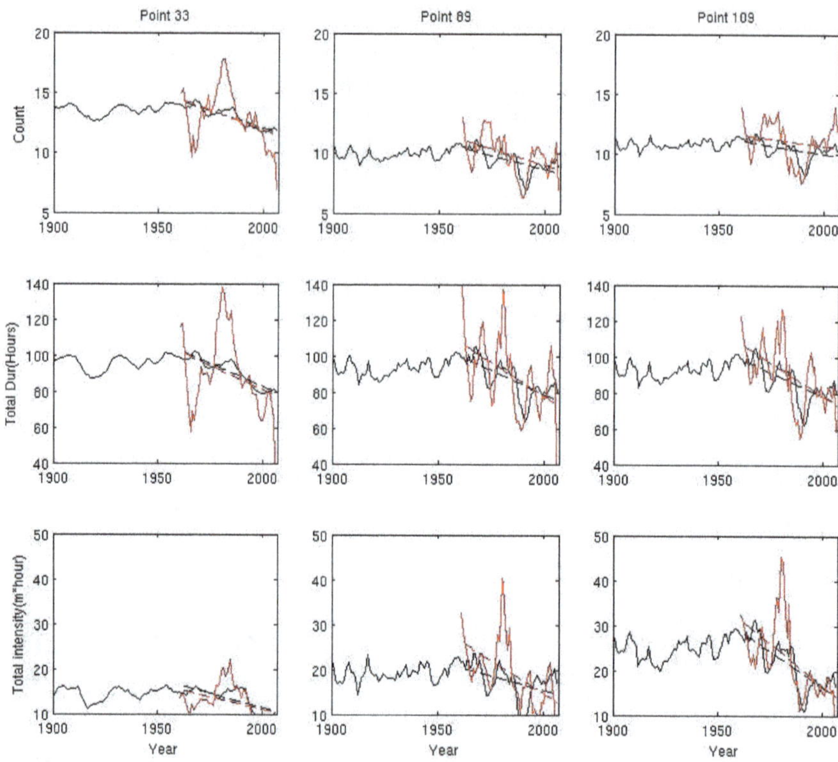

Fig. 12. Linear regression analysis results for the indices of the storm surges, results from the regression equation (black line), linear trend of the results from the regression equation (black short dotted line), model results (red line), linear trend of the model results (red short dotted line), and the indices of the storm surges at points 33, 89, and 109 from 1900 to 2006 using the linear regression analysis (black line).

Table 1. Coefficients of determination (R^2) and p values in the linear regression analysis method.

	A1	A2	A3	B1	B2	B3	C1	C2	C3
R^2	0.1640	0.1436	0.3345	0.2896	0.2482	0.1885	0.2036	0.3157	0.4197
p	0.0463	0.0797	0.0003	0.0014	0.0043	0.0250	0.0135	0.0005	<0.0001

5. Summary

As one of the greatest risks in coastal areas, storm surges are attracting more and more attention at present. The growing concerns about climate changes have added to the risk, and stakeholders and the public ask if storm surges have become more serious over the past several decades. Long-time datasets which can be used to assess such changes are in great need. Unfortunately, such data are unavailable for the coast of the Bohai Sea, but analyses of wind conditions together with numerical models allow generating long-time regional high-resolution data without using observational data. In this study, the reconstruction was done for the storm surge conditions in the Bohai Sea between 1961 and 2006.

The comparison between model results and observations, four storm surge cases, and long-time data from four tide gauges demonstrate that the numerical storm surge model (ADCIRC) is capable of reproducing storm surge conditions in the past few decades in the Bohai Sea.

Given this encouraging assessment, we examine the annual cycle, the interdecadal variability, and the long-time trend in the simulation at 127 points along the coast of the Bohai Sea. Results show that in three areas, that is, the Liaodong Bay, the Bohai Bay, and the Laizhou Bay, the storm surges are more serious than in others. Storm surges in the Bohai Sea undergo clear seasonal variations in that they are more serious in spring and autumn. The results from the analysis of storm surge indices show downward trends in storm surge count, duration, and intensity during the whole period.

Storm surges in the Bohai Sea are mainly caused by cold air outbreaks and extratropical cyclones, both of which are highly affected by the AO and the SH. In the past 50 years, the AO show an increase trend and the SH show a decrease trend, thus the cold air outbreak decrease in this area. In this way, the storm surges in Bohai show decrease in the past years. The correlations between the surge indices and the AO and the SH are calculated, and results show that in most of the Bohai Bay and the Laizhou Bay, the AO and storm surges are negatively correlated. And around the Liaodong Bay, the SH and storm surge indices have positive correlations, and the count of storm surges at some points in the Bohai Bay and the Laizhou Bay are positively correlated to the SH. Linear regression analysis between the storm surges and the AO and the SH shows that the results from the linear regression analysis correspond well with the interdecadal variations and the long-time trend. Based on this result, the conditions of the storm surges from 1900 to 2006 were estimated using the linear regression analysis. Thus, with the help of this method and the prediction of the AO and the SH, the conditions of storm surges in the future can be predicted.

6. Appendix: The Calculation of the Intensity of the Siberian High

Following the work by Gong (1999), an index I to represent the intensity of Siberian High is used here:

$$I = \frac{\Sigma_{n=1}^{N} P_n \delta_n \cos \Psi_n}{\Sigma_{n=1}^{N} \delta_n \cos \Psi_n} \tag{4}$$

where P_n is the sea level pressure at point n, and Ψn refers to the latitude of point n. When $P_n \geq 1028$ hPa, $\delta n = 1$, and when $P_n \leq 1028$ hPa, $\delta n = 0$. The selected area is bounded by 30°N, 60°E in the southwest and by 70°N, 120°E in the southeast.

Funding

This work is supported by the China Scholarship Council (No. 201306330027) and Public science and technology research funds projects of ocean (201305020-4), and we also really appreciate the support from Shanhong Gao for providing the wind fields.

Acknowledgments

We sincerely express our thanks to the developers of the ADCIRC model. JF also thanks for the Helmholtz-Zentrum Geesthacht for hospitality during a 24-month visit and for the provision of computing facilities during the time this work was carried out.

References

Bacopoulos, P., Dally, W. R., Hagen, S. C. and Cox, A. T. (2012). Observations and simulation of winds, surge, and currents on Florida's east coast during hurricane Jeanne (2004), *Coastal Engineering* **60**, pp. 84–94, doi:10.1016/j.coastaleng.2011.08.010.

Barcikowska, M., Feser, F. and von Storch, H. (2012). Usability of best track data in climate statistics in the western North Pacific, *Monthly Weather Review* **140**, 9, pp. 2818–2830, doi:10.1175/MWR-D-11-00175.1.

Bernier, N. B. and Thompson, K. R. (2006). Predicting the frequency of storm surges and extreme sea levels in the northwest Atlantic, *Journal of Geophysical Research-Oceans* **111**, C10, p. C10009, doi:10.1029/2005JC003168.

Blain, C. A., Westerink, J. J. and Luettich, R. A. (1998). Grid convergence studies for the prediction of hurricane storm surge, *International Journal for Numerical Methods in Fluids* **26**, 4, pp. 369–401, doi:10.1002/(SICI)1097-0363(19980228)26:4<369::AID-FLD624>3.0.CO;2-0.

Bromirski, P. D., Flick, R. E. and Cayan, D. R. (2003). Storminess variability along the California coast: 1858-2000, *Journal of Climate* **16**, 6, pp. 982–993, doi:10.1175/1520-0442(2003)016<0982:SVATCC>2.0.CO;2.

Church, J. A. and White, N. J. (2006). A twentieth century acceleration in global sea-level rise, *Geophysical Research Letters* **33**, 1, p. L01602, doi:10.1029/2005GL024826.

Conte, D. and Lionello, P. (2013). Characteristics of large positive and negative surges in the Mediterranean Sea and their attenuation in future climate scenarios, *Global and Planetary Change* **111**, pp. 159–173, doi:10.1016/j.gloplacha.2013.09.006.

Ding, Y. (1990). Buildup, Air-Mass Transformation and Propagation of Siberian High and Its Relations to Cold Surge in East-Asia, *Meteorology and Atmospheric Physics* **44**, 1–4, pp. 281–292.

Ding, Y. and Krishnamurti, T. N. (1987). Heat budget of the Siberian high and the Winter Monsoon, *Monthly Weather Review* **115**, 10, pp. 2428–2449, doi:10.1175/1520-0493(1987)115<2428:HBOTSH>2.0.CO;2.

Ebersole, B. A., Westerink, J. J., Bunya, S., Dietrich, J. C. and Cialone, M. A. (2010). Development of storm surge which led to flooding in St. Bernard Polder during Hurricane Katrina, *Ocean Engineering* **37**, 1, pp. 91–103, doi:10.1016/j.oceaneng.2009.08.013.

Egbert, G. D., Bennett, A. F. and Foreman, G. G. (1994). TOPEX/POSEIDON tides estimated using a global inverse model, *Journal of Geophysical Research-Oceans* **99**, 24, pp. 821–852, doi: 10.1029/94JC01894.

Feng, J., Jiang, W. and Bian, C. (2014). Numerical prediction of storm surge in the Qingdao area under the impact of climate change, *Journal of Ocean University of China* **13**, 4, pp. 539–551, doi:10.1007/ s11802-014-2222-4.

Feng, S. (1982). *Introduction to Storm Surge.* Science Press. Beijing, p. 241 in Chinese.

Feng, X., Yin, B. and Yang, D. (2012). Effect of hurricane paths on storm surge response at Tianjin, China, *Estuarine Coastal and Shelf Science* **106**, pp. 58–68, doi:10.1016/j.ecss.2012.04.032.

Francis, J. A. and Vavrus, S. J. (2012). Evidence linking Arctic amplification to extreme weather in mid-latitudes, *Geophysical Research Letters* **39**, p. L06801, doi:10.1029/2012GL051000.

Gönnert, G., Dube, S. K., Murty, T. and Siefert, W. (2001). Global storm surge: theory, observation and modeling, *Die Küste* **63**, p. 623.

Gong, D. and Wang, S. (1999). Long-term variability of the Siberian high and the possible connection to global warming, *Acta Geographica Sinica* **2**, p. 125.

Gong, D. Y. and Ho, C. H. (2002). The Siberian High and climate change over middle to high latitude Asia, *Theoretical and Applied Climatology* **72**, 1-2, pp. 1–9, doi:10.1007/s007040200008.

Gregory, J.M., Martyn, P.C. and Mark, C.S. (2001). Trends in Northern Hemisphere surface cyclone frequency and intensity. *Journal of Climate,* **14**(12), pp. 2763–2768. doi:10.1175/1520-0442(2001)014<2763: TINHSC>2.0.CO;2.

Hallegatte, S. (2007). The use of synthetic hurricane tracks in risk analysis and climate change damage assessment, *Journal of Applied Meteorology and Climatology* **46**, 11, pp. 1956–1966, doi:10.1175/ 2007JAMC1532.1.

Hansen, J., Ruedy, R., Sato, M. and Lo, K. (2010). Global surface temperature change, *Reviews of Geophysics* **48**, p. RG4004, doi:10.1029/2010RG000345.

Jeong, J., Ou, T., Linderholm, H. W., Kim, B., Kim, S., Kug, J. and Chen, D. (2011). Recent recovery of the Siberian High intensity, *Journal of Geophysical Research-Atmospheres* **116**, p. D23102, doi:10.1029/2011JD015904.

Jones, P. D. (1995). Land surface temperatures — Is the network good enough? *Climatic Change* **31**, pp. 545–558, doi:10.1007/BF01095161.

Jr, R. A. L. and Westerink, J. J. (1991). A solution for the vertical variation of stress, rather than velocity, in a three-dimensional circulation model, *International Journal for Numerical Methods in Fluids* **12**, 10, pp. 911–928, doi:10.1002/fld.1650121002.

Kang, S. K., Cherniawsky, J. Y., Foreman, M. G. G., So, J. and Lee, S. R. (2008). Spatial variability in annual sea level variations around the Korean peninsula, *Geophysical Research Letters* **35**, 3, p. L03603, doi: 10.1029/2007GL032527.

Karl, T. R., Quayle, R. G. and Groisman, P. Y. (1993). Detecting climate variations and change — New challenges for observing and data management-systems, *Journal of Climate* **6**, 8, pp. 1481–1494, doi:10.1175/1520-0442(1993)006<1481:DCVACN>2.0.CO;2.

Kulkarni, A. and von Storch, H. (1995). Monte carlo experiments on the effect of serial correlation on the Mann-Kendall test of trend, *Meteorologische Zeitschrift* **4**, 2, pp. 82–85, doi: 10.1127/metz/4/1992/82.

Langenberg, H., Pfizenmayer, A., von Storch, H. and Sündermann, J. (1999). Storm related sea level variations along the North Sea coast: Natural variability and anthropogenic change, *Continental Shelf Research* **19**, 6, pp. 821–842, doi:10.1016/S0278-4343(98)00113-7.

Li, X. (1955). A study of cold waves in East Asia, *Offprints of Scientific Works in Modern China — Meteorology (1919–1949)* (in Chinese). Science Press, Beijing.

Marcos, M., Tsimplis, M. N. and Shaw, A. G. P. (2009). Sea level extremes in southern Europe, *Journal of Geophysical Research-Oceans* **114**, p. C01007, doi:10.1029/2008JC004912.

Mattocks, C. and Forbes, C. (2008). A real-time, event-triggered storm surge forecasting system for the state of North Carolina, *Ocean Modelling* **25**, 3-4, pp. 95–119, doi:10.1016/j.ocemod.2008.06.008.

McCabe, G. J., Clark, M. P. and Serreze, M. C. (2001). Trends in Northern Hemisphere surface cyclone frequency and intensity, *Journal of Climate* **14**, 12, pp. 2763–2768, doi:10.1175/1520-0442(2001)014<2763:TINHSC>2.0.CO;2.

Mendez, F. J., Menendez, M., Luceno, A. and Losada, I. J. (2007). Analyzing monthly extreme sea levels with a time-dependent GEV model, *Journal of Atmospheric and Oceanic Technology* **24**, 5, pp. 894–911, doi:10.1175/JTECH2009.1.

Menendez, M. and Woodworth, P. L. (2010). Changes in extreme high water levels based on a quasi-global tide-gauge data set, *Journal of Geophysical Research-Oceans* **115**, p. C10011, doi:10.1029/2009JC005997.

Meng, X., Wu, Z. F., Du, H. B. and Wang, L. (2013). Spatio-temporal characteristics of cold wave over northeast China during 1961–2010,

Journal of Arid Land Resources and Environment **27**, 1, pp. 142–147. Chinese with English abstract.

Mudersbach, C., Wahl, T., Haigh, I. D. and Jensen, J. (2013). Trends in high sea levels of German North Sea gauges compared to regional mean sea level changes, *Continental Shelf Research* **65**, pp. 111–120, doi:10.1016/j.csr.2013.06.016.

Palumbo, A. and Mazzarella, A. (1982). Mean Sea-Level Variations and Their Practical Applications, *Journal of Geophysical Research-Oceans* **87**, NC6, pp. 4249.

Pascual, A., Marcos, M. and Gomis, D. (2008). Comparing the sea level response to pressure and wind forcing of two barotropic models: Validation with tide gauge and altimetry data, *Journal of Geophysical Research-Oceans* **113**, C7, p. C07011, doi:10.1029/2007JC004459.

Pawlowicz, R., Beardsley, B. and Lentz, S. (2002). Classical tidal harmonic analysis including error estimates in MATLAB using T-TIDE, *Computers & Geosciences* **28**, 8, pp. 929–937, doi:10.1016/S0098-3004(02)00013-4.

Ratsimandresy, A. W., Sotillo, M. G., Carretero Albiach, J. C., Alvarez Fanjul, E. and Hajji, H. (2008). A 44-year high-resolution ocean and atmospheric hindcast for the Mediterranean Basin developed within the HIPOCAS Project, *Coastal Engineering* **55**, 11, pp. 827–842, doi:10.1016/j.coastaleng.2008.02.025.

Singh, O. P. (2001). Cause-effect relationships between sea surface temperature, precipitation and sea level along the Bangladesh coast, *Theoretical and Applied Climatology* **68**, 3-4, pp. 233–243, doi:10.1007/s007040170048.

Thompson, D. W. and Wallace, J. M. (1998). The arctic oscillation signature in the wintertime geopotential height and temperature fields, *Geophysical Research Letters* **25**, 9, pp. 1297–1300.

Tsimplis, M. N. and Woodworth, P. L. (1994). The Global Distribution of the Seasonal Sea-Level Cycle Calculated from Coastal Tide-Gauge Data, *Journal of Geophysical Research-Oceans* **99**, C8, pp. 16031–16039, doi:10.1029/94JC01115.

von Storch, H. and Reichardt, H. (1997). A scenario of storm surge statistics for the German bight at the expected time of doubled atmospheric carbon dioxide concentration, *Journal of Climate* **10**, 10, pp. 2653–2662, doi:10.1175/1520-0442(1997)010<2653:ASOSSS>2.0.CO;2.

von Storch, H. and Zwiers, F. W. (1999). *Statistical Analysis in Climate Research,* Cambridge University Press, doi: 10.1017/CBO9780511612336.

von Storch, H., Jiang, W. and Furmanczyk, K. K. (2015). Storm surge case studies, in *Coastal and Marine Hazards, Risks, and Disasters,* Elsevier, Amsterdam.

Wang, Z. and Ding, Y. (2006). Climate change of the cold wave frequency of China in the last 53 years and the possible reasons, *Chinese Journal of Atmospheric Sciences-Chinese Edition* **30**, 6, pp. 1068–1076 in Chinese with English abstract.

Weisse, R., Bellafiore, D., Menendez, M., Mendez, F., Nicholls, R. J., Umgiesser, G. and Willems, P. (2014). Changing extreme sea levels along European coasts, *Coastal Engineering* **87**, pp. 4–14, doi:10.1016/j.coastaleng.2013.10.017.

Weisse, R., von Storch, H., Callies, U., Chrastansky, A., Feser, F., Grabemann, I., Günther, H., Pluess, A., Stoye, T., Tellkamp, J., Winterfeldt, J. and Woth, K. (2009). Regional meteo–marine reanalyses and climate change projections: Results for Northern Europe and potential for coastal and offshore applications, *Bulletin of the American Meteorological Society* **90**, 6, pp. 849–860, doi: 10.1175/2008BAMS2713.1.

Winterfeldt, J., Geyer, B. and Weisse, R. (2011). Using QuikSCAT in the added value assessment of dynamically downscaled wind speed, *International Journal of Climatology* **31**, 7, pp. 1028–1039, doi: 10.1002/joc.2105.

Woodworth, P. L. and Blackman, D. L. (2003). Evidence for systematic changes in extreme high water since the mid-1970s. *Journal of Climate*, **17**, pp. 1190–1197. doi:10.1175/1520-0442(2004)01 <1190:EFSCIE>2.0.CO;2.

Woodworth, P. L. and Blackman, D. L. (2004). Evidence for systematic changes in extreme high waters since the mid-1970s, *Journal of Climate* **17**, 6, pp. 1190–1197, doi:10.1175/1520-0442(2004)017 <1190:EFSCIE>2.0.CO;2.

Woodworth, P. L., Flather, R. A., Williams, J. A., Wakelin, S. L. and Jevrejeva, S. (2007). The dependence of UK extreme sea levels and storm surges on the North Atlantic Oscillation, *Continental Shelf Research* **27**, 7, pp. 935–946, doi:10.1016/j.csr.2006.12.007.

Woth, K., Weisse, R. and von Storch, H. (2006). Climate change and North Sea storm surge extremes: An ensemble study of storm surge extremes expected in a changed climate projected by four different regional climate models, *Ocean Dynamics* **56**, 1, pp. 3–15, doi: 10.1007/s10236-005-0024-3.

Wu, D., Gao, S., Wang, Y. and Chen, X. (2011). The atlas of monthly-average wind and air temperature in the Bohai Sea, Yellow Sea and East China Sea (1960–2007) China Ocean University Press, Qingdao.

Yang, G. (2000). Historical change and future trends of storm surge disaster in China's coastal area, *Journal of Natural Disasters* **9**, 3, pp. 23–30.

Yin, B., Hou, Y., Cheng, M., Su, J., Lin, M., Li, M. and El-Sabh, M. (2001). Numerical study of the influence of waves and tide-surge interaction on tide-surges in the Bohai Sea, *Chinese Journal of Oceanology and Limnology* **19**, pp. 97–102.

Zhang, H. and Sheng, J. (2015). Examination of extreme sea levels due to storm surges and tides over the northwest Pacific Ocean, *Continental Shelf Research* **93**, pp. 81–97, doi:10.1016/j.csr.2014.12.001.

Zhang, H. M., Reynolds, R. W. and Bates, J. J. (2006). Blended and gridded high resolution global sea surface wind speed and climatology from multiple satellites: 1987-present, in *American Meteorological Society 2006 Annual Meeting*, p. 2.23.

Zhang, K., Douglas, B. C. and Leatherman, S. P. (2000). Twentieth-century storm activity along the US east coast, *Journal of Climate* **13**, 10, pp. 1748–1761, doi:10.1175/1520-0442(2000)013<1748:TCSAAT>2.0.CO;2.

Zhang, X., Walsh, J. E., Zhang, J., Bhatt, U. S. and Ikeda, M. (2004). Climatology and interannual variability of Arctic cyclone activity: 1948-2002, *Journal of Climate* **17**, 12, pp. 2300–2317, doi:10.1175/1520-0442(2004)017<2300:CAIVOA>2.0.CO;2.

Zhang, Y., Ding, Y. and Li, Q. (2012). Cyclogenesis frequency changes of extratropical cyclones in the northern hemisphere and East Asia revealed by era40 reanalysis data, *Meteorological Monthly* **38**, pp. 646–656 (in Chinese).

Zhao, P. and Jiang, W. (2011a). A numerical study of storm surges caused by cold-air outbreaks in the Bohai Sea, *Natural Hazards* **59**, 1, pp. 1–15, doi:10.1007/s11069-010-9690-7.

Zhao, P. and Jiang, W. (2011b). A numerical study of the effects of coastal geometry in the Bohai Sea on storm surges induced by cold-air outbreaks, *Journal of Ocean University of China* **10**, pp. 9–15.

Zhu, Q., Lin, J., Shou, S. and Tang, D. (2000). *Principles and Methodology of Synoptic Meteorology, 3rd edn.*, China Meteorological Press.

Zou, X., Alexander, L. V., Parker, D. and Caesar, J. (2006). Variations in severe storms over China, *Geophysical Research Letters* **33**, 17, p. L17701, doi:10.1029/2006GL026131.

Author Index

B

Bisling, Peter, xxiv, 213

C

Cavicchia, Leone, xxiv, 3
Chen, Fei (陈飞), xxi, 169, 193
Chen, Xueen (陈学恩), viii, x, xvii,
xix–xxi, 57, 99, 113, 137

D

Du, Yan, 193

F

Feng, Jianlong (冯建龙), xxi,
285, 311
Feng, Shizuo (冯士笮), xix
Feser, Frauke, xxiv, 3
Fu, Gang (傅刚), xxii

G

Geyer, Beate, xxiv, 206,
213, 277
Guo, Donglin, 137, 247

H

Hasselmann, Klaus, xxix, 56, 100,
114, 132, 140

J

Jiang, Wensheng (江文胜), viii, xix,
xxii, 285, 311

K

Kanamitsu, Masao, 16

L

Li, Delei (李德磊), viii, xxi, 3, 29, 57,
213, 247
Li, Huajun (李华军), xxii
Lin, Lin (林璘), xxi, 137
Luo, Yongming (骆永明), xxii

P

Penduff, Thierry, 116, 132

Q

Qi, Jifeng, 247

Subject Index

www.ingramcontent.com/pod-product-compliance
Lightning Source LLC
Chambersburg PA
CBHW070757100725
29074CB00020B/35